D1600509

THE TECHNOLOGICAL INDIAN

The

Technological Indian

ROSS BASSETT

 Harvard University Press

Cambridge, Massachusetts
London, England   2016

Library of Congress Cataloging-in-Publication Data

Bassett, Ross Knox, 1959–
    The technological Indian / Ross Bassett.
        pages cm
    Includes bibliographical references and index.
    ISBN 978-0-674-50471-4 (alk. paper)
    1. Technology transfer—India—History.   2. Massachusetts Institute
of Technology—Foreign students.   3. India—History.   I. Title.
    T27.I4B37 2015
    338.954'06—dc23

                                                              2015012424

To Debbie, who shows me what courage is

Contents

THE TECHNOLOGICAL INDIAN

Introduction

O N MAY 30, 1884, M. M. Kunte, the headmaster of the Poona High School, addressed a group of middle-class Hindus in that western Indian city's spring lecture series. His speech was a remarkable piece of proselytization for a new global faith that he called "the art of mechanization"—what we would today call technology.[1] The Marathi-language periodical, the *Kesari*, recounted Kunte's speech, which had been given in English. Kunte said that if Hindustan wanted its glory to be revived, then the Hindu people needed to suspend the activities of all their religious, social, and sporting groups for a century and a quarter and then Hindus needed "to travel from village to village, taluka to taluka, district to district and start the activities of blacksmithy with frantic haste and zeal."[2] His speech reached a climax: "For eradicating the undesirable and establishing the desirable in society, there is no option but to follow and spread widely the art of mechanization. If you want to eat, be a machinist. If you want to win freedom, you have to learn to be a machinist. If you want to live as luxuriously as our rulers, you have to run the machines. If you want this country to progress like that of England, all of you have to become blacksmiths." He concluded that the art of mechanization "has become a Kalpavriksha," referring to the mythological Hindu wish-granting tree, stating that "achieving our desired progress would be next to impossible without its shadow and shelter."[3]

Kunte was part of a small group of English-language-educated Indians who in the late nineteenth century sought to respond not only to British control of their country, but also to a world increasingly dominated by machines. India had an ancient and rich technical tradition, where for centuries its manufactured goods, most notably cotton textiles, were sold and prized in Asia, Europe, and Africa. When the Europeans (increasingly British) came to India they took control of India's system of making and selling cotton textiles before they seized control of the country. Then in Europe, the British, through a series of transformations of the processes of production known today as the Industrial Revolution (as well as through tariffs and laws blocking Indian imports), captured cotton textile manufacture for their own country. The work formerly done by Indian spinners and weavers was now done in huge steam-powered factories in Manchester. India, under the control of a foreign power, became an increasingly poor and agricultural society left out of the benefits of what British writer Thomas Carlyle called "the age of machinery." What was to be done? Kunte saw India's salvation in mechanization.[4]

Kunte's speech contained a hint of unreality, even absurdity. Kunte had never been abroad, and his reference to starting village blacksmith shops, at a time when the United States was making 4 million tons of pig iron a year, suggested that he had not fully grasped the tenets of the religion he was preaching. Kunte had originally made his reputation as a student of Sanskrit and had no formal qualification for the role that he adopted as "apostle of industry."[5]

Implicit in Kunte's exhortation was that Indians were not machinists—in today's language they were not technological. Many British would have agreed with Kunte. Indeed, in 1881 the former Governor of Bombay, Sir Richard Temple, asserted that "the Hindus are not a mechanical race."[6] As Kunte spoke to this middle-class audience, whose background was more literary than technical, he implied that they could no longer leave making things to the thousands of crafts people in Poona—they had to step in.

In the early twenty-first century, the Indian middle class has no need for Kunte's exhortations. Under an expanded definition of the terms "blacksmith" and "machinist," his city (now known as Pune) is full of blacksmiths who build a wide range of machines, most notably automobiles. And today indeed all of India is full of those who "run the machines" through the programs they write that control the twenty-first century's quintessential machine, the computer.

Today, young middle-class Indians (in either India or the United States) so commonly seek engineering careers that the stereotype is now of a people who avoid liberal arts in preference to engineering—the exact opposite of what Kunte faced in 1884. In the spring of 2014, 1.4 million young Indians sat for the Joint Entrance Examination (JEE), which serves as the primary entrance point to an engineering education. In recent years the city of Kota in Rajasthan has become famous as a center of the JEE coaching industry, with over 100,000 students annually flooding the city to take classes in the hopes that they can gain entrance to the elite Indian Institutes of Technology.[7] Perhaps because the competition is less fierce, wealthier middle-class Indians increasingly come to the United States to enter undergraduate engineering programs, while Indians account for a large portion of the graduate student population in some American engineering programs. Second-generation Indian Americans also show a similar technological bent, swelling the ranks of engineering programs throughout the United States.

Indians' focus on engineering has led to remarkable achievements in India and America. Indian firms produce a wide array of technologically sophisticated products sold in a global market. India's information technology (IT) industry has become famous worldwide, generating over $100 billion of revenue and employing 3 million people. In 2012 a trade journal noted that IBM was on track to have more Indian than American employees. In the United States, the role of Indians in Silicon Valley is widely recognized. In 2015 Indian engineers occupy the presidency of major American universities and hold or have recently held deanships at Harvard, MIT, Penn, Berkeley, Carnegie Mellon, and UCLA. Indians' position in the American technological workforce is so central that the appointment of Satya Nadella as CEO of Microsoft was hardly noteworthy.[8]

How did this change happen? How did "Indian" and "technological" go from being mutually exclusive to being practically synonymous for the Indian middle class? And how did Indians fit so well into the American technological system, which was so foreign to

Kunte in 1884? This book argues that beginning in the late 1800s, a small group of middle-class, English-language-educated Indians began to imagine a technological India, and to do so, they looked to the United States and the Massachusetts Institute of Technology (MIT) in particular. For more than a century, Indian elites have gone to MIT, but perhaps more importantly, their technological imaginations have been shaped by MIT.

The fact that MIT served as the model behind the original conception of the Indian Institutes of Technology (IITs) between 1944 and 1946 is widely known. But the role that MIT has played in shaping technical education, and indeed technology, in India is far deeper than that. The first Indian attended MIT in 1882 and the first suggestion that MIT had something to offer India came in 1884 from Indian nationalist Bal Tilak's newspaper the *Kesari*.

After independence, the government of India explicitly sought MIT's help in developing the Indian Institute of Technology at Kanpur, resulting in a ten-year program anchored by MIT and supported by nine American institutions. Later four other IITs, ostensibly designed to showcase other nations' models of technical education, increasingly converged on the IIT Kanpur/MIT model. While the IITs were designed to make India self-sufficient in technical education, from the time of their founding, they were increasingly integrated into an American system of technical education, where MIT stood at the apex.

MIT's hold on the technological imaginations of the Indian middle class might be seen in Godavarthi Varadarajan, who in 1972 was an eight-year-old boy in a Tamil Brahmin family. That summer, his mother took her children from Ahmedabad, where her husband worked as a geophysicist for the state-run oil company, to her family's home in Madras. There, she and her father sat the young boy down for a serious talk about his future, laying out the path they thought he should aspire to. He should aim to be one of the winners of the National Science Talent Search Exam, given after tenth grade. After high school, he should secure entrance to one of the Indian Institutes of Technology. After IIT, he should win admission to MIT for graduate study. Finally, he should earn an MBA at Harvard Business School. These goals, part of a "joint journey," were continually reinforced throughout young Varadarajan's schooling. Varadarajan followed this path (except for the Harvard MBA), earning a doctorate in engineering from MIT.[9]

This book had its origins over thirty years ago when, as an engineering undergraduate at the University of Pennsylvania, I met a group of Indian graduate students from the Indian Institute of Technology at Madras. Over the course of our friendship they told me about the history of the IITs, four of which had a foreign nation as a patron. I traveled to India in the early Reagan years when the Soviet invasion of Afghanistan had drawn the United States closer to Pakistan, leading to an official distancing of relations between India and the United States. If official Indo-U.S. relations were marked by tensions, what I saw on the ground level was different. Indians approached me on the street telling me of their adult children in the United States. I was particularly struck by meeting a young man at a bus stop in Bangalore carrying a programming manual for the PDP-11, an American minicomputer made by Digital Equipment Corporation. When I returned to the United States and began work as an engineer, I had a number of Indian colleagues. Years later, as a professor studying the history of technology, those images seemed increasingly relevant as Indian engineers were becoming more prominent in the United States and American businesses were increasingly having software written in India.

I sought to understand when this technological connection between the United States and India began and how it developed. IIT Kanpur, because of its connections with the United States, seemed like a promising starting point. During a trip to Cambridge to examine MIT's holdings on IIT Kanpur, I discovered a collection of MIT commencement programs listing the name and hometown of every MIT graduate. This made it possible to develop a database of Indians who had gone to MIT—a reasonably small traceable sample. I imagined that this sample would include people who had gone to IIT Kanpur, but also some who had gone to the other IITs.

My assumption was that Indians had started going to MIT after the development of the IITs in the 1950s and 1960s, and surely no earlier than Indian independence in 1947, for in the colonial period they must have gone to British universities. But for the sake of completeness I grabbed earlier commencement programs off the shelf. Surprisingly, programs from before 1947 revealed a small but significant group of Indian MIT graduates. The first Indian attended MIT in 1882, far earlier than I would have imagined. In all, roughly 1,300 Indians earned degrees from MIT between its founding in 1861 and 2000.[10]

I then spent years trying to find out who these people were, how they got to MIT, and what they did afterward. Some of this work was done online, on the telephone, or through a collection of MIT alumni directories, but I also traveled the length and breadth of India, from Pilani to Chennai and from Bhavnagar to Jamshedpur, conducting scores of interviews and seeking information. As I did this research, I found that the individual stories of how these people got to MIT was a bigger part of Indian history than I could have imagined.[11]

This book is both more and less than a history of Indians who went to MIT. An attempt to include everyone who went to MIT would result in a directory, not a book. And because the focus is on technology, it concentrates on those who went to MIT to study engineering, resulting in the omission of many important Indian MIT graduates, ranging from Charles Correa, India's most important architect since independence, to Raghuram Rajan, as of this writing (2015) the governor of the Reserve Bank of India, albeit with an undergraduate degree in engineering from IIT Delhi. It focuses on those who came from India and so leaves out the many second-generation Indian Americans who have gone to MIT, starting with Amar Bose, of the eponymous audio company. (After the 1965 American immigration reforms, Indians began coming to the United States in greater numbers. By 1991, the number of MIT graduates who gave a hometown in the United States, but whose name suggested South Asian ancestry, was greater than those who gave a hometown in India.) It deals primarily with the period before India's liberalization of its economy in 1991.

This book uses an experimental approach to history, seeking to explore what a cross-sectional slice of roughly 850 Indian engineering graduates of MIT and the broader context of their lives can tell about

India and the United States. One might think that such a small group
of people in such a large country, chosen on such specific grounds,
would not be very revealing about Indian history, akin to trying to
understand a person's anatomy by taking an MRI of the big toe. But
remarkably, starting with these people and then expanding the image
to include the broader context of their lives pulls into the picture
major figures and themes in Indian history, often letting us see them
in a new perspective.

While today's high-tech India is often associated with Bangalore,
in the years before independence, Indians who attended MIT came
primarily from the western part of India: Bombay, Poona, and Gujarat. The first Indian to attend MIT was part of a movement for
the industrial development of India located in Poona, in which a central role was played by Bal Tilak's newspapers, the *Kesari* and the
Mahratta. Although Poona has long been seen as an important point
of origin of the Indian nationalist movement, the origins of India's
technological nationalist movement are often located in the swadeshi
movement in Bengal. This book both emphasizes important precur-
sors in western India and links them to the United States.[12]

Mahatma Gandhi enters this story by virtue of the fact that several of his close associates either went to MIT themselves or sent
family members to MIT. Given Gandhi's position as the outstanding
symbol of opposition to modern technology, this would seem to be
a great paradox. Historians have noted that one of Gandhi's great
strengths was his ability to draw from many traditions and speak in
ways that could be interpreted differently by different groups. This
work takes Shahid Amin's statement that there was "no single authorized version of the Mahatma" in a different direction. Contrary to
what we might think today, Gandhi's Satyagraha Ashram could be
one step on a path to MIT.[13]

A major theme in the study of Indian technology has been the role
of the state, both in the colonial and the postcolonial periods. While
the state is clearly visible in this study, at the same time a variety of
non-state actors can also be seen who combined nationalism, business, and technology in a variety of ways. A variant of the state in
the colonial period was the princely state. And while the princely
states of Mysore and Baroda are most often associated with efforts

at industrial development, this study brings out the role of the small princely state of Bhavnagar. An important thread throughout the book is the tension between those who sought technological development within the state and those who in varied degrees sought to operate outside the state.[14]

At the same time this study does not support a strict dichotomy between the state and private enterprise. This study introduces Indian entrepreneurs with a range of views, and even those whose approaches were less statist saw areas where state involvement was necessary. One of the ironies of this story is that while today (and particularly in contrast to Nehru's India), some have tried to cast the United States as an exemplar of a market-driven economy, MIT was to a significant degree an institution of the American state, and its role as an agent of technological development came about to a large degree through funding from the state.

This book captures an important thread in Indian business history. The House of Tata is often seen as exceptional among Indian businesses for a global outlook, which stretched back to J. N. Tata in the nineteenth century. Throughout this narrative, the careers of Indians who went to MIT frequently crossed with the Tata businesses. Tata's international perspective, combined with a group of young engineers exposed to computing at MIT, led in the 1960s to the establishment of Tata Consultancy Services, India's first global IT firm. But the story of globally oriented Indian business is more than the Tatas. The Birlas, often seen as a more conservative business family, had a closer connection to MIT than the Tatas, sending Aditya, a chosen heir, to MIT and winning MIT's support for the development of the Birla Institute of Technology and Science in Pilani. This study highlights a group of Indian business families, including Kirloskar, Lalbhai, Godrej, and Chauhan, who sent their heirs to MIT to develop competencies in engineering and management.[15]

It is tautological that a cross-section of Indians going to MIT would say something about Indo-U.S. relations, but this cross-section again puts those relations in a new perspective. Throughout the period covered by this book many things separated India and the United States. A virulent anti-Asian racism in America was codified into law, reaching its height in 1917 legislation that made it illegal for Asians to even enter the United States, except as temporary students. Only with the passage of immigration reform in 1965 did Indians have the ability to come to the United States on something like equal terms with Europeans. But even while American laws were at their most discriminatory, Indians continued to look to and come to the United States for technical education.[16]

After Indian independence in 1947, formal Indo-American foreign relations were held hostage to the Cold War. The United States sought allies in its struggle with communism while Jawaharlal Nehru committed India to remain nonaligned. At the same time America's alliances with Pakistan, India's chief rival and a country that was more willing than India to take sides in the Cold War, proved to be a constant irritant to India. Historians who have written about Indo-American relations during this period have often focused on the apparent paradox of the world's two biggest democracies not getting along, writing books with such titles as *The Cold Peace* and *Estranged Democracies*.[17]

But from the perspective of a small Indian elite with a desire for technical education and technological development, things look dramatically different. In fact their actions justify the use of the term "empire." This was not the political empire of the British Raj, but a different kind of empire, whose nature might be best understood by analogy to Harvard historian Charles Maier's definition of empire as a "particular form of state organization in which the elites of differing ethnic or national units defer to and acquiesce in the political leadership of the dominant power."[18] By looking to the United States and to MIT, a group of Indian elites accepted not American political leadership, but American technological leadership. This was not an empire imposed from outside. The ties between India and the United States were weak enough that Indian technical elites had a great deal of flexibility in the world they fashioned. Indians who looked to the United States saw a number of possible advantages to doing so: building up India, their family businesses, or themselves. Indians made this empire operational.

This is not to say that the United States played no active role in the creation of this empire. Americans ranging from professors to businessmen to government officials to librarians sought to build connections that would extend American influence in India. The United States government spent millions of dollars for programs to establish an American model of technical education in India. But all these efforts would have been for naught had it not been for the technological dreams of a group of Indians. The British, Germans, and Soviets who contributed to the Indian Institutes of Technology with the hope of extending their countries' influence in India often found to their dismay that their efforts were coopted and the institutions they helped create were in significant ways more closely connected to the United States than to their own countries.

Given American technological developments in the late nineteenth century and onward, it would have been more remarkable if Indian elites had been immune to American technology. And they were not alone. In varied ways, large parts of the world were part of an American technological empire in the twentieth century.[19]

While this book argues that the dominant model of the technological Indian today is the engineer trained to American standards, whether at an IIT, MIT, or some other institution, throughout the period covered by this book, there were other models available. Kunte's plea for Indians to be machinists was made in the belief that industrial technologies had outmoded the Indian craftsman and artisan. While Karl Marx, arguing for the effects of British textile machinery, once quoted a report claiming that "the bones of the cotton weavers are bleaching the plains of India," recent studies have shown that industrial technology left a space in which a dynamic group of Indian artisans could survive well into the twentieth century. Innovations in artisanal technology continued even as some British and middle-class Indians wrote off such "craft-based" technology.[20]

The British colonial ideal of the Indian engineer was that of a civil engineer subordinate to British engineers, working to build irrigation systems and other infrastructure necessary to sustain an agricultural economy. The British established colleges throughout India to train such Indians. Mokshagundam Visvesvaraya was an Indian civil engineer who reached the pinnacle of achievement allowed under that system. After successfully engineering a number of projects for the Bombay Presidency, Visvesvaraya resigned when it became clear that he would never be allowed to occupy the position of chief engineer. Instead he moved to the princely state of Mysore, where he served as dewan, the equivalent of prime minister, and tried to advance a program of industrialization. Ironically, in 1921 Visvesvaraya was called back by the state of Bombay to lead a commission studying technical education. He proposed the establishment of an institution modeled on MIT, but could not win the support of the British members of the committee, who viewed it as too expensive and unrealistic for an agricultural society.[21]

Although Mahatma Gandhi had a vision of India and the Indian people that was too holistic to be reduced to one dimension, he saw hand spinning on a charkha as central to transforming the Indian masses. He envisioned training them in the industrial values of a work ethic, time consciousness, and quantitative thinking, introducing the personal disciplines of industrialization without its hardware. Gandhi's vision of industry in India is often seen as losing out to Nehru's. This book looks at it differently.

As I have described my project, I have been frequently asked, "Why MIT? Why not a British university or some other American university?" And of course during the colonial period and after, Indians did go to a variety of British and American universities for technical training. As a small group of English-language-educated Indians saw their country's position and the process of industrialization taking place in Europe and America, they rejected the role that the British assigned them and sought something more. As these Indians looked at Britain, they saw concern there about British industrial decline, often focused on the weakness of the British system of technical education. They looked elsewhere. MIT began in 1861 as an institution that offered a science-based professional education in industry in contrast to the existing literary and classical education, British in origin, offered by Harvard and other American colleges. An MIT-type education offered the Indian middle class an entry into the world of technology on middle-class terms. Their attraction to MIT was a rebellion against the British system.

From its founding, MIT's supporters had national and international aspirations, seeking wide recognition for their institution and its approach to technical education. While the nineteenth century saw the creation of a number of American universities, none combined

MIT's focus on technology with its global reputation. As early as 1883, MIT's annual report boasted of how European authorities considered at least some specific aspects of MIT's program and facilities the best in the world. MIT's reputation would only rise in the years to follow. As information spread, albeit slowly in the late-nineteenth and early-twentieth centuries, if Indians were aware of one American engineering institution, it was often MIT.

Even as MIT changed in the twentieth century in ways that made it less appropriate as a model for India, first through an alliance with large corporations and then through an alliance with the U.S. government, where it received millions of dollars for research on advanced technologies with military applications, MIT remained for middle-class Indians the pinnacle of the American system of technical education, a source of aspiration. And it was a pinnacle that Indians could reach, helping them develop technical competencies and making them players in the world of high technology. But giving MIT such a key role in the Indian technological imagination, particularly in modeling the IITs after MIT, had its costs. It was an orientation that was more global than local, and paid less attention to India's specific technological possibilities than it might have. It was an orientation that focused on what a few brilliant Indians might do at the top, and paid less attention to empowering the Indian masses. The Kalpavriksha would be very selective in whom it favored, granting the wishes of some but not of others.

1

The Indian Discovery of America

IN 1884 THE *Mahratta*, a Poona-based newspaper controlled by Indian nationalist Bal Tilak, ran a three-part series under the title "Model Institute of Technology," in which it printed excerpts from the annual report of the Massachusetts Institute of Technology and noted its relevance for India. In its analysis of this twenty-three-year-old institution, which was still struggling to establish itself, the *Mahratta* somewhat prematurely asserted that "for ourselves, we are quite convinced that this Institute is the best conducted institute in the world."[1] Shortly thereafter, the publication of the report of the British Royal Commission on Technical Instruction gave the *Mahratta* occasion to compare systems of technical education in England and the United States. The *Mahratta*, recognizing British anxiety about whether it was ceding leadership in science and industry in an increasingly global environment, mischievously and subversively suggested the report's inadequacy to remedy Britain's lagging position in technological education. The *Mahratta*'s editors first allowed that the commission's recommendations were "highly useful at the present stage of technical education in England." However, the condescension contained in that statement became clear in the next clause, which asserted that the suggestions "cannot fairly compare with the masterly regulations of the great Massachusetts Institute of Technology, U.S. America."[2] If machines were to the

British "the measure of men," as historian Michael Adas has suggested, here was the periphery, itself behind in technical education and technology, using its global knowledge to ridicule the metropole on those very grounds.[3]

Britain's colonization of India occurred in stages. In the seventeenth century the British had come as traders, largely content to operate within Indian society as it was. But by the nineteenth century, the British increasingly sought to transform India, introducing British practices into India in ways that were designed to serve Britain's interests. Britain sought to make India an agricultural state that purchased its manufactured goods. The British started English-language schools to train Indians so they could work within the colonial system as clerks and other subordinate positions under British supervision. They built telegraph systems to strengthen their control over the land. They built irrigation systems to open more land to agricultural production. They built railroads, with components imported from Britain, to enable India to more effectively ship its agricultural products to ports.[4]

The way that the British sought to connect India and Britain created tensions, some of which can be seen in the small group of Indians who had received an English-language education. English gave this group access to a vast array of information available worldwide, which they used to understand the world and India's place in it. They saw India's poverty and understood it in a global perspective. They saw Europe and America developing industrially, while India did not. They saw characteristics of Indian society that they saw as incompatible with modern industrial society. But they also saw a Britain that was at variance from the Britain their rulers wished them to see. It was not an omnipotent Britain, but a Britain with a strain of anxiety about its own position in the world, fearful that Germans and Americans were catching up, wondering if British technical institutions were adequate to face global challenges.[5]

In the western Indian city of Poona, Bal Tilak and a group of English-language-educated Brahmins sought to interpret this world to their people and provide them a way forward. His newspapers, the English-language *Mahratta* and the Marathi-language *Kesari*, informed their readers of global trends, such as the increasing role of mechanization in the West, the rise of global trade, and the growing role of science in industry. But it did so in a way that sought to undermine the ideas of English technological and economic superiority that lay at the heart of empire. It framed news in a global rather than a colony-metropole perspective. Britain's status as an advanced country seemed far less secure once the United States and Germany were in the picture. Their economic rise suggested that India's position was a problem to be addressed rather than a fate to be accepted.

The *Mahratta* did not just report on global events: it called for changes in India consistent with what it saw around the world. Specifically, it called for India to establish industrial enterprises based on the latest developments in science and technology to enable it to compete in the world economy. While the *Mahratta* saw many examples throughout the world that India might follow, the *Mahratta* increasingly looked not to Britain but elsewhere, especially to the United States.

To industrialize, the *Mahratta* argued for the necessity of technical education, calling it "India's greatest need." In fact, when Queen Victoria died, the *Mahratta* wasted no time in asserting that a proper tribute to her would be a technical institute. Furthermore, the *Mahratta* called for the reform of Indian society, asking Indians, and Hindus in particular, to develop global bourgeois values. It called for them to interact more broadly with the rest of the world. It claimed that Indians needed to be more enterprising and self-reliant. In a paradoxical way, it saw individualism as the key to building an Indian nation.

Scholars now see the nineteenth century as a distinct period of globalization, both with respect to trade and the flow of information. The pages of the *Mahratta* drew attention to a rich informational environment coming into place in the late nineteenth century. The telegraph was merely one component of an information distribution infrastructure that included steam ships, railroads, postal services, printing presses, books, journals, and newspapers—all revealing an increasing information intensity of European and American societies. Information was frequently accessible through a global network, at times making its way to the *Mahratta* and other Indian newspapers. Once the *Mahratta's* editors got this information, they could interpret it in their own way.[6]

Getting MIT's annual report was one thing; taking advantage of MIT in India was something else. Even though the *Mahratta* could appreciate MIT, the career of Keshav Bhat, the first Indian to attend MIT, would show how difficult it could be to use knowledge from the West to develop industrial enterprises in India. Bhat studied sporadically at MIT beginning in 1882. But his efforts at establishing technology-based industries back in India failed. A cursory, skills-based education in the hands of a single individual provided precious little leverage in the effort to bring industrial enterprise to India. The complexities of technological enterprise were further shown by the fact that the meanings of Bhat's failure were contested. Did it show the futility of sending Indians to America? Was the problem with Bhat? Was it a problem of Indian opposition to an Indian entrepreneur? There could be no definite answer, and while Bhat was soon forgotten, the questions his failure raised would remain.

Poona, the *Mahratta*, and the World

Poona (now Pune) is a city in western India that in 1881 had a population of 130,000. Its original importance came from its status as the seat of the Peshwas, the last major Indian polity in western India, subjugated by the British in the early 1800s. Not coincidentally, thereafter Poona was the site of a large military garrison. After being brought under British control, Poona was economically and politically a tributary of Bombay, the capital city of the British administrative unit known as the Bombay Presidency, which encompassed much of western India. The two cities, separated by one hundred miles and the mountains of the Western Ghats, were joined by rail with the 1863 opening of the Bhore Ghat Incline. Later Poona became the summer residence of the governor of Bombay. In 1885 Poona employed 25,000 people in crafts, with the largest contingents being in metalworking and weaving. Poona's schools and colleges made it a regional education center, attracting young Indians from its hinterlands.[7]

In the late 1800s, a small group of English-educated Brahmins made Poona one of the most activist cities in India. To call their activism "political activism" would not be untrue but would underestimate its breadth. They started a political society. They started a private English-language school. They started newspapers. They started an organization for the promotion of industry. They started businesses. They wrote.[8]

One of the leading figures in this movement was Bal Gangadhar Tilak. Tilak is best known for gathering public support through a series of provocative acts, such as the revival of the cult of Shivaji, the seventeenth-century Maratha warrior, and the establishment of the Ganapati festival. He challenged the British more openly and more aggressively than his contemporaries in the still-nascent nationalist movement, serving multiple prison terms and making an early assertion of India's right to independence. Historian Richard Cashman has claimed that a key part of Tilak's success as a leader was his ability to "speak in different social and religious terms to various groups."[9] One of the terms in which he spoke, not so clearly heard by historians, was technological.

Tilak, born in 1856, was the son of a schoolteacher turned minor educational inspector for the British. Tilak's great grandfather had a more exalted post serving the Peshwas as the administrator of a town, but had to relinquish it when the British came to power. When Bal was ten, he came to Poona with his father, who had accepted a job there. Bal entered the Deccan College in 1873 earning a BA in 1876. At a time when very few Indians had a college education and when the most obvious way to use that education would have been in service to the colonial state, he and a small group of fellow Brahmins chose another path. In 1880 they established a private high school, the New English School. The next year, they founded two newspapers, the English-language *Mahratta* and the Marathi *Kesari*.[10]

The two papers had different audiences and, to an extent, different purposes. The *Mahratta* was aimed at English-speakers, the British as well as Indians in other parts of the country. The *Kesari* had a much more regional audience. The *Mahratta* had the tiny circulation of 460 copies a week in 1885, while the *Kesari* throughout the late nineteenth and early twentieth century had the largest circulation of any Indian-owned periodical in the Bombay Presidency (4,350 in 1885, rising to 17,500 in 1906). The *Kesari* carried large amounts of advertising that subsidized the money-losing *Mahratta*. Both papers consistently argued for the need for the industrialization of India and for technical education.[11]

Even though it was an English-language newspaper with a small circulation, the *Mahratta* could indirectly reach both English-reading and vernacular-reading Indians throughout the country by being read and excerpted by other Indian newspapers. Indian papers, like many newspapers worldwide, operated in a system where they exchanged news among themselves. In its early years the *Mahratta* had a feature titled "Our Contemporaries," in which it briefly commented on reports from other newspapers.[12]

Another regular heading of the *Mahratta* in its early days was "What the World Says," a section of global news. The *Mahratta* regularly carried stories from both British and American periodicals, but the exact provenance of those stories is unclear. It is possible that the system of exchange worked here, but the *Mahratta* specifically mentioned in 1881 that "London friends" had sent it copies of periodicals that had information of relevance to India.[13] The *Mahratta* occasionally wrote of British or American publications coming in the latest mail. Many stories were likely secondary or tertiary references, with sources coming from another paper within a paper the *Mahratta* had received. The Madras-based *Hindu* had a London correspondent and occasionally reports from British papers were explicitly attributed to him. Whatever their source, the *Mahratta* regularly cited articles from a wide range of English publications, including popular publications such as *Nineteenth Century* or *Fortnightly Review*. It occasionally cited pieces from more technical journals, such as *Nature* or *Engineering*. The American journals it cited most regularly were the *North American Review* and *Scientific American*. The range of periodicals the *Mahratta* had access to through this system of secondary and tertiary citation is evidenced in its 1881 reproduction of a column of business advice, which originally appeared in the Philadelphia-based *Confectioner's Journal*.[14]

Rising America, Declining England

In the late eighteenth and early nineteenth centuries, Britain was the home of what historians now call the Industrial Revolution. Technological change came to the fore. While not immediately or uniformly, factories increasingly replaced small-scale production, steam engines replaced animate power sources, machines displaced human skill, and iron displaced wood. Up through the first half of the nineteenth century, Britain had been the unquestioned leader in this process. But by the late nineteenth century, there were signs that Germany and the United States were catching up. Partially this was due to the fact that, as economists and historians have noted, following a path blazed by the leader is easier than establishing a new one. But the United States and Germany had other advantages. Their systems of education were in some ways superior to the British system, which grew increasingly important as science became more central to industry. The United States had vast natural resources and could take advantage of its large internal markets. Whether a true British decline in fact occurred and its causes, if it did, are debated by historians to this day.[15]

What is not debatable is that throughout the second half of the nineteenth century a group of Britons raised the alarm about what they saw as their country's declining position. The editors of the *Mahratta* were particularly attentive to these stories and reported them to their Indian readers. The very first issue of the *Mahratta* carried this prophecy, attributed to the London-based *Contemporary Review:* "'Yankee Doodle' is to be the anthem of the race on whose possessions the sun never sets; and 'Rule Britannia' will survive as an old remembrance of the Queen of the Sea. Chicago will supersede London."[16]

The second issue of the *Mahratta* carried as its lead article "The Gokulas of the New World," a commentary on an article on American cattle ranches from the *Fortnightly Review*. The *Mahratta* stated that "in the economy of the world, what is one man's gain, is often another man's loss. What America gains from her inexhaustible natural resources, is often times so much loss to England and other countries. The unlimited food-producing capabilities of America threaten the agricultural interests of the United Kingdom with ruin."[17] The *Mahratta*'s editors no doubt took pleasure in seeing the United States doing to England economically something similar to what England had done to India.

Most of the *Mahratta*'s sources for describing English economic decline vis-à-vis the United States came from English sources, but in 1883 the *Mahratta* reprinted a letter in which an Indian claimed to have seen American ascendancy with his own eyes, although with a formulaic language suggesting that he was picking up on existing tropes. The letter, carried under the title "A Bengali in New York," which had originally appeared in the *Indian Mirror*, a Calcutta newspaper, stated that "every European feature of life is here intensified in degree, and varied in kind. Activity, courage, enterprise, devotion, aspiration, are the main springs in every action, big or small. When the European walks, the American runs. Where the former hesitates, the latter seizes at once, and is more than half sure to carry off the prize through mere pluck. Wealth is literally rolling in the streets and flowing in the sidewalks. Machinery and science are utilised to every service of man."[18]

The Internationalization of the World

The *Mahratta* examined a wide variety of stories in Indian, British, and American publications covering global events and trends and excerpted them with its own commentary. While the range of stories over the period between 1881 and 1901 was great, a primary theme was what the *Mahratta* called "the internationalisation of the world," describing what today would be called globalization.[19] It used this phrase in 1883 in reporting on how Russians bought flowers in Italy and then shipped them in refrigerated boxcars to Moscow for display in the Kremlin. The *Mahratta* saw this process as a worldwide trend, but was most alive to its effects on India. And to that point "the internationalisation of the world" had been a disaster for India.

The *Mahratta's* most poignant and expansive statement on the effects of internationalization on India came in July 1881 in response to a story in the Melbourne *Argus* on plans to import fruit into India from Australia. This report drove the *Mahratta* to the brink of despair. In a continuation of the process that began with the British displacement of Indian weavers, the *Mahratta* saw Australian apples, apricots, and nuts displacing Indian products. The *Mahratta* half-facetiously feared this was only the penultimate step in the process: the conclusion would come when "New York will undertake for the convenience of natives to supply us with cooked food after taking proper assistance from the refrigerating chamber." The Australian fruit invasion would cause Indian farmers to "take shelter in the gloomy regions of Pluto." The *Mahratta* bitterly and sarcastically concluded that the loss to Kabul and Kashmir of the fruit trade was justified "since the Australians, Americans and others of the civilized world are the fitter races, they only deserve to survive the Cabulis, Kashmeres and ignorant Indians."[20]

Australian fruit was but one case of internationalization affecting India, and the *Mahratta* reported on many others. German dyes were threatening Indian indigo. American kerosene was displacing Indian vegetable oil for lighting. New leather tanning techniques developed in America were supplanting traditional methods used in India. Typewriters were displacing copyists. Australians were making ghee and shipping it to India. In perhaps the ultimate indignity, in 1894 the *Mahratta* reported that the Cashmere shawl, "once the most coveted article of luxury alike in Asia and Europe has now been replaced by worthless stuff." Threats were everywhere. In 1888 the *Mahratta* carried a warning from another paper of the possibilities of English manufacturers applying industrial metalworking machinery to the process of making brassware and capturing this market from the Indian artisan.[21]

Even as the editors of the *Mahratta*, along with other early nationalists, recognized a number of ways that British colonial policies were economically disadvantaging India by draining wealth from the colony to the metropole or denying Indians higher-level positions in government service, the chief problem that the *Mahratta* saw was that India was being left out of a fundamental global transformation. This transformation, mediated by an aggressive capitalism, was privileging industry over agriculture, large-scale operations over small-scale operations, and scientific methods over traditional ones. The *Mahratta* accepted that the world, India included, was living in the machine age. In 1881 it wrote that "a purely agricultural country such as India is, cannot hope to win the race of life" in what it called "the mechanical age of modern times."[22] That same year, in advice to the newly installed Gaikwar (Maharaja) of the princely state of Baroda, the *Mahratta* wrote that "the people should be taught how improved European mechanical contrivances are used to facilitate every little work. They should be taught, in a word, that henceforth they have to live by the products of machinery and not of mere unassisted manual labor."[23]

Given this view, the editors of the *Mahratta* put a very high value on technical education. In 1885 the editors commented on a list of Indians studying in England and were unpleasantly surprised to see that a majority of them were studying law or medicine. The *Mahratta* then laid down a stark statement of India's situation: "Our connection with England has disclosed to us one fact, namely that if we ever mean to rise in this world, we must study science and the industrial arts. We must beat England at her own weapons. England is superior to India, not so much because there are eminent lawyers or eminent physicians, as because there are mills and handicrafts and skilled men to work in both. As long as we do not make any real appreciable advance in these, we must not expect to improve India materially."[24] In 1891, even as the *Mahratta* expressed gratitude for Parsi entrepreneur J. N. Tata's offer to fund scholarships for Parsi youth to study abroad, it expressed disappointment that his generosity was not more focused. "We wish the donor would confine the beneficiary to the study of some industry," the editors wrote, suggesting that careers in medicine, law, or the Indian Civil Service were less worthy of support than careers in industry.[25]

In 1884 the *Mahratta*'s editors wrote that in reviewing a government report on native publications from the previous year, "it was principally the publications under the classes Arts and Science that we looked for with interest and curiosity." However they were disappointed, with only a few publications in these areas. The editors then exhorted their readers, saying, "If we are to rise, we can do so not by reading novels, dramas or poetry. We must be familiarised with the elements of the Sciences. We have to compete with the western nations who have secured so vast an advantage over us in this field."[26]

India in the Technological World

The *Mahratta* expressed great hope in science and technology for India's future. And its content over the years demonstrated that hope. It worked to keep its readers up to date on the latest developments in science and technology around the world. Although Tilak and his movement are not usually seen in these terms, the editors of the *Mahratta* were ardent students of technology who sought to educate their readers on the increasing role that scientific technology was playing in the economic life of the world. In its third issue the *Mahratta* carried on its front page a lengthy description of the newest creation from Alexander Graham Bell, the inventor of the telephone. Bell had first described his photophone, a device that transmitted speech through light, at a scientific meeting in Cambridge, Massachusetts, in August 1880. The provenance of the *Mahratta* article was unclear, but it contained no evidence that it had been modified by the Poona editors. While doubtlessly the editors of the *Mahratta* were experimenting with the material appropriate to its readership, this article presumed a high degree of scientific literacy. The article asserted that the photophone was the next step beyond the telephone, the phonograph, the audio phone, or the microphone, assuming the readers were familiar with each of these devices. In its detailed description of the photophone's operation, the article casually asserted that "every student of science" knows that the electrical conductivity of a metal was a function of the metal's temperature. The article's appearance in the *Mahratta* was an implicit statement that Indians should be familiar with the world of science.[27]

Thomas Edison was a regular figure to the reader's of the *Mahratta*, appearing in its pages at least twenty times between 1881 and 1897. The paper reported on the opening of Edison's Pearl Street Station in September of 1882, where he inaugurated his system of incandescent lighting using centralized power. The *Mahratta* followed American and British papers in repeating far-fetched schemes connected with Edison, as when it reported on a transatlantic railroad tunnel, with Edison ostensibly providing the electric locomotive, and asserted Edison was working on "aerial navigation."[28]

With the death of Louis Pasteur, the *Mahratta* published its most expansive statement of the role of science and technology within the modern nation-state and the world. Paraphrasing and expanding on a statement made by Pasteur, the *Mahratta* wrote, "Men like Pasteur and Edison belong to no one country or people. The whole world claims them as its own and looks to them with eager expectancy for the means of increasing their physical and material well being."[29]

Even as the *Mahratta*'s editors saw the effects of internationalization with alarm, the paper expressed a universal hope in the work of Edison. And while the work of Pasteur could easily be imagined to benefit "all mankind," to state the same about Edison's work, particularly in a country where the peasantry lived in extreme poverty, suggests that "all mankind" really meant a global middle class.

The *Mahratta* accepted the idea of technological progress and wanted to participate in it. In its comments on an 1882 article from the *Times of India* on the electric light, the *Mahratta* stated, "It has always been observed that recent scientific developments give us commodities which combine cheapness and convenience," noting how kerosene was an improvement over coconut oil and gas an improvement over kerosene. It called for an economic analysis of the electric light, seeking its general use if it were cheaper than other means of illumination. It closed its piece calling for an entrepreneur to give Indians (middle-class Indians) the chance to show how forward-looking they were: "Would that some enterprising company undertake to convince the natives of the great utility of electric light and we are sure the labor will not be lost."[30]

Eight years later, the *Mahratta* took the occasion of receiving from "a friend of ours who is now in America" catalogs from Thomson-Houston, one of the leading electrical manufacturers in the United States, to assert that "there is no doubt that there is a great future before electricity," further noting that American papers seemed to announce daily new electrical inventions. In commenting on rumors that Thomson-Houston was going to establish an agency in India, the *Mahratta* expressed its confidence that if it did so, it would have a good business.[31]

Creating Bourgeois Indians

A consistent theme of the *Mahratta* was the need for Indian society to reform itself so it could compete more effectively in the world of global capitalism. Specifically, the *Mahratta* called for Hindus to be willing to travel overseas, and for Indians generally to show more enterprise and to demonstrate more pluck. While the *Mahratta* noted colonial policies it saw as unjust and damaging to Indian interests, such as the tariff or civil service appointments, its editors consistently argued that Indians had to take things into their own hands and help themselves rather than wait for changes in policy. "Will the Natives Be More Enterprising?" published just six weeks after the *Mahratta's* launch, opened by quoting the proverb "Man is the architect of his own fortune." It lamented that it had "notoriously become known to all the civilized countries of the world that India is the poorest nation on the surface of the earth." To change this situation, rather than complaining about the drain of resources from India, the newspaper asserted that Indians needed to become "capitalists and enterprisers." The *Mahratta* put forth a vision of India as a "nation of traders, machine makers and shopkeepers."[32]

The *Mahratta* stood clearly and consistently for the fundamental necessity of Indian engagement with the industrialized world. In 1881 the *Mahratta* rhetorically asked "where is the nation on the surface of the earth, which has made any progress in material industries without keeping large intercourse with the nations in which those industries are found in a developed form." India could not industrialize without contact with the West. However, the strictures on foreign travel accepted by much of the Hindu community made this sort of contact difficult. The *Mahratta* acknowledged that a wealthy man would not send his children abroad for training "unless the social opinion of his community sympathizes with his act and even admires it."[33]

While the *Mahratta* supported overseas travel, it did so in the context of orthodox Hinduism. In 1890 it noted with approval a report of a Bengali man who was planning on traveling to England, all the while maintaining orthodox Hindu practices. He was taking his own Brahmin cook and planning on wearing Indian clothes the whole time. The *Mahratta* observed that the minds of orthodox Hindus who had never traveled abroad "stand in need of opening out," and there was no better means of doing this than travel abroad. Two weeks later the *Mahratta* gave evidence from traditional Hindu texts, including the Ramayana, the Rig Veda, and the writings of Manu, that showed that ancient Hindus had made long-distance sea voyages.

Thus, the prohibition of sea voyages was a later corruption of Hinduism, and so in arguing for foreign travel, the *Mahratta* was not abandoning or even in a sense "modernizing" Hinduism.[34]

In 1889 the *Mahratta* published an extraordinary excerpt from Amrit Lal Roy's reminiscences, which could be seen as a cross-cultural Horatio Alger story to encourage self-help among Indians, both individually and as a nation. Roy was recounting his time in America, when he, though highly educated, was schooled by a New York City boot-black. Roy was penniless and looking for a job. The boot-black, although uncouth, had pluck and that "democratic spirit" that made an American feel "that he is good as anybody." Roy and the boot-black, Jim Martin, could be seen as ideal representatives of their civilizations. Roy was a grown man, well educated, but poor and without a clue about how to get along in the world. Martin was an uncouth teenager, but full of energy and optimism, willing to do whatever it took to raise himself up in the world.[35]

When the boot-black asked Roy if he wanted a shoe shine, Roy said he did, but couldn't afford it. The boot-black generously offered it gratis. When Roy told the boot-black of his troubles in finding a job, the boot-black urged Roy to simply walk into a restaurant, ask for a broom and a shovel, and start cleaning. The fact that Roy had not done so already raised the possibility in the boot-black's mind that Roy's problem was aversion to manual labor. He asked, "Per'aps you want something fine? You don't want to dirty your fingers? Are a dude, eh? But mind you this ain't no country for dudes, not for suckers neither. . . . Every one gits along in this country who's got spunk in him. Take hold of what'e'r you can git and stick to it till you find something better." Roy said that he thereafter thought of the boot-black as his guardian angel. The lesson Roy took from this incident was explicitly one for the entire Indian nation: "Sunk low as we are to-day in the scale of nations, the materials are profuse, and all around us, by taking advantage of which we may rise as high as it is legitimate for men and nations to rise." However, if Indians were to wait for help to come from outside sources, either from "some kindly disposed Englishman, or perhaps the man in the moon," he said, "we shall only go down lower and lower."[36]

Another example of the consonance between late nineteenth-century American culture and the values that the *Mahratta* sought to encourage among middle-class Indians came from an excerpt that it ran in 1881 from the *Confectioner's Journal*, a Philadelphia-based trade publication. This piece had almost surely come to the attention of the *Mahratta*'s editors by being excerpted in *Scientific American*. In the article, Chicago-based confectioner C. F. Gunther, trying to introduce both his products and modern methods of business into the world of the rural American drug and candy store, gave eighty-three pithy reasons "Why Some Confectioners Do Not Make Money." They included:

> They are lazy.
> They hope for fortune to drop in their lap.
> They talk politics too much.
> They make no changes in goods.
> They fail to invent or have new ideas.
> They fail to give loafers the cold shoulder.
> They always stay home and travel not.
> They fail to remember that their art is a science.
> They must wake up to the role of improvement.[37]

Gunther provided this advice in the hope of changing small businessmen, perhaps in rural towns, into modern managers. In republishing Gunther's piece, the *Mahratta*'s editors showed that they thought these lessons in American capitalism would have applicability among middle-class Indians.

The Movement for Industrialization in Poona

In Bombay a diverse group of Indians was building a textile industry; meanwhile, Poona, with little industry to speak of, became the home to what was arguably the largest campaign for industrialization in India in the late 1800s. Paradoxically, this campaign was led by Brahmins who had little experience in business, adapting themselves in a variety of ways as they held on to both their caste and religion.[38]

The most remarkable figure in this movement was M. M. Kunte, whose 1884 speech introduces this book. Kunte, born in 1835, earned a BA from Bombay University. He was the headmaster at high schools in Karachi and Kolhapur before being appointed the headmaster of the Poona High School, the first Indian to hold the position. He was a Sanskrit scholar, producing poetry and studies of Hindu philosophy and Indian civilization, which won him recognition in Europe.[39]

In the mid-1870s Kunte began a program aimed at the industrialization of India. He collected a wide variety of plant and mineral specimens, carrying out experiments on them in search of ones that could be the basis of Indian industries and making them the basis for a private museum. He started a pencil manufacturing operation. Although he was a regular speaker in Poona's summer lecture series, one writer claimed that the goal of his talks was always the same: "to point out to the intelligent classes of his countrymen, not merely by words but by deeds, the supreme importance and prime necessity of starting up industries such as have been, in modern times, developed in Europe and America."[40] Kunte had what the *Mahratta* called "grand visions" of "factories and like depots of trade," covering the entire Poona region, even out to "the Bhambarda and Chatursingi grounds [roughly four miles from the Central rail station]."[41] One writer called Kunte "the Morning star of Science and the Apostle of Industry." Although the *Mahratta* was largely supportive of Kunte, his views were so strong that the newspaper caricatured them as "wanting to see every child hammering away at a piece of iron."[42]

The great European and American international exhibitions of the nineteenth century, ranging from the Crystal Palace Exhibition in London of 1851 to the Paris Exhibition of 1889 and the Chicago World's Columbian Exposition of 1893, were grand spectacles showcasing technology and promoting international trade. In parallel with these exhibitions were less noted ones pursuing similar goals on a smaller scale, including a series of exhibitions held in India. In Poona, citizens formed a committee, which privately organized exhibitions. The grandest one up to that time was held in 1888, with support coming from the Bombay and Bengal Presidencies as well as from many princely states. It was opened on October 6 by the Governor of Bombay, Lord Reay, and ran for two months.[43]

Like other exhibitions of the time, a central part was the display of modern technology. The exhibition hall was lit by electricity, the apparatus having been loaned by the princely state of Kolhapur. After the fair had been opened for two weeks, the organizers tried to draw crowds through a magic show accompanied by a display of electric light.[44]

The exposition's stated purpose was to encourage Indian industries by bringing them to the attention of the public. (Although the *Mahratta* noted its disappointment at the few entries in the First Class division—machinery and tools.) The *Mahratta* claimed that the exhibition vindicated Poona's right to be seen "as one of the few cities that have adapted themselves readily to the changed circumstance of modern times." It also saw the exhibit as a sign "of the industrial era which is dawning just over the horizon."[45]

Lord Reay's speech inaugurating the exhibition demonstrated that he understood his primary audience and what they wanted to hear. He put particular emphasis on the more sophisticated technologies and the role of Brahmins in developing them. Reay praised the fact that those Indians devoted to "the old studies" were beginning to see that they should not neglect "the new ones." In observing Poona's leadership in "literature, philosophy, and older studies," he expressed confidence that Sanskrit scholars "will hail the advent of native Edisons and Armstrongs." If Poona Brahmins were taking a social risk by their advocacy of industry, Reay moved to ensure them that they could transfer their elite status, stating that "leaders in arts and manufactures have a right to a foremost position in all countries."[46]

In 1890 a group composed largely of Brahmins from Poona established the Industrial Association of West India, which that year began an annual conference. By the next year the *Mahratta* was explicitly drawing parallels between the Industrial Conference and the National Congress, the nationalist organization founded in 1885. The *Mahratta* took solace in the fact that while Poona had been denied its planned role as host of the initial meeting of the Congress (because of an outbreak of plague in the city), it had inaugurated the Industrial Conference. The *Mahratta* called the Industrial Conference a "necessary coordinate" to the National Congress, quoting approvingly an unnamed English economist's assertion that "a nation which is not great industrially cannot become so politically." Even though participants from the Industrial Conference mainly came from western India, the *Mahratta* hoped that it would gain the Congress's national audience. The *Mahratta*'s editors, who had been pushing industrialization in India for years, asserted that the advent of this conference marked "the real beginning of the movement."[47] The Industrial Conference, like the National Congress, included both Indians and Europeans, but not, by and large, members of the Indian business community.

A letter in the *Times of India* challenged the Poona Brahmins' efforts to claim authority in this area, mocking their enterprise in the most scathing terms. The correspondent, "Mirror," who wrote after the 1892 Industrial Conference, asserted that the mind of the Brahmin was not "an industrial mind." The Brahmin was too proud to labor, and instead of being like Edison and using his ingenuity to help industry, the Brahmin used his ingenuity to make a living off other people's work. The author asserted that the only products Poona Brahmins had a global advantage in was "bags of wind" and "industrial platitudes."[48]

But David Gostling, a prominent Bombay architect who presided over the conference, came to the defense of the Poona Brahmins. Gostling, who admitted that he had initially resisted the offer of the presidency because of the "curious air of unreality about the whole thing," reported a different picture of the Poona Brahmin than "Mirror" had suggested. Gostling described how he saw a factory, managed by a Poona Brahmin, producing thousands of pieces of brass, copper, and white metal using a powerful machine press to make "articles suited to Hindoo ideas." On the last day of the conference, Gostling met with a group of Brahmins who were completing plans to start various enterprises—soap-making, silk-dyeing, match-making. Gostling asserted that the Poona Brahmin had changed from the stereotype presented by "Mirror"; now he wanted to visit Europe to see the latest industries and he wanted to "develop the commercial instinct."[49]

In 1892 a paper at the conference advocated sending a delegation to the next year's Chicago World's Columbian Exposition, suggesting meeting there with representatives of other nations to discuss developing commercial ties. Although the planned group did not materialize, the 1893 Industrial Conference featured two nights of talks by a Mr. Nagpurkar, who had attended the Chicago exposition. The *Mahratta* praised Nagpurkar for the light he threw not only on the exposition, but on "American and European civilization in general." The *Mahratta* further claimed that Nagpurkar's talk must have led "in not a few cases to the formation of something like accurate ideas with regard to the Americans, and the farthest limit of civilized comfort and convenience to which they have undoubtedly attained, leaving behind even their parent nations of Europe."[50]

MIT and Technical Education in Two Lands

The editors of the *Mahratta* and the *Kesari* were passionate about the need for India's industrialization. As these educated men, who themselves had started a school, thought about the process, not surprisingly they accorded education a central role. Over the first ten years of its existence, the *Mahratta* may well have devoted more space to the question of education, particularly technical education, than any other single subject.

In April 1884 the *Kesari* carried an editorial on "The Need of an Industrial School." It began by giving its diagnosis for why India was in the impoverished state that it was: the decline of Indian craftsmanship and the flooding of the market with cheap foreign goods. The industrial school was the *Kesari*'s proposed remedy. This school would provide children of craft workers with a scientific knowledge of craft skills while spreading Western skills widely. The editorial sought to encourage private donations for the establishment of the school, which would honor a recently deceased professor at Deccan College. Although the authors claimed not to want to prejudge the form the new school should take, they stated their plan to introduce their readers to "the renowned industrial school of Massachusetts in America, which is famous world-wide and has been successfully running for more than two decades."[51]

Following up on the *Kesari* editorial, over the next four weeks the *Mahratta* published excerpts from MIT's annual report along with the newspaper's own commentary on it. How it got the MIT report is unclear. Perhaps it was through Keshav Bhat (who I will discuss in the next section); perhaps through another source. The *Mahratta*'s piece on MIT stated that technical education in India should be developed by examining schools that existed in "France, Russia, Holland, United States in [*sic*] America &C, and adopting the best of them all for our guidance." But the *Mahratta*'s editors were convinced that MIT was "the best conducted institute in the world."[52]

The presumption behind the editors' statement was so great as to be laughable. How could the Indian editors of a newspaper in a small city in India, with no technical background, no training in technical education, and no firsthand knowledge of a technical institute in Europe, the United States, or elsewhere, consider themselves qualified to judge the world's best technical institute? What was this institute that the *Mahratta* was so enthusiastic about?

MIT was founded in Boston in 1861, out of a decades-long vision of William Barton Rogers to establish a school that would provide professional training based in science to those seeking an industrial career. Rogers, who had been a professor of geology at the University of Virginia, moved to Boston in 1853, which he believed would provide more fertile ground for his school, with its textile industry and the city's scientific culture.[53]

At the time, three main streams of collegiate education existed in the United States. The largest was a liberal arts education, grounded in the knowledge of Greek and Latin, originally developed for the training of ministers. This was the education offered at Harvard College, which had itself been developed based on English models. The next were a very few technical institutes, such as Rensselaer and West Point, which offered practical hands-on training for those seeking to work as engineers. Finally there were a few scientific schools, such as the Lawrence Scientific School at Harvard and the Sheffield School at Yale. These schools were small, with the students often "private pupils" of a particular professor, with whatever scientific education the student received being of an unsystematic kind typically geared away from the practical. Compared with the coordinate classical school, the scientific school was, in the 1869 words of MIT chemistry professor and later Harvard president Charles Eliot, "the ugly duckling."[54] Rogers proposed to start a school that would occupy new intellectual territory. It would largely dispense with the classics and offer a rigorous systematic scientific training in the service of industry.

Rogers was able to get enough support from Boston area businessmen and the Massachusetts government to start the school in 1861, but it still faced a number of challenges. First, its founding coincided with the outbreak of the Civil War, hampering fund-raising, so that instruction did not begin until 1865. Harvard provided another challenge. In spite of its weaknesses, the Harvard scientific school became the vehicle for Harvard's attempts to swallow the new school.[55]

America's existing "shop culture" posed a more implicit challenge to MIT. At the time of MIT's founding, many young men got their start in an industrial career by working in factories and workshops, where they got practical on-the-job experience but little formal training. This was the dominant mode of training for industry, received by Thomas Edison and a majority of the men who built American industry in the late nineteenth century. Even with the development of steam engines and telegraphs in the nineteenth century, the utility that a scientific education would provide to a career in industry was far from clear.[56]

It required ambition to attempt to establish a new model of education in the face of well-entrenched incumbents, and Rogers's aims were imperial. In his 1846 proposal for what would lead to MIT, he asserted that his institution "would soon overtop the universities of the land in the accuracy and the extent of its teachings in all branches of positive knowledge."[57] Rogers sought not to build a local technical school, but by claiming this intellectual territory, to achieve preeminence in the United States.

New models are not established overnight, and by the 1880s, if MIT had gained a foothold, it had not brought Rogers's vision to fulfillment. In the late 1800s the boom and bust cycles in the American economy driven by railroad construction and associated financial chicanery at times made it difficult to find students willing and able to pay for an education at the institute. The Morrill Act of 1862 provided each state with support for establishing colleges for agriculture and the mechanical arts. These schools, not as ambitious as MIT, were local and required little or no tuition.

In spite of the fact that the original *Kesari* article had spoken about a school for the children of crafts workers, the excerpts from the MIT annual report made it clear that MIT was not such a school. A primary source of the attraction of MIT to the editors of the *Mahratta* was that it occupied an educational space that was vacant in India. The principle education offered by the British to their caste was essentially a classical and literary one. As they looked at the technological developments around the world, the poverty of their country, and the few jobs it prepared them for, they had seen its inadequacy. To advance through shop culture was not an option for them. The motto of MIT was "Mens et Manus" (Mind and Hand), but it clearly privileged the mind in a way that would have been attractive to the editors of the *Mahratta* and the *Kesari*. The MIT education was technical, but not merely technical. It provided an education that was befittingly broad for those who were to lead their society.

As the editors of the *Mahratta* went through its curriculum, they compared it to Indian universities. MIT required constitutional history, while Indian history curricula "hardly ever went beyond dates and battles." MIT required political economy, which was not required in Indian universities. The students in the chemistry lab were encouraged to do original work, which the *Mahratta*'s editors thought should be introduced in India. The laboratory facilities seemed far better than what was available to Indian students.[58]

The editors of the *Mahratta* saw the integration of science and practice. They quoted from the annual report on how the chemistry program had a laboratory of industrial chemistry that taught such subjects as "the production of dyes, colors, paints and glazes; tanning; glass making; fermentation; distillation; destructive distillation." The report also stated that the school provided facilities for the students' own special investigations "having reference to their future work in life."[59]

The editors of the *Mahratta* believed that this education was what was needed, but also that the British were denying it to them, saying, "At present when the Indian industries have all ceased and it is desirable to revive them under the existing wretched conditions of the Indian nation, it is of the utmost importance to give a strong impetus to practical sciences like the above. It is not the study of English language or of Mathematics so much as that of Industrial Chemistry that we really want. But to our extreme regret we find that we are kept back from it—the very necessary of life—with what purpose we know not, or knowing do not like to disclose."[60]

One of the ironies of the *Mahratta*'s complaints was that the British had in fact not kept Indians back from all technical education. In 1847 the British had established Roorkee College (later Thomason College of Civil Engineering, now IIT-Roorkee). By the 1860s engineering colleges had been established in Bengal, Madras, and Poona. Roorkee was established in conjunction with the construction of the Ganges irrigation canal and its purpose was to train Indians who would occupy lower-level positions in the Indian Public Works Department, particularly in its irrigation department. These engineering colleges were precocious, with Britain at the time relying more on a system of apprenticeship for civil engineering, and only a handful of engineering colleges anywhere predated Roorkee. At the same time the Indian engineering colleges were limited in their scope, designed to produce Indian engineers or engineering assistants who would work on public works projects in service to the Indian state. From the *Mahratta*'s point of view, these colleges were of some utility, producing Indians who could occupy positions in the Public Works Department, but as their discussion of MIT shows, ultimately the *Mahratta*'s editors sought something much more: technical institutes that would produce trained men capable of starting businesses and industries.[61]

The *Mahratta* continued its emphasis on technical education, calling it in February 1886 "the most important question of the day," and in October of that year a subject "of the highest importance to the country." In 1886 the weekly *Mahratta* carried forty-three articles on technical education, more articles than written on just about any other subject. This was not merely an issue to the editors of the

Mahratta. Historian Deepak Kumar has argued that "in the last few decades of the nineteenth century technical education overrode almost every other educational issue" and was "hailed as the panacea for all ills."[62]

The Indian National Congress began issuing resolutions concerning the need for technical education in India in 1887 and passed such resolutions almost every year until well into the twentieth century. Just like the editors of the *Mahratta*, the members of the Indian National Congress had seen that the growth of technical education had seemed to be related to the rise of industrialization around the world and believed that it had the power to transform India from an agricultural to an industrial country.[63]

While the *Mahratta* had since its beginning championed technical education, several factors gave the issue special salience in the Bombay Presidency in the mid-1880s. Viceroy Lord Ripon had retired in 1884, and proponents of technical education seized upon the idea of a technical institute as a means of honoring him. In 1885 Lord Reay was appointed governor of Bombay. Reay had served as the president of the 1884 International Education Conference, where he had emphasized the importance of technical and scientific education over classical education. It was possible to believe (and perhaps the editors of the *Mahratta* did) that Reay had been appointed governor of Bombay primarily to introduce a modern educational system.[64]

After several government reports on technical education that largely endorsed the status quo and enraged the *Mahratta*'s editors, in 1887 Reay provided a concession to advocates of technical education in Bombay. The Queen's Jubilee came in 1887, and Reay announced that he was supporting the establishment of a technical institute in Bombay in honor of the Jubilee. The Bombay Presidency would provide an annual contribution, but the rest of the funds would come from citizens' private donations and from the municipal government. Furthermore, the aims of the institute would be limited, focused mainly on training technicians working in Bombay cotton mills. Although the *Mahratta* was glad that in its words "something will now be done" in an area that it had long been agitating for action, it was unhappy with the scale of the program. It noted sarcastically that Lord Reay's statement that "the most pressing want" of the citizens of Bombay was technical education was belied by the government's allocation of only 25,000 rupees a year to the subject.[65]

In 1889, when Lord Reay formally opened the Victoria Jubilee Technical Institute, both the *Mahratta* and Reay emphasized their areas of accord. Reay suggested a three-level scheme for technical education that had the Poona College of Engineering at the top, implicitly acknowledging Poona's claim to be the center of learning for at least the Bombay Presidency. The *Mahratta* suggested that this marked the beginning of a partnership between India and England, where India would combine its technical skills and natural resources to help England compete with European nations. It wrote, "[I]f India can be given what she wants—and by this we mean technical scientific knowledge—we can show that we are more than a match to England's rivals."[66]

Keshav Bhat

As the *Mahratta* pushed its campaign for technological education and industrial development both in Poona and in India as a whole, two men, Keshav Bhat and Bhaskar Rajwade, became known in the Bombay Presidency for having traveled to America for technical education. And Bhat had the precise kind of education and experience it had been calling for. Keshav Bhat had enough pluck and enterprise to travel across the ocean, attend MIT, and work in a Boston dye house to master the art of dyeing cloth the shade called Turkey red. Then he had returned and opened a dye works in Poona under the pretentious name "Indo-American Dye House Company." However, if the writings of the *Mahratta* on technical education and industrialization had been models of clarity about what needed to be done, the career of Bhat was full of ambiguity. His efforts to start industries in India clearly failed, but why and what those failures meant was uncertain.

The basic outline of Bhat's life can be pieced together from a variety of sources. He was born in 1855, making him a year older than Bal Gangadhar Tilak. Whether Bhat was born in Poona itself is unclear, but he seems to have lived most of his life in the Poona region.

He attended Baba Gokhale's school in Poona. Gokhale, who was one of the first Indians to graduate from the University of Bombay and was recognized as a brilliant scholar, ran a private school that was at the time noted for being outstanding and "able to stand the competition of the missionary schools."[67] Tilak attended this school briefly, and so the two men may have known each other.

One of India's first nationalist institutions was the Poona Sarvajanik Sabha, established in 1870 as a shadow Indian representative institution. In the late 1870s and early 1880s, the Poona Sarvajanik Sabha had a campaign to encourage Indians to go abroad for technical training. A number of people volunteered to go, but the Sabha was unable to raise the money to support them. Bhat, who was in his mid-twenties at the time and not as well educated as some of the others who had applied, was, in the words of a later recounting, "crazy to go to England or America," so he borrowed several thousand rupees, mortgaging his house.[68] Bhat's zeal to go abroad, apparently in defiance of the strictures against overseas travel common in the Hindu community, showed a high degree of individualism.

Bhat first went to England, carrying with him letters of introduction from Bombay merchants. However, these letters were of no avail, and British manufacturing firms' refusal to let him enter their shops foiled his efforts to learn a trade. Bhat seems to have gone to England without knowing anyone, and his later accounts did not mention any Britons who assisted him while he was there. He later said he was seen by Britons "as a spy prying into the secrets of their industries." Frustrated in England, Bhat left for America, landing in Boston in July 1882.[69]

Ironically, Bhat was far better able to take advantage of personal connections with Americans than he had been able to with Britons. He came to Boston with a letter of introduction from Henry Ballantine, the U.S. consul to Bombay. Ballantine's brother, a minister in a Boston area church, received Bhat. Bhat began studying at MIT and simultaneously sought admission into American dyeing works in order to learn the secret of dyeing cotton Turkey red. Bhat failed to gain entry to a dyeing works for five months, a failure he attributed to the fact that he "had no influential friends." It is suggestive of the lack of connection MIT had with the textile industry that it was only through connections back in India that he was able to get training in textile dyeing. A Bombay merchant, B. V. Shastri, wrote to a Boston merchant engaged in the East Indian trade, J. H. Woodford, on Bhat's behalf. Through Woodford, Bhat gained entry into a John Cochrane's Turkey Red dye works in Malden, where he learned the art of Turkey red dyeing.[70]

Bhat's sojourn in America shows the limitations of MIT, both for himself and for America. Bhat was classified by MIT as a "special student," a category that made up 40 percent of the student body. A large portion of the students at MIT were not getting the systematic education MIT's founders had envisioned. And Bhat's association with MIT appears peripheral. At the time MIT did not have a formal program in textiles or textile chemistry, and although MIT offered organic chemistry, everything about Bhat suggests that his purpose was to learn a specific skill, rather than to master a discipline. When Bhat returned to India, his primary claim to fame was that he knew the technique for dyeing cloth Turkey red, rather than that he was educated at MIT. And learning the Turkey red process clearly seems to have come more from working at Cochrane's dye house than from MIT. (Cochrane himself had no college education.) While Bhat was in the United States, the Sabha spent 4,500 rupees on him, redeeming his home from the mortgage and sending funds.[71]

Several ironies exist in Bhat's efforts to learn Turkey red dyeing. By the time Bhat learned the process, the dye could have been made one of two ways. The first was the traditional plant-based dye, made from the madder root. The second would have been dyeing using synthetic coal tar-based dyes developed by German research chemists, which were increasingly rendering the older plant-based dyes obsolete. If Bhat learned the plant-based method of dyeing, it was being outmoded. If Bhat had been using the coal tar-based dye, he would be dependent on the German dye industry—he had not learned the complex chemistry behind synthesizing dyes. What Bhat took technologically from the United States was done ad hoc, rather than based on a strategic understanding of the technologies that were likely to be important in the future.[72]

In 1884 Bhat came home. He quickly gave a report of his sojourns to the managing committee of the Sarvajanik Sabha, stirring controversy. At the meeting, seemingly in an attack on the Sabha, he implied that he had not been properly supported during his time in America and that his pleas for assistance had gone unmet. The *Mahratta* strongly took up his case, asserting that it was only through Bhat's "pluck and determination" that he persevered through his many difficulties. Although the *Mahratta* generally tried to destigmatize manual labor, it bitterly complained that due to the failure of his supporters in India, Bhat had to "work like a common coolie to earn his bread."[73]

Over the next three years in India, Bhat's primary achievement seems to have been convincing doubters that he had really learned something in the United States. Bhat prepared samples of colored cloth and had them tested by chemists and a team of three judges. The judges found that Bhat had mastered the art of dyeing.[74]

In 1887, taking advantage of the enthusiasm for industry, Bhat began a dyeing venture, based in Poona. Its name, the Indo-American Dye-House Company, was an obvious attempt to capitalize on Bhat's tenure in the United States, given that the operation had no other connection to America. This was an early swadeshi factory. It planned to issue a thousand shares of stock, each valued at 20 rupees. The *Mahratta* included an economic analysis in its discussion of Bhat's venture, which purported to show an expected profitability of 15 percent. The *Mahratta* also emphasized that except for a boiler and a hydraulic press, which would be imported from England, all the equipment for the venture would be made in India. By early 1888 the *Mahratta* announced that all the stock had been subscribed, and expressed hope that the factory would soon be operational.[75]

It was not to be, and the problems appear to have been more personal than technical or economic. In September 1889 the *Mahratta* announced that due to disagreements between Bhat and the company's secretary, Bhat had left the company and gone back to America. By December the *Mahratta* was sounding the death knell for Indo-American, stating that the firm's closure would result in the loss of 3,000 rupees, which would be felt by the subscribers, most of whom "were not rich or even well-to-do persons."[76]

By 1890 Bhat was back at MIT, now studying "the principles of electricity." In May, he addressed the Free Religious Association of

Boston, making a spirited defense of Hinduism as well as a statement of India's need for industrial development and America's special role in meeting that need. He asserted that Poona once had 2,000 men working in copper, but the importation of machine-made sheets of copper from England had cost many jobs. Fifty Indian families had made their livelihood by dyeing the color red, but they too had been dispossessed by British technology. Bhat argued that because of British interest in keeping India down, the country's only hope for industrial development came from turning to America.[77]

Bhat argued that the question before his arrival in America was whether Americans would assist Indians to learn technical arts and whether the cold American climate would be tolerable to Indian vegetarians. Bhat claimed that his stay in America had answered both questions in the affirmative and that in India "hundreds of young men are trying to come here, dreaming of America in their sleep." Bhat noted some of the difficulties that he had had in America, including the fact that boarding housekeepers seemed afraid of the color of Indians and suggested the formation of a society that would aid Indian students. Bhat took ill at some point on this stay and only through the help of two Americans was he able to return to India. Whether he had to return to India before he had mastered electrical technology or whether he tried to start an electrical business in India is unclear.[78]

In May 1891 the *Mahratta* reported that Bhat, back in India, had spent the previous half year studying processes for canning fruits and vegetables. He gave the *Mahratta* correspondent a piece of guava that had been canned in an "air tight tin" a month previously. The correspondent found the guava as good as fresh fruit. Bhat's plan was to send samples of canned mangoes to England and America in the hopes of establishing a business. The *Mahratta* suggested "there was every reason to believe" Bhat would develop "a prosperous business," but nothing came of it.[79]

By this time the *Mahratta* could state that Bhat's story was "well known," and it appears to have been an exceptional one. He was one of two Indians who had gone to the United States for technical education mentioned in the Bombay and Poona press. (Bhaskar Rajwade, from Bombay, had been supported by a group of Indians to go to the

United States and study glassblowing.) Both men's lives stood as test cases for the effectiveness of American training for establishing industry in India. In 1885, less than a year after Bhat had returned, the Anglo newspaper the *Bombay Gazette* published an article using Bhat to contest the wisdom of sending Indians abroad for training. It noted that it had cost 10,000 rupees to support Bhat and Rajwade on their ventures to America, but questioned whether India had received any gain. The author claimed that Bhat and Rajwade came back to a "public that is unable to appreciate their craft and a market unprepared to receive their wares." The author called for a more systematic movement for the promotion of industry within India, and the establishment of technical training schools in India, with the teachers, for an initial period, "imported from Europe or America."[80]

A more biting letter to the *Bombay Gazette* ended with the same suggestion. If the *Mahratta* saw America as an example to be followed, some Britons saw America as a land of charlatans. The *Times of India* had regular articles about bogus American medical degrees being peddled in India. A correspondent taking the moniker "Industry and Commonsense" wrote to the *Gazette* ridiculing Rajwade and calling him a "modern day Don Quixote." Industry and Commonsense suggested that the reason Rajwade went to America was that "a shrewd American missionary wanted his passage paid to America" and that this missionary then abandoned Rajwade when they "reached the shores of go-aheadism." He mocked Rajwade for thinking he could learn a trade in months and for the way he went about trying to start a business. In this analysis, Indians needed the paternalism of the British to keep them on a realistic course and protect them from American hucksters.[81]

One of the oddest chapters of Bhat's career came in 1896, when he was the object of a large public meeting in Poona. The stated purpose of the meeting was to express thanks to two Americans who had helped Bhat when he was ill in America five years earlier. Bal Tilak introduced the meeting. Bhat was allowed to tell his story and he attributed his previous failures to a lack of appreciation for his skills and his inability to raise capital. The peripatetic Bhat had settled down and spent the previous two years dyeing cloth in a laboratory under the patronage of the chief of the princely state of Miraj. Bhat's patron expressed total satisfaction with his work, several samples of which were on display. More than anything, the meeting seems to have been an attempt to rehabilitate Bhat.[82]

Bhat next appeared in 1899, this time speaking in his own words to an audience of his fellow alumni of MIT in a letter that was included in the first volume of MIT's *Technology Review*. What can be drawn out of the ambiguity of Bhat's letter is that he saw his time at MIT and Boston positively, but whether it had had a beneficial effect on his career was far from clear. His letter suggested a warmth and depth of connection that one might have expected to come from an extended stay in Boston. He wrote of his reverence for the "very kind American friends and factory owners" who had helped him, citing a Vedic injunction to respect teachers as a father. Bhat further wrote that he had periodically informed his classmates of his doings in India—these were lasting friends.[83]

Bhat wrote that people had accused him of ill-spending his time and money in America—understandable given that he had so few substantive accomplishments to show for his sojourn in the United States. However, Bhat credited MIT with helping him clear his name from these accusations. He did not mention his previous failures. Instead he wrote that he was considering returning to plans to start a dyeing business that had been shelved due to famine and plague in India. Furthermore, he hoped to initiate his son into the business.

His uncertain situation limited his ability to praise his alma mater. He wrote that "the credit of the success that I might or rather wish to achieve" in his new venture would be due to MIT. And he testified that without the "invaluable practical knowledge" he got from MIT, "I should not have been where I am." But his exact position was unclear. Was he a success? Given how coveted an MIT education was to be in India in the late twentieth century, his praise for MIT in double negatives was ironic. "I don't hesitate to call myself a son of Massachusetts Institute of Technology," he wrote.[84]

From this distance it is impossible to make any definitive assessment of Bhat's career. But Bhat clearly shows the limits of what a single person with an American technical education could do in India. Bhat clearly had the gifts of a showman and a promoter, but his career does not show signs of a solid grounding that might have come from a business family (such as J. N. Tata, who will be discussed later) or from being part of a larger organization. He seems to have had the ability to see business opportunities, but not the persever-ance necessary to see them through the inevitable accompanying difficulties. In any case, by 1899 Bhat's career had not proved that MIT had anything to offer India.

A Memorial to the Queen

Given the importance of technical education to the editors of the *Mahratta*, the *Kesari*, and to other middle-class Indians, it was only fitting that the Victorian era in India literally ended with a battle over technical education. Immediately after Queen Victoria died on January 22, 1901, the question arose in India as to how to properly honor her memory. The Viceroy of India, Lord Curzon, favored a museum to be housed in a suitably grand building. The editors of the *Mahratta* met Curzon's proposal with derision, calling it "a huge pile of stone." They argued that rather than serving a sentimental pur-pose, Queen Victoria's memorial should be "thoroughly practical and utilitarian" and that "Technical Education alone in the present cir-cumstances is such a purpose." The *Mahratta* claimed, boldly and not a little presumptuously, "The public have clearly set their minds upon a great Technical Institute as the most fitting memorial to the Queen," and then proposed that the memorial should take the form of either "a great Central Technological College in the heart of India or a net work of Provincial and District Technical Schools."[85]

Curzon was already on record as viewing the Indian campaign for technical education with contempt. Nothing better represents the scorn and condescension with which Curzon met Indian aspirations for technical education than a 1900 speech where he called technical education "a phrase that seems to have an extraordinary fascination for the tongue in India."[86] Curzon's image suggested that Indians were children who liked the sound of a pair of words and repeated them over and over without having any idea what they meant. The United States, Japan, Germany, and other nations of Europe were adults, able to develop and profit from a system of technical educa-tion; India was not.

The *Mahratta* made further arguments in support of technical education as a fitting memorial to Victoria. Although previous proposals for technical education had been rejected with the government pleading poverty, now any funds that the government would contribute would be supplemented by public subscriptions: linking the desire to honor Victoria with the enthusiasm for technical education could be expected to multiply donations. The *Mahratta* implicitly put to the test the Indian government's claim that its previous lack of support for technical education was due to insufficient funds. The *Mahratta* then enlisted the words of Victoria herself in support of their cause, quoting her 1858 proclamation that it was "our earnest desire to stimulate the peaceful industry of India."[87]

Curzon acted quickly to implement his idea of a memorial to the late Queen, announcing pledges of over 3 million rupees in just fifteen days after her death. In a forceful statement of his plan for a memorial to Victoria in a speech in February 26, 1901, barely a month after Victoria's passing, Curzon felt obliged to disparage the idea of using technical education as a memorial. The contempt Curzon had for the enthusiasts toward technical education can be seen in his assertion that "the economic problem" of India "is not to be solved by a batch of Institutes or a cluster of Polytechnics. They will scarcely produce a ripple in the great ocean of social and industrial forces."[88]

Of course, Curzon won this skirmish, and his Victoria Memorial sits today in a prominent place in Kolkata, but it is hard to put into words how galling this episode was for nationalists. Indians paid taxes to support the British Indian government, which had the money to support larger British strategic interests such as launching a military adventure in central Asia, but claimed poverty when it came to technical education in India. Now Indians were expected to donate funds to honor Queen Victoria, above and beyond what they paid in taxes, but have absolutely no say in how those funds were to be used. Curzon's memorial may have been intended to remind Indians of British greatness, but for some it would have had another meaning.

The next year the *Mahratta* commented on British Prime Minister Arthur Balfour's speech at the opening of the Manchester School of Technology. In his speech, large parts of which the *Mahratta* reproduced, Balfour acknowledged that the day of Britain's unchallenged industrial preeminence had "passed never to return again." He asserted that industry had changed from a "prescientific" to a "scientific" stage, and other nations had grasped this change sooner than Britain. Although the *Mahratta* did not note it, the principal of the new Manchester School had visited MIT and claimed to have used it as a model.[89]

The *Mahratta* stated that the opening of the school showed how England was "regenerating its economic industries," the very process that England seemed to be hindering in India. It then went on to suggest a reason for the difference in response between the two countries by quoting an article published in *The Nineteenth Century* and written by Sir John Gorst, a Tory member of Parliament and a leading figure in British education. Gorst asserted that in future global rivalries in "commerce and manufacture," "the uninstructed nations will have to reconcile themselves to being menial servants of the rest of the world, and to perform the lower and rougher operations of modern industry, while all those which require taste, skill, and invention gradually fall into the hands of people who are better taught." The implication was so obvious that the *Mahratta* did not state it directly: the lack of support for technical education in India was a cynical attempt to keep India down, as Britain herself struggled to keep up with the United States and Germany.[90]

At a time when few Indians had traveled abroad, the *Mahratta* presented its readers with an idealized portrait of the world, based largely on reports. That picture had a message for India that was simultaneously pessimistic and optimistic: pessimistic because it saw India as being impoverished by being excluded from a fundamental transformation of the world, and optimistic because it claimed that the United States and other countries provided examples that India could follow to quickly end its poverty. A formula that "technical education" and "pluck" would equal prosperity was a simple one that could inspire action. Even though MIT had done very little in the United States, the *Mahratta*, in searching for a route to industrialization, gave MIT a position far beyond what it warranted.

Keshav Bhat, the one Indian who had traveled abroad, studied at MIT, and then attempted to start businesses in India, belied much of the *Mahratta*'s optimism. He failed and no definite postmortem was possible. Bhat's failure quickly became forgotten to most, and in the new century many more Indians would look to America in pursuit of technical education and industrialization for their country.

2

American-Made Swadeshi

IN 1912 Upendra Nath Roy, a young Indian student of engineering at the University of California at Berkeley, wrote "Engineering Education in America" in the Calcutta-based *Modern Review*, an article containing a wealth of practical information for Indians considering coming to the United States. He provided a list of the leading American engineering colleges, the majors they offered, their admissions requirements, and their tuition fees. But in addition to this practical information, Roy made an idealistic plea for young Indians to come to America and study engineering. Roy stated, "America today leads the world in Engineering. America so to speak is a country made by Engineers." He noted how Chinese and Japanese students were coming to America to study, but "the number of Indians at engineering schools is distressingly small." He laid down a challenge to Indians, saying "we must come to this country, not by dozens but by hundreds."[1]

Although Roy's article was suggestive of the various articles carried by the *Mahratta* in the nineteenth century, it was grounded in a different reality. The new century brought a new phase in the Indian technological nationalist movement, where there was much more doing. Indians started businesses and voluntary organizations to support industrialization, and sent students abroad for industrial training.

These efforts at industrialization were highly decentralized and varied, dependent on local conditions and a wide variety of actors, ranging from India's leading business family to individuals to voluntary associations to princely states. Efforts at industrialization ranged from a small glass factory to a modern steel mill. An important subset of these efforts involved connections with the United States. Roy mentioned a few in his article, such as the assistance J. N. Tata received from American firms in building India's first modern steel facility and in hydroelectric plants built near Bombay. He and other Indians were now studying in American universities in numbers unknown in the nineteenth century. But connections to the United States also existed in cases of small-scale industrialization, likely unknown to Roy. An aspiring entrepreneur used American journals to help launch an agricultural implements business, while an American-educated engineer led a nationalist glassmaking venture. These connections to the United States are not surprising, given its status as a rising industrial nation. However, connecting to the United States inevitably meant bypassing the imperial metropole.

This industrial activity was significant, but hardly of the sort to suggest that a new India was at hand. It did however raise questions: Could industrialization happen apart from the state? Could nationalism and capitalism coexist? Was there a contradiction between a cry for greater individual entrepreneurship and a cry to build up the nation? Could Indians use the United States, a rising technological power but one whose institutionalized and legalized racism extended to India, to aid in their country's technological development? These questions were implicit, when the nationalist movement seemed to move in a different direction.

The Global Indian Entrepreneur: J. N. Tata

As the Poona Brahmins argued for a more industrial, more entrepreneurial, and more global Indian, there was in fact one Indian who served as a prototype of what they called for: J. N. Tata. While Indians—Tata included—had traveled to China to sell opium, and Indians had traveled throughout the Indian Ocean region, Tata had gone from China to England to France, then to Australia, to Japan, and to America in search of business opportunities. Tata, a long-term strategic thinker with the resources to try his ideas, was India's one-person industrial policy throughout the late nineteenth century, working to integrate India into the world of global capitalism.[2]

Tata, born in 1839, was a Parsi, a small religious community that had migrated to western India from Iran hundreds of years earlier. When the Europeans came to western India, the Parsis developed a role as intermediaries between them and the inland Indians. The Parsis proved extraordinarily adaptable to the new circumstances brought by the British and profited greatly from them. As the British moved from the western Indian port of Surat south to Bombay, the Parsis moved with them, becoming Bombay's leading business figures. Unlike Hindus, their religion did not discourage overseas travel and in the nineteenth century a number of Parsis had traveled to Britain to gain business and technical knowledge.[3]

J. N. Tata's father was a trader, and the younger Tata joined him, traveling first to China to manage the opium trade and then to Britain in 1864 when the American Civil War had opened the British cotton market to India. But the younger Tata was determined to move from trading into manufacturing. He was not the first to do so, but what differentiated him was his systematic, long-term strategic thinking, the breadth of his vision, and his insistence in looking beyond India to take in best practices throughout the world.[4]

In 1877 Tata opened his Empress Mills textile factory. Previous cotton mills had been located in Bombay; after much study Tata decided to locate his in Nagpur, in central India, so as to be closer to cotton fields. He hired British experts and, after a false start, bought not the cheapest production equipment but the most efficient. His businesses were marked by continual expansion and introduction of new equipment. He made a mark for himself in India by being an early adopter of the ring spindle, which was capable of operating at a higher number of rotations per minute than existing machinery. And Tata's British experts further improved the ring spindle so it could operate at higher speeds still.[5]

Frank Harris, Tata's biographer, observes that it was "Mr. Tata's practice to visit every important exhibition," and Tata attended at least the 1878 Paris Exhibition, the 1893 World's Columbian Exhibition in Chicago, the 1900 Paris Exhibition, and the 1902 Dusseldorf Industrial Exhibition. The display of the latest technological developments was central at each of these exhibitions, and Tata would have been able to see and evaluate many new technologies from throughout the world. At the 1878 Paris exhibition, Tata would have seen Thomas Edison's phonograph, Alexander Graham Bell's telephone, and electric arc lamps lighting the streets. While many of the visitors to these exhibitions were awed by the spectacular displays of technology, Tata's career suggests that he saw things differently, carefully examining the technologies and considering the business possibilities for applying them in the Indian environment. His consistent presence at exhibitions would have allowed him to judge how the technology had progressed over time.[6]

Tata not infrequently wrote to the *Times of India* proposing new courses of action that would have wide benefits to India. In 1886 he wrote to note that Indian cotton manufacturers' existing market was limited to coarse yarns. He proposed spinning mills based on new technology—the throstle—which would enable Indian mills to spin higher count yarns and thus win new, higher value markets, where Indian mills had previously been unable to compete.[7]

In 1896, in his continuing quest to keep India competitive in cotton globally, he wrote a thirteen-page memorandum urging Indians to conduct experiments on producing Egyptian cotton in India. Tata saw this as another necessary step to produce finer yarn and compete with Japan and Europe. He asserted that efforts by the Indian government had failed, and his memo proposed an independent large-scale, decentralized program that sought to get the educated throughout India to make experiments growing Egyptian cotton. He offered advice to would-be cotton growers and sent out forms to gather data.[8]

In the midst of the drought and famine of 1899, Tata wrote a letter to the *Times of India* showing that for years he had been conducting a global search for methods to improve the supply of water to Indians through artesian wells. The *Times* reported that Tata had "personally examined the famous well at Grenelle, near Paris, the wells used on sheep runs in Australia, some of the wells in America, and a few of those existing in London." Twelve years earlier he drilled an artesian well to a depth of 500 feet at his ancestral home in Navsari in Gujarat. Tata drilled three more artesian wells in the area of Navsari, which demonstrated to local citizens the superiority of his methods. During the then-current drought, he had loaned his apparatus to others interested in drilling wells and sent to London for further materials.[9]

As mentioned in Chapter 1, Tata also dedicated funds to support Indian education. In 1891 he announced a scholarship program, funded by 500,000 rupees, to enable Parsis to study in England or Europe. The next year he announced a similar program that would be available to Indians of any religion.[10]

In the latter years of his life, Tata conceived of and began planning three large-scale projects that had the potential to fundamentally change the technological environment of India: a hydroelectric plant, a steel plant, and a research university. Each involved the transfer of an innovation already transforming the world of Europe and America. Each would necessitate Tata looking increasingly to the United States. But none would be completed before his death in 1904.

As Chapter 1 showed, some English-educated middle-class Indians had seen technical education as India's fundamental need and had been pushing for its establishment with almost no success. From the late 1880s, Tata began considering the possibility of establishing a university in India that would conduct research at the highest level. Beginning in 1896, B. J. Padshah, a key lieutenant of Tata, went on an eighteen-month study trip, examining institutions of higher learning throughout the world, in preparation for the establishment of Tata's university. In 1898 J. N. Tata made a proposal to establish an Indian research university, which would include a branch devoted to scientific and technical teaching and research. Padshah claimed to have gotten the basic idea of a research university from Johns Hopkins and Clark Universities in the United States. The *Mahratta* saw Tata's institute as a virtual declaration of scientific and industrial independence for India: now Indian scientists, engineers, and agricultural workers would be able to develop independent capability to meet Indian needs and to compete in a global market.[11]

Tata did not propose to provide all the funds for the institute, but the very fact that he was promising significant funding meant that the Indian government, which had been so reluctant to establish new ventures in technical education, had to consider this one. The seriousness with which Tata's offer was taken even by the British government can be seen by the fact that the incoming Viceroy, Lord Curzon, met with Tata and a group of supporters of his project almost immediately, the day after his ship from England landed in Bombay and even before he had gotten to Calcutta and formally taken up his position. As shown in Chapter 1, Curzon did not share Indians' enthusiasm for technical education. After the meeting, Curzon praised Tata's generosity and, while denying any hostility to the plan, raised questions about whether there would be sufficient students for such an institution, and then whether graduating students would be able to find employment.[12]

In spite of the speed with which Curzon took up Tata's proposal and his seeming open-mindedness toward it, J. N. Tata would die and ten years would pass before Tata's institute was realized, and then in a shell of its original form. After J. N. Tata's death, wrangling continued between the government and his heirs, with the Indian Institute of Science finally established in Bangalore in 1908. The Institute was continually engulfed in controversy over its management and in its first decades failed to live up to the overly optimistic expectations of the nationalist press.[13]

It was characteristic of Tata's long-term and strategic view that another of his points of focus was iron, a fundamental material of Victorian industrial society. Nothing so clearly showed India's abject industrial prostration than its failure to produce iron or steel. This failure was particularly humiliating since in the preindustrial age, iron produced by India had surpassed European iron in quality and quantity. But in the late 1800s, although India had an abundance of iron ore, it had to import almost all the iron that it used.[14]

J. N. Tata had begun his methodical exploration of iron production in 1882. He spent decades searching for adequate supplies of the proper quality coal and iron ore along with transportation networks to bring them together. In 1902 Tata had reached a point where he traveled to the United States to make arrangements to move forward on an iron and steel plant. The visit also showed his range of contacts in the United States and the breadth of his industrial interests.[15]

Tata visited Washington, where he met President Theodore Roosevelt and those at the highest level of government. Tata was welcomed at a reception whose attendees included Admiral George Dewey and Elihu Root, the Secretary of War, as well as the Japanese and Chinese ministers. Former president of Johns Hopkins Daniel Coit Gilman's attendance would have been of particular interest to Tata.[16]

The purpose of Tata's trip was business, not sightseeing, but Tata had such a wide range of interests that he traveled extensively throughout America during his six-week stay. He saw textile mills in Pelzer, South Carolina, and Atlanta, and iron and steel plants in Birmingham, Alabama, and Chattanooga, Tennessee. He went to Grand Rapids, Michigan, to pick up furniture samples to explore whether similar pieces could be made in India. In Cleveland, he saw hoisting machinery used to unload ships and thought they could be used to advantage in Bombay.[17]

In 1902 the place in the world to go to obtain expertise in steel-making was Pittsburgh, Pennsylvania. In the last three decades of the nineteenth century, Andrew Carnegie had perfected a method of large-scale steel production based on the use of the latest technologies and the relentless pursuit of cost efficiencies. Carnegie had retired from the steel industry in 1901, but many of his men were still in Pittsburgh. Tata visited the steel works at Homestead, which a contemporary said stood "at the head of the steel works of the world."[18]

Tata not only saw Homestead, but he secured the services of Julian Kennedy to introduce the latest American techniques into India. Kennedy had worked for Carnegie, serving as superintendent of the Homestead works and developing and implementing new technologies that had greatly expanded production. After leaving Carnegie, he started his own consulting firm, which built steel mills in the United States and throughout the world. (Kennedy would later win contracts for modernizing British steel plants on American lines.)[19]

While he was in Pittsburgh, Tata also met with George Westinghouse, whose firm pioneered the development of alternating current electric power plants. Tata had a vision of developing a hydroelectric station in the western Ghats of India to provide power for Bombay. Westinghouse arranged for Tata to visit the hydropower plant at Niagara Falls that his company had built.[20]

Like the Institute of Science, a modern iron plant was a reflection of Tata's long-term perspective, taking years to come to fruition. After Tata's death in 1904, responsibility for the steel plant passed to his son Dorab. Finding iron and coal with suitable qualities and in locations where they could be economically transported took years of effort. In 1907, after being rebuffed by the British capital markets, the Tata interests raised 23 million rupees from investors in India, mainly western India, who combined pride in an industrial effort led by Indians with confidence in the Tata's moneymaking abilities. A significant amount of money came from Indian princely states, with the dewan of Bhavnagar earning a spot on the board of directors.[21]

With the capital raised, the Tatas signed a contract with Julian Kennedy for the construction of the iron works, which produced its first iron in 1911. It was the most sophisticated industrial facility in India, and it had been built using Indian capital and was controlled by an Indian board of directors. By 1913 Tata was exporting pig iron to San Francisco, a symbolic triumph for a country many considered to be hopelessly backward but which was now selling an industrial product to one of the world's leading industrial nations.[22]

Even in the midst of this triumph there was tension. The superintendent of the plant was an American and the senior managers were either American, British, or German. A 1913 article in the *Modern Review* applied to the Tata Steel plant a complaint long made against British-controlled enterprises in India: that Indians were excluded from higher-level positions in spite of their qualifications. The author bitterly complained that the hope that the Tata plant would provide an opening for Indian scientific intellect had not been fulfilled. The Indian nationalists' complaint that the Tata plant was still employing foreigners in senior positions continued into the 1920s. Ironically, the Tatas, who had done so much to build up India's technological capabilities, had, in their drive for profits and stability in the iron works, seemed to deny Indians' individual technological capabilities.

Not until 1938 did an Indian, Jehangir Ghandy, educated at Carnegie Institute of Technology in Pittsburgh, serve as the general manager of the plant.[23]

Small-Scale Industrialization in Western India

While the Tata undertakings attracted notice throughout the world, two industrial ventures in western India, the Paisa Fund and Kirloskar Brothers, received no such attention. In contrast to government programs or efforts of business families or other elites, these ventures represent an industrialization from below that suggests the campaigns of the *Mahratta* and *Kesari* found supporters. Moreover, both of these campaigns involved links to the United States, one through its chief engineer, the other through magazines and journals.

The Paisa Fund Glass Works was one of India's first modern glass works. It represented a fusion between an economic nationalist movement and one of India's first American-trained engineers, which held together for nearly a decade before breaking. A Maharashtran school teacher, Antaji Kale, had the idea of promoting the industrialization of India through a Paisa Fund, a fund raised by seeking minute contributions from the masses. (The paisa was one sixty-fourth of a rupee, a very small amount of money.) Initially he planned for the money to go to provide scholarships for students to study abroad. In 1901 he began traveling from one small Maharashtran town to another, giving lectures and seeking to start local committees to collect subscriptions. His lectures traced India's poverty to its industrial decline, asserting that only by a "revival" of Indian industries could the country be regenerated. By 1904 Kale claimed to have visited over 200 villages to promote his scheme.[24]

Although Kale's message was consistent with the campaigns of the *Kesari* and the *Mahratta* to promote industry, it was only after years of work, largely alone, that he won the support of leading figures in the nationalist movement in Maharashtra, most notably Bal Tilak. In a 1904 meeting in Poona, Tilak noted that Kale had continued with his idea even though he had been publicly ridiculed. By 1907 (the year the Tatas raised 23 million rupees for their iron venture), Kale had succeeded in raising 7,000 rupees.[25]

Kale's focus was collecting the money; he had less to do with how the money was raised. Eventually the Central Committee reoriented the program from sending men abroad to setting up ventures in India. The committee decided to make glass its main focus. Glass was a widely used material that had not been successfully made in India using modern methods. Furthermore, it did not require a large capital investment. A nationalist school in Talegaon, a town near Poona, offered the Paisa Fund land.[26]

The fund then hired Ishwar Das Varshnei to lead the glassmaking effort. In hiring Varshnei, the organization hired one of the Indians with the widest industrial training outside India. Varshnei would come to be called the "father" of the modern Indian glass industry. Varshnei, the son of a prominent cloth merchant in Aligarh, was born in 1880. After two years of college he left India around 1902 for Japan, where he studied for two years at Tokyo's Higher Institute of Technical Training while also spending six months working in a factory. He first began working on sugar technology, but then he switched to glassmaking in response to a claim that indigenous glass-making was impossible in India. In 1904 he came to the United States, giving his destination to immigration officers as St. Louis, suggesting an intention to visit the exposition there. He traveled across the United States, visiting several industrial locations along the way, before enrolling at MIT in chemical engineering for the fall 1904 term, enrolling as, what he would call in 1911, "a special post graduate student in glass." During his time in Boston, Varshnei also worked in a factory for three months.[27]

Varshnei is recorded in the MIT records as being a student only in the 1904–1905 school year. In June 1905 Varshnei left the United States for England as part of a delegation of American chemists, attending the annual meeting of the British chemical society and touring the country. He came home after touring factories in England, Belgium, Germany, Italy, and Austria. He spent six months working in a glass factory in Germany.[28]

By 1906 at the latest, he was back in Aligarh. In 1905 he started a glass works in the nearby city of Sikandra Rao, which seems to have failed. The next year he reported to the MIT alumni magazine, *Technology Review*, that he had established a glass factory that was

operating successfully in Aligarh. Varshnei was a serial entrepreneur, and throughout his career he had remarkable, although not
complete, success both in his specific factories and in training others
to enter the field.[29]

It was at this point where Varshnei was recruited to operate the
Paisa Fund glass works. The Paisa Fund had ambitious goals, aiming
not just to start a single business, but to aid in the industrialization of
India. To that end, it sought not only to produce glass, but also to
produce glass workers. Here Varshnei had access to resources that he
never would have had in Aligarh. Varshnei recruited four glassblowers
from Japan, who would initially run the factory and train Indians.
The Paisa Fund ran advertisements in the *Kesari* and other newspapers seeking applicants for its training program. One hundred and
sixty applicants led to sixteen young men selected, which dwindled to
eight actual trainees, with one cause of the attrition being that the
works had taken the unusual step of not observing caste, requiring
everyone to use the same glassblowing implements.[30]

The hybrid nature of the operation as both a factory and as a
training school caused difficulties. How was the factory supposed to
produce a reasonable quantity of goods when its small staff was busy
training workers? Ultimately the solution that the Paisa Fund and
Varshnei agreed to was that Varshnei would lease the plant for five
years and be responsible for production and the practical training of
workers. Someone else would take care of the workers' theoretical
training.[31]

The technological nationalism of the Paisa Fund posed a challenge
for Varshnei and the operation of the glass works. Bal Tilak, one of
the chief supporters of the Paisa Fund operation, was mistrusted by
the British and was jailed from 1908 to 1914 for sedition. Furthermore the Paisa Fund operation was co-located with a nationalist
school, the Samarth Vidyalaya, which was under suspicion by the
British for promoting sedition. (Although the Samarth was in some
ways a Hindu school, its curriculum included Benjamin Franklin's
Autobiography.) In 1910 the Indian government closed the school. The
Paisa Fund was supported by donations, and these donations were
often raised at festivals associated with Hindu nationalism. The Paisa
Fund, with its combination of training and production, never seems

to have been expected to work as a self-sustaining enterprise, and so needed continual contributions. Nationalism and association with Hindu movements were important to winning broad support in Mahrastra: people were unlikely to support it for purely economic reasons. However, the Paisa Fund also needed the support of the colonial government.[32]

The Paisa Fund did a reasonable job of maintaining that balance. Its official reports asserted that its funds would not be used in controversial religious, political, or social matters. Newspapers abstracted in the Bombay government's *Report on Bombay Newspapers* constantly maintained that the aims of the Paisa Fund were solely economic, not political. In 1910 three officials from the government of Bombay visited the glass works and thereafter issued a memorandum encouraging its patronage by the government. The 1911 visit to the glass works by the governor of Bombay, Sir George Clarke, was interpreted by supporters as sanctioning donations to the Paisa Fund. In 1911 the Paisa Fund sent a contingent to the Delhi Durbar that made a glassblowing demonstration before King George. But newspaper reports show the Paisa Fund continued to operate under a cloud of suspicion: one paper ran a letter inquiring if government servants could contribute to the Paisa Fund, while other papers continually protested the nonpolitical nature of the Fund.[33]

Furthermore, the hybrid nature of the operation, where it was a business needing continual public support, encouraged second-guessing of its work by the public. Tilak's *Kesari* showed through satire the kind of constant criticism the Paisa Fund faced. "Why was the Paisa Fund Factory started at place X and why not at Y? Why Glass of all commodities, and not sugar, cloth or matches was selected for manufacture? Why was Ishwardas appointed? Could the furnace not be built facing West instead of East?"[34]

In 1915 Varshnei's contract with the Paisa Fund ended. Varshnei left the works at Talegaon to take over a glass works at Ambala, in northern India, nearer to his hometown. Varshnei had fulfilled his lease of five years and had trained people who would take over the works' operation. The Paisa Fund's mixture of technology, business, education, politics, and religion seems to have taken a toll on Varshnei, and his later ventures were more focused. One of the main accomplishments of Varshnei's time at the Paisa Fund was in training workers, and by 1922 a dozen glass factories in India had been started by Paisa Fund alumni.[35]

A 1916 article described Varshnei's Ambala factory as having a difficult time keeping up with demand, selling its products to railroads and government agencies. Varshnei subsequently established another glass works in the northern Indian city of Bhajoi. In each of his operations, Varshnei combined the factory with a training operation.[36]

MIT per se did not play a critical role in Varshnei's career. His knowledge of glass technology appears to have owed more to Japan than to MIT. When Varshnei's son, B. D. Varshnei, went abroad for training in preparation for entering the glass industry with his father, he went not to MIT, but to Sheffield University in Britain. He then went to the United States and worked for seven months in a glass works.[37]

Varshnei appears to have steered clear of the politics of independence throughout his career. Although Mahatma Gandhi made the practice of "noncooperation" famous throughout India, a large glass works inherently required cooperation with the government. A 1916 article asserted that Varshnei was promised government support to establish a glass works. In 1919, as discontent was spreading throughout India, Varshnei hosted the lieutenant-governor of the United Provinces, who after a tour of the factory offered the hope that his "officers will assist it in any way that is possible." In 1921, as Gandhi's movement swept through India, Varshnei was floating a company to manufacture window glass on "the most improved and modern lines." Whatever Varshnei's exact politics were, he clearly made a decision to concentrate his energy on business rather than politics.[38]

Varshnei's businesses achieved a measure of success, but not such that it made him widely known outside the glass industry. His career illustrates the development of the kinds of skills and businesses necessary for an industrial nation. Varshnei was intent on following global practices and moving into more complex and higher value glass products. His specific goal was windowpane glass. Beginning his efforts in 1923, it was only five years later, with experts brought in from Belgium and with European equipment, that Varshnei's U.P.

Glass Works became the first Indian operation to successfully manufacture windowpanes. By 1932 Varshnei's operation employed a
thousand people, producing about 30 percent of the window glass
used in India.[39]

Another western Indian business, that of Laxmanrao Kirloskar,
got even less attention than the Paisa Fund, but it too was an example
of industrial enterprise from below. Kirloskar's business suggests how
calls for industrialization resonated. Kirloskar was a remarkable individual, with a strong mechanical ability and a willingness to break
with tradition. But his success also owed much to the broader movement for technological and scientific development in western India
and technical information from the United States.

Kirloskar was a Brahmin born in 1869 in Gurlhosur, a small
village roughly 240 miles south of Poona in the Belgaum district
of the Bombay Presidency. His father had retired from a job as a
government surveyor. The younger Kirloskar had a passion for me-
chanics, drawing, and painting. In 1885 he left home to take courses
in Bombay at the J. J. School of Art, where he took two years of
painting and continued on with mechanical drawing. By impressing
the principal of the newly established Victoria Jubilee Technical
Institute (VJTI) with his ability to accurately copy drawings, he got
a job there teaching industrial drawing at the salary of 45 rupees a
month.[40]

Kirloskar's years at VJTI were central to his formation in several
ways. He got access to resources that he would have been unlikely to
have had elsewhere. He began reading American technical publications, such as *American Machinist*, *Scientific American*, and *Foundry*,
subscribing to these publications himself when he had the money. He
took advantage of the Institute's workshop and learned how to operate and repair its machines. He was later appointed by the principal
to teach the course on steam engines.[41]

Kirloskar's interests were in business and he attempted a variety
of undertakings. Some quickly failed, such as an effort to make shirt-
buttons and an effort to make paper and envelopes for physicians'
bills and medicines. He had more success with bicycles, first buying
them in Bombay and shipping them to back to Belgaum for resale
by his brother. In 1897 he and his brother received a patent for a

water filtration system, but there is no evidence that it was ever marketed.[42]

That same year, Kirloskar resigned from VJTI after being passed over for promotion in favor of an Anglo-Indian. He left Bombay and returned to Belgaum. In the more remote Belgaum the direct presence of the British would be smaller and his interactions would be more with Indians. But his move to Belgaum was not a retreat into localism. Perhaps one of the most valuable resources Kirloskar had was his collection of American technical information. Kirloskar continued to subscribe to the American journals even when he had returned to Belgaum, and his brother "meticulously filed and indexed" them. They supplemented these with mail-order catalogs from the United States.[43]

Kirloskar wrote to Stover Manufacturing Company in the United States and secured a franchise for selling windmills, which he sold to wealthy Indian farmers. Kirloskar then saw a fodder cutter advertised in an American catalog that claimed to convert previously unusable plant matter into feed for animals. Kirloskar thought it could be a successful product in India. He bought one, tested it, copied it, and it became a moderately successful product. He then made a plow based on American technical specifications. The plows went unsold for two years, when a wealthy Indian farmer bought them all. He reported back that while they were good in general, their metal tips wore out faster than those of British plows. Kirloskar himself made comparison tests, which bore out the farmer's complaint. He then found information in back issues of *Foundry* on how to harden steel, which led to plows that proved to be as good as the English ones.[44]

Kirloskar's products were able to tap into a larger movement for an improved scientific agriculture in the Bombay Presidency. In 1909 the inaugural meeting of the Deccan Agricultural Association, established to modernize agricultural practices in the area, heard of the increasing popularity of Kirloskar plows and chaff-cutters. In 1910 Harold Mann, a leading advocate of scientific agriculture in India, gave a lecture to the Deccan Agriculture Society. Present at the lecture were "large numbers of the leading landholders" of the Deccan as well as "a considerable number of cultivators," which led to much of the meeting being held in Marathi. Mann argued that for effective dry land cultivation, wooden plows needed to be replaced with iron plows, and he mentioned Kirloskar as making one such plow.[45]

In 1912 Mann wrote an article for *Indian Industries and Power* claiming that among cultivators in western India, farmers' strong desire for improvements would lead them to adopt improved farm implements. He then went on to describe the operation of Kirloskar, stating that the number of plows made had grown from two in 1906 to 906 in 1911, while the number of employees had increased from seven in 1905 to eighty-five in 1911.[46]

Kirloskar was pragmatic. Kirloskar's brother had been a subscriber to Tilak's *Kesari* since its founding in 1881. Although Bal Tilak visited the Kirloskar operation in 1906, the brothers kept their business relatively free of politics. In a 1908 ad in the *Kesari* for a threshing machine, Kirloskar highlighted its endorsement by the Bombay government. For all the talk in Indian nationalist publications about India's need to move beyond agriculture into industry, Kirloskar's business strategy was a wise one that recognized the centrality of agriculture in the Indian economy. Here Kirloskar was likely to find products with a substantial market. And Kirloskar made simple products, relying on hand or bullock power.[47]

Kirloskar not only tried to promote agricultural products, but also a larger program of reform. A 1923 article on Kirloskar in a Mysore agricultural journal noted his efforts to promote social reform within his employees, particularly in refusing to acknowledge caste differences and creating an environment where upper and lower castes comingled. Although nominally a Brahmin, Kirloskar was described as "at his best when wearing his suit and oil smeared clothes and chatting freely with the boys in the shops."[48] In 1916 he started a magazine called *Kirloskar*, and while it was originally started to publicize the company's products, in 1920 it was reoriented toward the general public. A 1961 brochure put out by the company marking its history claimed that the magazine's editorial policy stressed "rationalism, scientific outlook, rapid industrialisation and social progress."[49] One such example was a special issue in 1927 celebrating Charles Lindbergh's flight across the Atlantic, saying "the whole world has an equal claim on this hero" who "has inaugurated a new era when the whole world will be made easily accessible for every individual."[50]

Indian Students in America

Although this work has so far concentrated on efforts to promote industrialization and technical education in western India, a parallel movement existed in Bengal. In 1878 the *Statesman* in Calcutta announced a proposal to form an "Albert Temple of Science," a school devoted to practical science that would be a memorial to Victoria's husband. In 1886 at the same time the *Mahratta* and *Kesari* were agitating for technical education, Pramatha Nath Bose, who had earned a geology education in England and was working for the Geological Survey of India, wrote a pamphlet arguing that if Indians were to develop the new class of science-based industries, they had to have a science-based technical education. He advocated the creation of a science and technological institute. Bengalis also discussed sending Indians abroad for technical training. As in western India, plans and proposals were abundant yet evanescent, likely to vanish before anything concrete had been accomplished.[51]

That changed in 1904. A group of Bengalis formed an organization for the promotion of technical education, the Association for the Advancement of Scientific and Industrial Education of Indians. It sought to raise 100,000 rupees a year, roughly a third of which would fund thirty Indians a year to study abroad. The rest would be used to establish a science-teaching laboratory, to prepare Indians for going abroad, and also to enable those Indians who had returned from abroad to start businesses.[52]

Although the founders of the Association were prominent Bengalis, they had hopes of creating a mass organization. While it began in Calcutta, it sent out delegates to the outlying areas in hopes of establishing committees in every district of Bengal. Membership required payment of four annas a year, a figure that the Allahabad *Indian People* (itself with an annual subscription of 112 annas) said could be readily afforded by the "poorer middle class," who "frittered away" that much every month or possibly every week. The *Indian People* further went on to recommend that "millions of leaflets in the vernaculars" be spread throughout the country announcing the organization and that a team of volunteers be established to canvass for subscriptions.[53]

In a certain way the Association did get broad support. Maharajas from princely states were prominent among the attendees at early meetings and by August had committed to contribute to scholarships totaling 550 rupees a month. By that same time delegates had gone out to outlying areas and established committees in twenty-nine districts. One village school proposed a tuition surcharge that would then be contributed to the Association.[54]

The organization emphasized economic rather than political goals, achieving these goals through self-help rather than government assistance. Particularly striking in this regard is a speech at a May 1904 meeting given by Bipin Chandra Pal. Pal later would become famous, along with Lala Rajpat Rai and Bal Gangadhar Tilak, as one of the troika of political revolutionaries "Lal, Bal, and Pal" seeking independence for India. But here he focused on economics and industry.[55]

Pal noted the efforts that Indians had previously put into literary and intellectual pursuits but claimed that these would amount to nothing without the advancement of industrial education and an increased focus on economics. Although some had previously claimed that the "new civilization" of Europe led to physical decay, Pal asserted just the opposite. From his travels in the United States and Europe, he observed that Europeans and Americans were physically larger than Indians and seemed to live longer. Although this was not the infamous doggerel that Gandhi had grown up hearing, about how "the mighty Englishman He rules the Indian small" because the English ate meat, it was a relative of it, linking the European economic system to physical strength and health. He described hearing Harvard economist Edward Cummings speak in Boston and quoted in agreement Cummings's assertion that nations had their basis on material and physical advancement.[56]

Perhaps an inevitable accompaniment of the movement's ambition was a naïveté about the process of industrialization. While speakers asserted that India needed to lessen its dependence on agriculture, and that too many Indians were absorbed in the world of literary, legal, and clerical work, they were often speaking about their very own backgrounds: their own wealth may well have been based on agriculture and they may have worked as lawyers, clerks, or authors. While many had doubtless read about the process of industrialization, few understood it in any systematic way. Few had traveled outside of India, but even seeing an industrial society was no guarantee that one understood it in any detail or what it would take to industrialize India. These limitations can be seen in Pal, perhaps the most well-traveled and cosmopolitan of the speakers. He expressed the hope that if they could find the means of sending six students a year to New York, "they would learn all the industries of the place and then return to teach the people the same."[57]

Exactly what Pal meant by "all the industries" of New York was unclear, but he appears to have been thinking of modern industry in a preindustrial way: that an industry involved a narrow set of skills that an individual could master. It was still several years before American engineer Frederick Taylor would utter his famous dictum, "in the past man has been first, in the future the system must be first," but the development of complex technological systems had made the statement increasingly true in the United States. While even many Americans would have thought of industrialization in terms of individuals such as Carnegie, Edison, or Westinghouse, it was increasingly about the hundreds and thousands of skilled workers, university-trained engineers, scientists, and managers working within a complex technological system. The Association saw its goal as to emulate the success of its European and American brethren, but in 1904, that was not something that could be done by sending a few people to the United States.[58]

The Bengali events were propelled to a significant extent by the proposed and then announced partition of Bengal. Beginning in 1903 British officials had publicly considered the partition of Bengal into two states, a plan officially announced in 1905. One of these states would be primarily Hindu, the other primarily Muslim. Bengalis, Hindus in particular, saw this as a violation of the integrity of their homeland and an effort to weaken Hindu influence in the state. Hindus in Bengal responded with the swadeshi movement, a program of boycotting British goods in favor of merchandise manufactured in India. The swadeshi movement was a highly amplified version of the preexisting drive for industrial development that had existed in Bengal and western India. It gave new and more powerful political meaning to the act of seeking industrial education abroad. A school for technical education was established in Calcutta. A range of publications and institutions supportive of industrialization were established as part of the movement. However, the swadeshi movement opened up a wide range of avenues for activism, from education to political violence.[59]

One of the new institutions born in the swadeshi period that renewed the call for technical education was the *Modern Review*, a monthly journal edited in Bengal by Ramananda Chatterjee. Although it had a small circulation (5,000 in 1918), it came to be recognized as one of India's leading publications. It was the primary outlet for English-language translations of the great Bengali writer Rabindranath Tagore. Like the *Kesari* and the *Mahratta*, a significant part of the *Modern Review* was devoted to science and technical news as well as calls for the industrialization of India. From its content, the *Modern Review* argued that the modern Indian should be one who could appreciate the poetry of Tagore, but also articles about "Legumes as Nitrogen Gatherers," a scientific article written by Tagore's son, educated at the University of Illinois, or "Electricity in the Byproduct Coke Industry in the Tata Iron Works." What made the *Modern Review* different from the nineteenth-century *Kesari* and *Mahratta* was that it could draw on Indians who were studying abroad (usually technical subjects) as correspondents.[60]

In 1905 the Association for the Advancement of Scientific and Industrial Education of Indians supported seventeen students to study abroad, a number that increased to forty-four the following year, and then to ninety in 1907. One of the most important contributions of the Association was not in the specific numbers of students it sent abroad, but in establishing an atmosphere where going abroad was possible both physically and socially. The leaders of the Association understood that challenging the strictures against foreign travel held by some Hindus was fundamental to its success. In a meeting held six months after the Association's establishment, the president claimed to have solved "the much vexed problem of sea voyage" by getting the sanction for such travel by a variety of orthodox Hindu pandits. The Association's send-off for students in 1906 started at the Kali Temple in Calcutta, where a Hindu priest blessed the scholars and expressed the hope that "Hindu society" would accept the scholars on their return (a hope that suggested some uncertainty).[61]

This first significant movement of Indians abroad for technical training took a question that had been discussed in hypothetical terms for years by the Indian nationalist press and made it a live one: Where should these Indians go for technical training? The Association's original materials had mentioned Europe, Japan, or America in an off-handed way, but there were political implications to these choices. England was an obvious choice, but it also had problems. Indians had established a path to England to study law and to prepare for the Indian Civil Service examination. Indians had a variety of ways of making connections with English schools. The movement to send Indians abroad was an implicit protest against the Indian government's lack of action, and it was paradoxical to make that protest by sending students to England. To those who followed technical education, it was a common complaint in the early part of the twentieth century that the English denied Indians access to factories for fear of competition. Furthermore, those with the most sophisticated knowledge would have known the increasingly widespread view that England lagged behind the United States and Germany.[62]

Japan was almost literally diametrically opposed to Britain. It was the Asian power that had maintained its own sovereignty and industrialized by adapting Western technology, but had kept the process under its own control. The year 1905 marked, at least symbolically, the end of Japan's tutelage, as it defeated Russia and demonstrated that Asian powers were not always destined to remain prostrate before European ones.[63]

In 1896, following Japan's defeat of China, the *Mahratta* said that "Japan and everything Japanese" had been "exerting a strange fascination on our minds." The *Mahratta* suggested that Indians could now use Japan as a training school instead of looking to Europe or the United States. A 1908 history of Indians in Japan gave 1898 as the date Indian students regularly began going to Japan. In 1900 one of the Indian students in Japan wrote a letter giving basic advice for

Indians considering coming to Japan for studies that they could apply to "industrial development of their own land." He suggested that the biggest challenge was the language, which he estimated would take four months to learn. He estimated living expenses at a modest 50 rupees a month, and asserted that "the Indian students of Tokyo will be very glad to receive their brethren from India."[64]

A young Indian student later recounted arriving in Japan in 1906, shortly after its victory over Russia. "Our admiration for Japan in those days was boundless. We looked upon every Japanese as a hero. Had they not helped to kill the spectre of the 'foreign devil' in the Orient for good? Therefore we were overjoyed to arrive in Japan at the moment when they were celebrating their victory."[65]

But for all this enthusiasm about Japan, what is striking is that the student, Rathindranath Tagore, Rabindranath's son, did not stay in Japan but continued on to the United States. Exactly why Tagore made the decision to forego Japan for the United States is unclear; a number of Indians did study in Japan, but Tagore's path was a common one, to the United States by way of Japan, and the United States became the more important center for Indians seeking technical studies. The main factors militating against going to Japan were the difficulty of the Japanese language and the fact that in going to Japan, Indians would not be going to the apex of technological knowledge. In 1902 an Indian wrote from Tokyo telling those who had the means to support themselves in America or Germany "not even to dream of Japan." He claimed that six months would be wasted learning Japanese; moreover, Japan, still in its "infancy," could not teach a student as much as Germany or America. A 1909 report on a meeting of Indian students in Japan concluded with a recommendation that Indians going abroad should go to Britain or the United States.[66]

In the early 1900s, Indians went to a variety of countries for technical study: Japan, Britain, Germany, and the United States, among others. However, in the pages of the *Modern Review* and other Indian periodicals, Indians, mostly those who were in the United States, made the case for going to that country for technical training with an unmatched ardor. In 1913 the *Kesari* wrote that Indians seeking a specialized education should go to the country that excelled in that specialty. It identified Germany as the place for producing scientists, but the United States as the place for producing technologists, stating that those "who want to secure good technical education should make it a point to go to America." In 1911 a correspondent for the *Modern Review* wrote, "We like our energetic young people to come in hundreds and thousands to American Universities. Nowhere is there more opportunity."[67]

Since the sixteenth century, America has had its boosters who have written tracts urging people to travel to this new land and make a new life for themselves. Sarangadhar Das deserves to be classified as one such booster, but what he promoted was not the American physical environment, but the American social environment. And his goal was not permanent Indian settlement in America, but for Indians to use America to re-create themselves and then to return and to create a stronger India. In 1911 Das, in two articles in the *Modern Review*, told in the most idealistic terms why Indians should come to study in America.[68]

Das, who had been a student at the University of California at Berkeley but was working at the time he wrote, based his article on the roughly sixty Indian students he knew on the Pacific coast. These young men were in many cases twice failures, having failed in their studies in Indian universities, then having gone to Japan, where they had encountered further difficulties. But they had redeemed themselves in the United States.

Das asserted that the failures of the students in Indian universities were a reflection on the poor environment of Indian universities— Das called it "enervating." But average students in India had met with great success in the United States. Das could only imagine the success that India's best students would have.[69]

Das's article implicitly referred to the British stereotype of effete and lazy Bengalis, but rather than claiming that the stereotype was not true, Das asserted that the American environment had allowed Indians to transform themselves, breaking free of the stifling restrictions that had been imposed both by Indian tradition and colonial oppression. In India, Das, a high-caste Hindu, had been an "idler" and had learned to "hate every kind of manual labour." When Das came to America and had to work to support himself, he was fired from several jobs because he lacked the simplest of manual skills—he did not even know how to use a broom. However, he grew to accept and even revel in manual labor. He cleaned toilets and served as a day laborer in a sugar mill, bragging about his muscles as "strong as iron bands."[70]

Das concluded with "A Call of Duty to Young India." Indians needed to emigrate to the United States "because we have been satiated with all kinds of servility and we long for manliness. . . . We emigrate to enable ourselves 'to develop the hidden resources of our continent.' We emigrate to learn to make the things 'right in India instead of importing them from foreign countries.' We emigrate in quest of knowledge." He urged Indians "to come here annually by hundreds," whether or not they had money.[71]

Das's article unwittingly suggested the tension between Indian nationalism and individualism. While the purpose of Indians going to school in the United States was to build up the Indian nation, most of the Indians who were there had gotten there by disobeying their elders. Das was narrating the collapse of the Association for the Advancement of Scientific and Industrial Education of Indians. Most of the Indians in the United States were largely self-sufficient, receiving little or no support from India. From Das's article, Indian students reveled in their self-sufficiency, even as they violated traditional Indian habits.

Das's own experience illustrates this. He was sent to Japan with scholarships amounting to 25 rupees a month from both the Association and from a maharaja. Das became frustrated with his education in Japan and asked his patrons for support to go to America, but was refused, being told he would get no money if he went to America. Das went anyhow but had to find work in America. In America, he met and married a Swiss woman, Frieda Hauswirth. Das had an inspiring story of personal transformation. But what was less clear was how Das and the other students would be able to apply in India the technical skills and personal lessons they had gained in America. Could this personal transformation, even if experienced by hundreds or thousands of Indians, change India?[72]

Das received several dozen queries asking for additional information and in response wrote a further article, published in the December

1911 *Modern Review*. In contrast to his previous one, this article, "Information for Students Intending to Come to the Pacific Coast of the United States," focused on the practical. Who should come? (Almost anyone, but they may need to attend high school in the United States.) What clothes should they bring? (Not the "English styles," which look "ludicrous" in "Yankeeland.") What is the best university? (To Das, there was no doubt that it was the University of California.) How should the immigration inspectors' questions be answered? (When asked if you believe in polygamy, say "no.") Das addressed himself to those students who planned to work to support themselves, claiming that for them, the West Coast universities were the best. They offered cheap tuition and abundant opportunities for work in an area not crowded with immigrants. Das offered to meet at the dock those students who sent him their arrival details.[73]

In 1912 a group of Indian students in the United States formed the Hindusthan Association of America, putting a formal structure around Das's work. It started off with fifty members in 1912, but was up to 150 by the end of 1913. One of its first initiatives was to write circular letters to Indian periodicals providing information and offering practical assistance to students who would come to America. Members from New York and the West Coast offered to meet incoming students from India.[74]

The students also sought at least the moral support of elites in both the United States and India. They wrote letters to dozens of prominent Americans, mostly educators, explaining the Association and seeking permission to make the addressees honorary members. By 1915 the Hindusthan Association had a remarkable roster of American honorary members, including Jane Addams and presidents, chancellors, or deans at MIT, Stanford, Chicago, Brown, Clark, and Purdue, among other schools. Several wrote endorsements, which no matter how generic were used by the Association in its publicity. Richard Maclaurin, president of MIT, asserted his willingness to do anything in his power to help the Association's work, while the president of Brown, W. H. P. Faunce, a Baptist minister, said that American students could learn much from Indian students, with the East and West being like the left and right hands, both being needed for completeness.[75]

The Association also had endorsements from prominent Indian nationalist leaders such as Dadabhai Naoroji and Sarojini Naidu. Its most powerful yet poignant endorsement came from the poet Rabindranath Tagore, who in 1913 became the first Indian to win the twentieth century's highest mark of global intellectual achievement, the Nobel Prize. Earlier that year he was in Chicago and gave a series of lectures, a portion of the proceeds of which went to aid the Association. He had come in part to support his son, who was hoping to complete a doctorate in agriculture at the University of Illinois.[76]

Tagore had a long-term vision of revitalizing India using Western science and technology while retaining Indian culture. Since the 1890s Tagore had been trying to start small businesses on his family's estates. He had initially been a supporter of the swadeshi movement, but had withdrawn when he saw it was not following his lead of creating a new distinctive system of Indian education and economic development, and instead had been captured by forces drawn to simplistic appeals to emotion, nationalism, and violence. He made a statement to the Hindusthan Association saying that Indians needed to "walk defiantly in the wide world in pursuit of modern scientific and industrial knowledge to modernize our ancient and glorious land."[77]

Asia's Mixed Welcome from America

In coming to America, Indians were coming to a country that was increasingly attracting foreign students. In 1912 an early survey of foreign students in the United States compiled by the U.S. Bureau of Education found almost 5,000 regular foreign students in American universities. The major global hinterlands were Canada, the Caribbean, and Asia. The large Canadian and Caribbean delegations could be explained by propinquity. Asian students often came to the United States for technical education. China sent 549 students and Japan sent 415 (with the two countries sending more students to the United States than all of Europe). India and Ceylon together sent 148 students.[78]

While Chinese students had come to study in the United States since the middle of the nineteenth century and continued on through the twentieth century through individual initiative and the support of American missionaries, it was ironically an antiforeign movement in China that helped bring about a systematic increase of Chinese students in America. In 1900 a group of Chinese nationalists, with support from the Chinese government, had attacked foreigners and foreign property in China, in what became known as the Boxer Rebellion. After Western countries quashed the attacks, they forced the Chinese to pay heavy reparations. Theodore Roosevelt signed a bill in 1908 calling for excess reparations due to the United States to be used to fund Chinese scholarships to study in the United States and also to establish a school in China to prepare Chinese students to come to America for further study. In 1905 there were 130 Chinese students in the United States. By 1914 there were 847.[79]

The Indian nationalist press regularly reported on Chinese efforts to send students abroad for technical training, with equal measures of praise for Chinese efforts at nation building and concern that India was lagging behind. In 1909 the *Modern Review* wrote that after a breakdown in relations between China and Japan, the Chinese were planning on sending 2,000 students to the United States, showing "it was not to be left behind in the race for modernization." In 1910 the *Modern Review* carried an article stating that the "exodus of a great number of Chinese students to foreign universities for acquiring knowledge in sciences and arts" was "one of the most remarkable events in contemporary history."[80]

Compared to Indian students, Chinese students were more often found in private universities on the East Coast. This was partially due to the fact that Chinese student had outside funding and could afford to pay higher tuition. The Chinese students were heavily but not exclusively concentrated in engineering, with that field comprising one-third of all majors in 1914. That same year, when MIT had 114 foreign students, forty-six of them were from China. (The next largest foreign contingent was fifteen Canadians.)[81]

MIT's attitude toward foreign students merits special consideration. MIT was from its start an institution with ambitions that stretched far beyond Boston. Given the location of his initial supporters, it would have been unseemly for William Barton Roger's institution not to have had "Massachusetts" in its name, but MIT's founder had a national vision from the start, predicting in his early conception of the institute in 1846 that it would "soon overtop the universities of the land."[82] MIT's special sense of its manifest destiny, that it offered something distinctive, kept it going through financial scarcity and Harvard takeover bids. As early as 1883, the MIT annual report was reporting comparisons that showed it aspired to compete with the best universities in Europe.[83]

The president of MIT's annual report often took notice of the distance from which it attracted students. In 1910 it stated that "the number of students coming from a great distance gives some measure of the reputation of a school and is specially significant for this Institute whose fees are relatively very high." MIT was competing with true land-grant colleges that charged very little tuition. The 1912 annual report included data on the geographic center of its student population, showing a movement westward that paralleled the country's as a whole.[84]

By the end of the twentieth century, after having subdued a space between the Atlantic and the Pacific, the United States became a global imperialist power by taking the Philippines. In an analogous fashion, MIT, if not having subdued the continental United States, at least having won students from much of the country, began looking abroad. MIT's global reach might have had the effect of serving American imperial interests, but those at MIT would have seen it as serving a vision of technological education.

In his 1901 annual report, MIT president Henry Pritchett addressed the question of why MIT, at the time facing both constraints of finances and space, should be educating those outside Massachusetts and the United States when such students might later prove to be competitors to those in the Commonwealth. Pritchett's response demonstrated the standards that he set for MIT, writing that no institution of higher education "which undertakes to limit its ministry to its own section or community is worthy of the highest devotion." Pritchett went on to state that an institution that "seeks to serve the purposes of widest education and of the highest training, best serves its own community when it serves best the citizens of the whole country and the whole world." On a more practical level, Pritchett claimed that the "attendance of students from abroad is the best barometer we have of our own alertness and fitness."[85]

While MIT had a few foreign students in the nineteenth century, by 1910 it was calling itself "one of the most cosmopolitan institutions in the world," and two years later bragged that its foreign student population of 100 out of 1,611 gave it a proportion of foreign students that was twice as great as any other American institution.[86]

To maintain and strengthen its position, MIT recruited globally. In 1910 it appointed a prominent alumni, Jasper Whiting, a "special commissioner" charged with examining ways MIT could attract foreign students and thereby "extend the range of its influence."[87] Whiting then proceeded on a nearly yearlong trip to Japan, China, and India on behalf of MIT. In Japan he worked to create a permanent MIT alumni association. In China Whiting had a range of American and Chinese contacts. The Chinese head of Ching Hua College, the college established with the Boxer reparations money (and later known as Tsinghua), was "fully acquainted" with MIT, and he and Whiting discussed what the Chinese wanted from an American education.[88]

Although Whiting went to India, the MIT archives contain no record of his activities from India and no evidence exists of any activity stemming from his trip. India was not the fertile field for Whiting as Japan and China had been, and the expectations there were much lower. India did not have an existing base of alumni that Japan had or the supporting infrastructure that China had. India's British rulers did not have the interest in industrialization that the Chinese and Japanese did, and Whiting seemed to have lacked Indian contacts.[89]

One of the ironies of Indians and other Asian students coming to America in the early twentieth century was that for all the Indian students' enthusiastic talk about America and for all their American allies spoke of the joining of East and West, America as a country did not love the Indian students back. America in the late nineteenth and early twentieth century was a white man's country, where racial discrimination took many forms, some subtle and some not. While anthropologists said that Indians were of the same race as Europeans, to most Americans and in law, Indians were not white. As Indians studied engineering in the United States, and as Indian periodicals reported back in India on their exploits, many Americans thought of Indians as fundamentally different from them, and certainly not a technological people.[90]

In 1917 the *Chicago Tribune* made a withering criticism of Indian society based on its upper classes, asserting that they were manifestly unfit for self-rule: "You see Japs studying engineering abroad, studying finance, studying trade, studying law; Hindus never. At Oxford and Cambridge they devote themselves to philosophy and belles-lettres and become exquisitely polished young highbrows. Not one of them understands railroads, or banking, or manufacturing, or commerce or government. An India run by such dreamers and swamis would perish miserably."[91] Of course, the *Tribune's* editorial contained many errors: many Indians did study engineering abroad as well as law. But the errors show how much more powerful stereotyped images were than facts to the editors of what styled itself "the World's Greatest Newspaper."

World War I led to a rise in anti-immigrant sentiment in the United States, which found expression in a 1917 law barring Indians and most other Asians from entering the United States. Students, who were not considered immigrants, were excluded from this ban. In 1923 the United States Supreme Court ruled that Indians were not eligible for American citizenship, as they were not white under a common understanding of the term.[92]

In June 1922 the MIT Executive Committee recorded its concerns with how U.S. immigration policy threatened its vision of American technological supremacy. It stated that the attraction of foreign students was "of great importance to the United States in extending and consolidating her newly acquired international leadership." However, overly harsh immigration laws annoyed incoming students and "lessened their enthusiasm" for America. The resolution therefore called for a modification in the immigration laws to exempt students from its provisions.[93]

In spite of the concerns of the MIT Executive Committee, an enthusiasm among Indians remained about coming to America for education. A 1925 article in the *Modern Review* by an Indian emphasized that in spite of the laws, there was "no difficulty" for Indian students to enter America. He again emphasized India's need for technically trained personnel and stated that America offered "the best field" for study and was the only place that offered "exceptional facilities" for research work.[94]

Back in India, the colonial government continued its lack of enthusiasm for technical education. In 1922 a committee headed by Mokshagundam Visvesvaraya, one of India's most eminent engineers, made its recommendations on technical education for the Bombay Presidency. Visvesvaraya's program for expansion of Bombay's technical education had as its centerpiece the establishment of a new College of Technology, modeled after either MIT or the younger Manchester College of Technology, which itself owed a debt to MIT. Visvesvaraya's proposals, which had the support of the other Indian members of the committee, were vetoed by the British majority on the committee, who claimed such an expansion of technical education was not needed.[95]

In the first decades of the twentieth century, the United States and India were unlikely partners for technical interactions. While the industrialized United States was a country of unprecedented wealth and popular education, agricultural India was a country of widespread poverty and illiteracy. However, a group of middle-class educated Indians, infinitesimally small compared to the larger Indian population, working largely outside of the colonial state, used American technical resources for the industrialization of India. Given the limited resources Indian students coming to America had available, it was the western land-grant colleges, not MIT, which were most attractive. By 1922, with the rise of Mahatma Gandhi and his mass movement seeking swaraj for India, not only MIT but the whole subject of technical education was, by all appearances, irrelevant for India. Or so it seemed.

3

Gandhi's Industry

THE TECHNOLOGICAL and economic bent of the Indian nationalist movement was increasingly eclipsed by a political nationalism in the early twentieth century, a process that seemed to be
complete when Mahatma Gandhi came to the fore in 1919 and built
a mass political movement. From that point on nationalist leaders
appeared to believe that political power took precedence over economic and technological development. However, these appearances
were deceiving. Between the years 1915 and 1948, two lawyers from
western India waged a tacit struggle over their very different visions
of India. The struggle between these lifelong friends and associates
would seem to have been a very unequal one: one was one of the
best-known figures in the twentieth-century world, able to enlist
millions of people in his movements, while the other, unknown
even to this day, spent most of his life in the small princely state of
Bhavnagar. But it would be the vision of Devchand Parekh, the
obscure lawyer, that would prevail, both in the official policies
of the government of independent India and in the aspirations of
the English-language-educated middle class, more than the vision
of Mahatma Gandhi.

Devchand Parekh's vision was simple: to renew India by sending
Indians to the Massachusetts Institute of Technology for training
and have them come back and start industries. His was a more clearly

enunciated version of what the *Mahratta* and *Kesari* had called for in the nineteenth century. He pursued it quietly, through his family and in his state. No manifestos were written or demonstrations made, but between 1927 and 1940, six of his sons, sons-in-law, or nephews went to MIT, while six other young men from his state also attended.

Gandhi stands as the twentieth century's foremost critic of industrial society, with his loincloth, his commitment to the archaic charkha, and his frequent fasts standing in opposition to societies based on mass production. And if Gandhi seemed to mock industrial society, industrial society returned the favor. Gandhi could be seen as the stereotypical untechnological Indian, the Hindu ascetic, who had renounced earthly things and was content to live in poverty while nurturing his spiritual side. Parekh's MIT and Gandhi's Satyagraha Ashram would seem to be perfectly diametrically opposed institutions, with one devoted to changing the world through the relentless application of technology and the other devoted to the application of "truth force."

But that view of Gandhi and his movement lets the symbols Gandhi created obscure the reality of the man and his movement, for while Gandhi opposed modern industrial society, at the same time he had a vision that was based on a different kind of industry— personal industry. In Gandhi's work promoting handspinning, he combined preindustrial hardware, the charkha, with the values of industrial society, including work ethic, time-consciousness, and quantitative thinking. In his efforts to transform India by seeking to voluntarily introduce industrial values to 300 million Indians while keeping them in their homes, Gandhi may qualify as the most audacious engineer in a century of audacious engineers.

For many a dream is an idle reverie or a vague hope, but this was not true for either Gandhi or Parekh. The passion and energy that each man put behind his vision made the inevitable unfolding of a different reality all the more difficult to endure. Gandhi had hoped to co-opt industrial values in the service of his nonindustrial nation. Instead, for some young Gandhians, it was Gandhian values that were co-opted, as they headed off to MIT as preparation for building an industrial society in India.

The Divergence of Parallel Lives

Mohandas Gandhi, better known to the world as Mahatma Gandhi, was born in 1869 on the Kathiawar Peninsula of India, where his longtime friend Devchand Parekh was born two years later. Among the factors historians have emphasized as being central to Gandhi's life are his birth and early history in Kathiawar, his family's background as ministers to princely states, and his education in the United Kingdom.[1] Each of these would be important for Parekh as well. Devchand Parekh was born in 1871 and grew up in the small town of Jetpur. In the nineteenth century the Kathiawar Peninsula of Gujarat was composed of several hundred princely states. The British directly controlled much of India, but not all of it. In some areas they allowed princely states to remain, ruled by a local prince or maharaja, but with a British resident in place as a monitor who would ensure that British interests were being served. In princely states the direct hand of the British was felt less keenly, local rulers had certain levels of discretion, and more opportunities existed for Indians to occupy administrative positions.[2]

Devchand's father, Uttamchand, was a lawyer who worked in the royal court of Jetpur. The princely state of Jetpur consisted of the city of Jetpur and 143 villages, ruled by sixteen talukdars or princes. Family history asserts that Uttamchand came into his fortune when one of the talukdars was unable to pay his fees. Instead Uttamchand received the revenue rights to two villages for twelve years, which happened to be years of good monsoons. Following this he built a medi, a mansion, the only two-story house in Jetpur not owned by talukdars.[3]

The lives of Uttamchand Parekh and his son Devchand uncannily paralleled those of Karamchand Gandhi and his son Mohandas. They lived in the same milieu, the small princely states of Gujarat. While both fathers had lacked formal education, each family saw that the son needed a formal English-language education to thrive. When Devchand enrolled at Cambridge, he gave his father's profession as that of a native State minister, which was the profession of Karamchand, and the two fathers may have known each other.[4]

The Gandhis had a longer tradition of service to the princely states than the Parekhs, and Karamchand Gandhi served the states of Porbandar and Rajkot, which were much larger than Parekh's Jetpur. Karamchand's death, when Mohandas was only sixteen, upset the family and thereafter Devchand consistently fared better materially than Mohandas, whose hard-pressed family situation limited his options. He went to the local Samaldas College in Bhavnagar, but left after the first term, whereas Devchand went to Elphinstone College in Bombay, where he would have been part of an elite group of students. Mohandas's family concentrated its resources on his education as a final effort to prevent the family's decline, whereas the Parekhs were moving up and many of Devchand's siblings received a college education. Mohandas had to struggle to raise money to go abroad, while Uttamchand appears to have been able to fund Devchand himself. Although both Mohandas and Devchand went to England to study, Devchand went to Cambridge, while Gandhi essentially studied on his own in London.

In 1893 Devchand entered Cambridge University. Gandhi, who had come to England five years earlier, wrote movingly in his autobiography of his difficulties in adjusting to life in England. However, several factors would have made Parekh's situation easier. Parekh, at twenty-two, was three years older than Gandhi when he came to England. Parekh, having studied at Elphinstone College in Bombay, would have doubtlessly been more comfortable with the English language and had more experience dealing with English people.[5]

While Parekh's family had the money to send him to Cambridge, he faced his own challenges in the foreign culture. In his first three terms Parekh studied as a "Non-Collegiate Student," meaning that he was not associated with any of Cambridge's constituent colleges and lived in a hostel, where he avoided paying some of the fees required by the colleges. Perhaps Parekh did this to conserve funds, but he may also have done so to have greater control over his diet. (As a Jain, he was a strict vegetarian.) After a year he was admitted into Peterhouse College where he graduated with a second-class honors degree in the Moral Sciences Tripos in 1896. He received his calling to the bar from the Middle Temple in 1897.[6]

In 1893 both Gandhi and Parekh, each now outside of India, would have encounters with modern Western industrial society that would shape the course of their lives. Gandhi's came through the railroad, a technology that would be a standardizing force throughout the world. While the railroad brought people together, when Gandhi was evicted from a train in Pietermaritzburg South Africa for presuming to sit in the first-class coach reserved for whites, the lesson he learned was of separateness. He could be educated in England, dress and act like the English, but he would not be accepted on European terms. In retrospect, Gandhi spent the rest of his life cultivating that difference, no longer trying to be like Europeans, but building an Indian identity.[7]

Parekh's transforming event took the form of a more positive challenge and carried with it not the message of Indian difference to be accepted and cultivated, but of sameness—that there was a single path to salvation to be followed by Indian, English, and American alike. Parekh's life-changing event was a meeting with the great English economist Alfred Marshall. According to Parekh's son, while Devchand was at Cambridge, Marshall counseled him that Indians should not be coming to Britain to study liberal arts and become lawyers; instead they should go to America—specifically to MIT—to study engineering and then return to India to set up industries that would improve the Indian standard of living. In response to this advice, Parekh went to the United States in 1893 to visit MIT and began a correspondence with MIT officials to receive catalogs.[8]

Although much of the testimony of Devchand Parekh's son (ninety-five years old in 2008, when interviewed by the author) cannot be verified directly, indirect evidence strongly supports the outline of his account. Students in Parekh's curriculum would have heard lectures from Marshall in political economy. Marshall was known for his openness—he set aside two afternoons a week in which any member of the university could call on him at home.[9]

Although no correspondence between Marshall and Parekh survives (and there may never have been any), a 1910 letter by Marshall is suggestive. In the letter Marshall wrote apparently to a B. B. Mukerji of Lucknow University: "For twenty years I have been urging on Indians in Cambridge to say to others: 'How few of us, when we go to the West, think of any other aim, save that of our individual culture? Does not the Japanese nearly always ask himself in what way he can strengthen himself to do good service to his country on his return?' "[10]

Earlier in the letter, Marshall had written in praise of J. N. Tata, the great Indian entrepreneur, saying the country could use a "score or two" of men like him. But Marshall maintained a pessimism about India, writing, "[S]o long as an Indian who has received a high education generally spends his time in cultured ease; or seeks money in Indian law suits—which are as barren of good to the country as is the sand of the sea shore—nothing can do her much good." In an earlier letter, Marshall wrote: "I do not believe that any device will make India a prosperous nation, until educated Indians are willing to take part in handling things, as educated people in the West do. The notion that it is more dignified to hold a pen and keep accounts than to work in a high grade engineering shop seems to me the root of India's difficulties."[11]

These all echo what Marshall is alleged to have said to Parekh. One might note multiple levels of irony in Marshall's words. First of all, Parekh need not have gone to Cambridge to get that advice; he could have gotten it in India by reading the *Mahratta* or *Kesari*. Furthermore, Marshall was complaining about a lack of industrial spirit in India at precisely the time when some historians have noted a decline in industrial spirit in England, which was increasingly being overtaken by the United States and Germany. Finally, an heir of Adam Smith and a colonizer was urging Indians to rise above individual self-interest and think of the good of the country when an appeal to narrow self (or group) interests had been one of the main strategies of the British in colonizing India. Marshall was no Indian nationalist, and he showed traces of racism in his dealings with Asians.[12]

Marshall was familiar with MIT and its approach to technical education was consistent with his way of thinking. In 1875 Marshall went to the United States, where he spent two weeks in Boston, hosted for part of the time by Charles Eliot, the president of Harvard and former professor of chemistry at MIT. Marshall was a regular correspondent with his fellow economist and the president of MIT,

Francis Amasa Walker. In 1886 Walker wrote Marshall telling him of the opening of MIT's school year and that he would send Marshall an MIT catalog so that he could see "how unlike an English University is a Yankee School of Technology."[13] In Marshall's *Principles of Economics,* he described a failed effort of his at Bristol to introduce a technical education program of several years duration based on alternate six-month periods of studying science and six-month periods of working in workshops, an approach consistent with MIT's.[14]

Although it is not possible to verify that Parekh visited MIT in 1893, strong indirect evidence exists to support this claim. Passenger manifests show that in June 1893, Devchand Parekh disembarked in New York City from the liner *Paris.* The presence of Indians in the United States was so unusual that an Indian visitor was often commented upon in newspapers. In July 1893 the *Washington Post* ran a story describing the visit of Parekh and a companion, Golkuldas Geria, to the American capital. Parekh's companion observed that the main difference he noted between America and India was the widespread use of machinery in America. The report suggests the two had a nationalist bent: they described conditions under British rule and then stated, "Now that is what you would call tyranny, would you not?" Parekh and his colleague had come to Washington from Chicago, where presumably they had been to the World's Columbian Exposition.[15]

Shortly after Parekh's return to India, he had an experience more in line with Gandhi's, showing the difficulties an Indian had in being accepted in a European-dominated society. It started out as the seemingly perfect opportunity for him. An eight-month furlough for O. V. Mueller, professor of history and political economy at Elphinstone College Bombay, had created a teaching vacancy, which Bombay officials intended to temporarily fill with Parekh. In writing back to the India Office in London for approval, the government of Bombay met resistance. First, a check of Parekh's credentials at Cambridge told little about him other than he was a "quiet man, was liked in college, and his conduct was in every way satisfactory."[16]

But the deeper problem was that some in the India Office believed that Indians were fundamentally not suitable to teach subjects such as history or political economy. The strength of this belief led to five months of correspondence between the India Office and Bombay officials over an eight-month appointment. Sir Charles Lyall, a longtime administrator in India who was now working back in the India Office, wrote "that the ordinary run of educated natives have very perverse notions of history, and still more so of political economy."[17] Lyall and others considered these positions ones which should be reserved for those from England. The secretary of state for India ultimately overruled these concerns and appointed Parekh to the temporary position. During his time at Elphinstone, Parekh would have likely come into contact with those who shared Lyall's ideas and were not afraid to put them into words. In December 1900, after holding the position for less than five months, Parekh resigned his position.[18]

Exactly why Parekh resigned is unclear, but by 1902 Parekh was back in the Kathiawar Peninsula, a smaller stage than Bombay, where his interactions with the British would be fewer. His largely private life can be traced through its intersections with Gandhi's more public one. In August 1902 Gandhi, back in India after nine years in South Africa, wrote to Parekh, vividly describing his difficulties in finding legal work in Bombay: "I am now free to lounge about the High Court letting the Solicitors know of an addition to the ranks of the briefless ones." In November, as Gandhi prepared to return to South Africa, an option that was made attractive by his lack of work in Bombay, he hurriedly invited Parekh to join him in South Africa. Parekh's more successful legal career in India is suggested by the fact that he did not go with Gandhi.[19]

While Gandhi's career would take him from the Kathiawar Peninsula, making him a champion of independence for all of India, Parekh's career would keep him in Kathiawar. And while Parekh worked to fulfill Gandhi's vision, he also had a vision of his own—of a nation industrially transformed and independent. Devchand had studied law and economics, but would encourage his brothers to study for careers related to industry. All his brothers stayed in India for their education, some earning degrees in chemistry at Bombay University. Brother Vallabhdas earned a master's degree and won a gold medal for his studies before teaching chemistry at a local college.[20]

In 1908 the brothers established a laboratory in their Jetpur home. Shortly thereafter Devchand met Prabhashankar Pattani, the dewan or prime minister of Bhavnagar. Bhavnagar was a much larger princely state, just thirty miles from Jetpur. Pattani offered the Parekhs twelve acres of land at a nominal lease for ninety-nine years, and based on that support, the Parekhs began a chemical works in Bhavnagar. The Parekhs began their business by canning mango juice using canning equipment purchased from America and by making papain from papaya fruit for use in India and abroad. A later history of the venture called Devchand "the guiding spirit" of the venture while Vallabhdas was the "scientific brain."[21]

The major themes of Parekh's professional life were support for Gandhi and support for technological development in India. When Gandhi returned from South Africa in 1915, he was more a son of Kathiawar than a son of India. Kathiawar was not only his home, but also the home of his closest allies. After being received in Bombay for a week upon his arrival, Gandhi spent the next two weeks in Kathiawar. His next stop after his hometown of Rajkot was Parekh's hometown of Jetpur, where he stayed with Parekh and the citizens of Jetpur honored him at a reception held by Parekh.[22]

In his first year back from South Africa, Gandhi spent a disproportionate amount of time in Kathiawar, and Parekh was with him on a number of occasions. British surveillance noted Parekh's presence with Gandhi in Bagasra on December 12, 1915, while Gandhi's diary for 1915 specifically notes Parekh's presence three times, designating him with the Gujarati honorific title "Devchandbhai."[23]

Given how much time they spent together, they must have talked about Parekh's passion for industrialization. Another associate of Gandhi, Shankarlal Banker, recounted that he and Parekh spoke with Gandhi about their ideas for the industrialization of India during the 1915 meeting of the Indian National Congress in Bombay. Parekh and Banker may have hoped that Gandhi would make their cause his cause, but Banker recounted that Gandhi said nothing, but listened patiently.[24]

In that first year back, Gandhi famously traveled by third-class rail throughout the length and breadth of India, getting to know the famous (Nobel Laureate Rabindranath Tagore) and the common, and making his first steps to becoming an all-India figure. Gandhi also moved his base of operation from Kathiawar to Ahmedabad, the large Gujarati textile center. Ahmedabad, unlike the princely states of Kathiawar, was directly controlled by the British and much more integrated with the rest of India, making it a more promising center of a national movement. In 1917, as Gandhi sought funds for his ashram in Ahmedabad, Parekh was one of the first few dozen people he sent an appeal to.[25]

A picture of Gandhi and Parekh together in Jetpur in 1915, with Parekh in English clothes while Gandhi wore traditional Kathiawar garb, suggested how far the two men's paths had diverged, with Gandhi increasingly giving up the British for the Indian, while Parekh had not. Gandhi and Parekh had been longtime friends and they would continue to be so. They shared much, including Indian middle-class sensibilities. But events after 1915 would increasingly draw them apart. Physically, they would be drawn apart by Gandhi's movement onto the national scene, as he moved his base of operations first to Ahmedabad and then later to Bombay, Wardha, and New Delhi. They also drew apart intellectually, for while one pillar of Parekh's professional life was support for Gandhi, the other was his vision of an industrial India. But their future paths notwithstanding, the two had things in common.

Gandhi as Engineer

Technology would be a central part of the India of each of their dreams. For Gandhi that technology would be khadi, cloth made from thread spun on the age-old charkha, the handspinning wheel. The charkha was such an archaic technology to Western eyes that it seemed to be an antitechnology. But Gandhi was not so antitechnological as he might have seemed.

Indian historian Sumit Sarkar has observed that "varied sections of the Indian people seem to have fashioned their own images of Gandhi."[26] Sarkar was referring particularly to illiterate villagers who might base their image of Gandhi on rumor or very limited information, but Sarkar's point holds more broadly. Gandhi drew his inspiration from such a wide variety of sources, and wrote so widely to such diverse audiences and for diverse reasons using a range of languages that it was possible for people from diverse backgrounds to find a Gandhi whose ideas matched their own. This accounts for the wide range of supporters Gandhi had: from Indian business tycoon G. D. Birla, to illiterate Indian peasants, to Indian Brahmin intellectuals like Jawaharlal Nehru, to English liberal ministers like C. F. Andrews.

Those who had close contact with Gandhi on a daily basis may well have developed different views of Gandhi and what "Gandhianism" was than those who experienced him from a distance only through his writings and media reports. In 1932 Gandhi admitted as much to a correspondent: "The writer of articles in *Young India* is one person, and the man whom the inmates of the Ashram know intimately is another. In *Young India*, I might present myself as one of the Pandavas, but, in the Ashram, how can I help showing myself as I am?"[27]

One example of the different guises Gandhi could appear in comes from Richard Gregg, an American lawyer, who in the late 1920s spent two and a half years in India loosely associated with Mahatma Gandhi's movement. In 1928 he wrote *Economics of Khaddar*, an engineering and economic analysis of Gandhi's program of handspinning. Gregg, who had once planned to be an engineer, wrote from the perspective of "good business men," citing Henry Ford as one of his authorities for sound principles. Gregg asserted that the mass of unemployed Indians were "engines" and that "Mr. Gandhi proposed to hitch them to charkhas and thus save a vast existing waste of solar energy." Gregg concluded, "Mr. Gandhi seems to be, in effect, a great industrial engineer."[28]

Suffice it to say, Gregg's has not been a common interpretation of Gandhi. Mahatma Gandhi, both through his words and the symbolic persona he created, stands as the twentieth century's foremost critic of modern industrial society. His 1909 *Hind Swaraj* contains an uncompromising critique of machinery, calling it "a great sin," "a snake hole," and just plain "bad."[29] More powerful than his writing, read both then and now by only a relative few, was his image, seen by those who attended his rallies as well as those far and wide who saw his picture or his image on a newsreel. By wearing a loincloth after 1920, Gandhi created a symbol that, as Emma Tarlo has written, showed his rejection "not only of the material products of Europe, but also of the European value system with its criteria of decency."[30]

Despite such statements and symbols, which seem to allow no room for compromise with Western civilization, scholars have noted that a central aspect of Gandhi was his hybridity, combining the Indian and the Western, in a dialogical fashion. In his program of civil disobedience, satyagraha, while Gandhi asserted its essential Indian nature, he also acknowledged debts to such thinkers as Thoreau, Ruskin, and Tolstoy. David Hardiman has asserted that Gandhi held to an "alternative modernity," while Robert Young has called it a "countermodernity."[31]

Gregg's assessment of Gandhi and Gandhi's connection with Parekh draws attention to another aspect of Gandhi's hybridity, in which he combined diverse elements into a new whole that could be recognized (at least in part) by people from a wide variety of backgrounds. Gandhi appropriated the language and, in some ways, the actions of engineering, a field that was increasingly coming to the fore among an Indian middle class.

In 1967 noted South Asian scholars Susanne and Lloyd Rudolph observed striking similarities in the personal characteristics of Gandhi and those of the prototypical American engineer and man of the middle class, Benjamin Franklin. In asserting the limits of such binary categories as modern/traditional, the Rudolphs noted that Gandhi shared habits with the paradigmatic modern man, including time-consciousness, a work ethic, a passion for organization, and attention to financial details. The Rudolphs assert that "neither man proposed to let the control and mastery of his worldly environment escape him."[32]

Historian Lewis Mumford's assertion that the clock is the key machine of the industrial age provides a suggestion to locating Gandhi, given that one of his few possessions when he died was a watch. The Rudolphs observe that he "employed his watch as a species of tyrant to regulate his own affairs and the lives of those associated with him."[33] In one of the most striking comparisons between Franklin and Gandhi, they reproduced side-by-side both men's daily timetable. Both men had routines they kept to throughout the day, with

Gandhi managing his day down to a finer level of detail. While Franklin only directly regulated his own life, Gandhi imposed his time-discipline on all who lived at his ashram, through a series of bells not dissimilar to the bells that regulated factory workers.[34]

Max Weber argued that fundamental to the economic rise of Europe and the West was the establishment of a work ethic. A regular British criticism of Indian society was that Indians, especially those in the upper strata of society, lacked a work ethic. In Gandhi's autobiography, he recounts reading John Ruskin's *Unto This Last* on a train in South Africa in 1904 and from it forming the idea that the "life of labour" was "the life worth living." Gandhi's life was from this point marked by a passion for manual labor, whether it was ironing his own collars, washing his own clothes, covering excrement, or spinning yarn.[35]

Gandhi's assertion of the positive value of manual labor for people of all social statuses had very strong resonances with British economist Alfred Marshall's prescription for India described earlier. In a 1925 speech to Indian science students, Gandhi said: "Those who go in for this class of education or for higher education are drawn from the middle class. Unfortunately for us and unfortunately for our country, the middle classes have almost lost the use of their hands and I hold it to be utterly impossible for a boy to understand the secrets of science or the pleasures and the delights that scientific pursuits can give, if that boy is not prepared to use his hands, to tuck up his sleeves and labour like an ordinary labourer in the streets."[36]

One of the fundamental transformations of Western society was the rise of the bureaucratic organization. In spite of the fact that Gandhi is often thought of as a follower of the light of his own personal conscience, he was very much an organization man. Gandhi was born into a hereditary caste; he spent most of his life as a member and shaper of voluntary organizations. When he had a new cause to advocate, he typically developed a new organization to support the cause. He refashioned organizations to make them operate more effectively. Gandhi was an organization builder.

With Gandhi's initial political involvement in South Africa, he created his first political organization, the Natal Indian Congress.

Although he consciously chose the name to suggest similarities to the older Indian National Congress, Gandhi's organization was more sophisticated and more effectively designed as a vehicle for action. The Indian National Congress met once a year to pass resolutions; the Natal Indian Congress met monthly or weekly and did things. Gandhi struck those off the membership rolls who failed to pay their dues or who missed meetings. He used the Natal Indian Congress as a vehicle to train South African Indians in the protocol and etiquette of operating as part of an organization.[37]

Although Gandhi's entry into all-India politics was most visibly seen through the Rowlatt Satyagraha and the Noncooperation Movement, he simultaneously modernized the organizational structures of the Indian National Congress through a new constitution. The Rudolphs detail a remarkable set of changes Gandhi initiated in the process of reengineering the Congress into an effective mass organization: restricting the size of the delegations, regularizing processes for selecting delegates, making provisions for a professional staff, and making divisions of the Congress contiguous with linguistic divisions in India, rather than British administrative divisions. This latter organization of the Congress had hobbled it by practically requiring that those in Congress know English to bridge linguistic barriers.[38]

Gandhi maintained a lifelong interest in the politics of Kathiawar, his home region in western India. But as Kathiawar was geographically isolated with a unique set of political issues, Gandhi could devote little time to it after he became a figure on the all-India stage. One of the ways that Gandhi and Devchand Parekh worked together after 1915 was that Parekh served as the secretary of the Kathiawar Political Conference and was Gandhi's emissary to Kathiawar politics. However, Parekh was tempted to rely on his famous friend for the organization's success. In 1929 Gandhi rebuked the conference for being overly dependent on him: "If you cannot do without my presence, it is better that the Conference is not held. . . . I am trying to free even Devchandbhai from this addiction and I wish to tell him that he should not abjectly believe that the Conference cannot be held without me. If we wish to organize the people, we should have the capacity to do so without any man, however great or talented he may be. . . . When Kitchener died, the Government machinery did not come to a halt, the Empire did not perish, the War did not end; only another man took his place."[39] Gandhi was enunciating a fundamental principle of modern industrial society that could have been affirmed by American engineers such as Frederick Taylor or Alfred P. Sloan.

Gandhi's numeracy is another fundamental congruence between Gandhi and industrial society. Particularly after 1924, his writings and those of his associates that appeared in *Young India* suggest an implicit acceptance of Victorian physicist William Thomson's maxim, "When you can measure what you are speaking about and express it in numbers, you know something about it."[40] The pages of Gandhi's periodical, *Young India*, at times look like a business or technical journal, with tables of numbers on many pages giving lists of donors, figures of khadi production, or almost anything else that could be reduced to numerical form. A typical and regular Gandhian heading was "An Instructive Table."[41] One might be tempted to call Gandhi's numeracy a fetish but for the knowledge that entire societies are based on such obsessions.

The Rudolphs largely confined their analysis of Gandhi to the political realm. However, the characteristics they described so pervaded Gandhi's actions that he can fairly be called an engineer, specifically the twentieth century's newest and most expansive type of engineer, an industrial engineer, whose ambit was every aspect of human life, seeking to engineer objects, human processes, and indeed humans for maximum efficiency. "Efficiency" was a word that Gandhi often used, and often not in a dramatically different way than it would have been used by Frederick Winslow Taylor, the American industrial engineer and father of scientific management.[42]

After a 1929 tour of Andhradesha to raise money for khadi work, Gandhi published in *Young India* a complete list of the collections and the expenses, with the expenses being detailed down to the anna (a sixteenth of a rupee, a very small amount). (The major expenses were renting cars, including a "new Ford," and keeping them supplied with petrol and oil.) Gandhi noted the expenses were 5 percent of the revenue collected, a fact he ascribed to the "businesslike" organizer of the tour, but Gandhi also suggested that with even better management, the expense ratio could be further reduced.[43] Gandhi and his retinue also thought of this efficiency in terms of the actions of people. In 1927 Gandhi's secretary, Mahadev Desai, noted with satisfaction the arrangement for Gandhi to collect donations from the people during a tour of Bihar. People were able to give their donations directly to Gandhi, but were prevented from touching his feet, which "always gives rise to a terrific rush and crash," with the result that "men at the rate of 14 per minute passed through his hands."[44]

In the pages of *Young India*, Gandhi would often write his assessments of the annual meeting of the Indian National Congress, including what might be called the industrial engineer's judgments on the meeting, commenting on the time management skills displayed, the discipline in the meeting, and the sanitation facilities provided. Gandhi's impressions of the 1924 Congress meeting in Belgaum were as "a triumph—not yet of swaraj—but certainly of organisation. Every detail was well thought out. Dr. Hardikar's volunteers were smart and attentive. The roads were broad and well kept. They could easily be broader for the convenience of the temporary shops and the easy movement of thousands of sight-seers. . . . The sanitary arrangements though quite good needed still more scientific treatment than what they had. The method of the disposal of used water was very primitive."[45]

Gandhi's engineering approach can be seen in his attempt to transform the cow protection movement. Cow protection had been a divisive Hindu-Muslim issue, but when Gandhi got seriously involved with it in 1924, he tried to make it less sectarian and more about scientifically managing cows so as to increase milk production to a level that could justify keeping older cows, thus avoiding religious conflict. His journal *Young India* reprinted "The Cow: Mother of Prosperity," a Taylorist tract produced for the use of American farmers by the American farm machinery company International Harvester. Gandhi began a model dairy at Satyagraha Ashram, managed by Dinkar Pandya, who had studied agricultural science at the University of California at Berkeley.[46]

The Charkha and Gandhi's Industrious Indian

The technology Gandhi most famously associated himself with is the charkha or the handspinning wheel. Its archaic nature has implicitly tended to reinforce the notion that Gandhi was not a man of technology; he was the opposite of an engineer and cared not at all about efficiency. The charkha might be seen as the ultimate proof of Gandhi's opposition to modern industrial society. Indeed scholars have tended to see the charkha as a symbolic stunt, an example of Gandhi's penchant for getting distracted into spending too much time on peripheral issues. Both in Indian historiography and biographies of Gandhi, his handspinning program is largely lost in the overtly political events of the period: the Rowlatt Satyagraha, the Jallianwalla Bagh Massacre, the Noncooperation Movement, the Salt March, and so on.

But perhaps it is not Gandhi but we who have been distracted. Indian history and Gandhi in particular have been held captive to 1947. Those events that can be understood on a path headed toward Indian independence receive more attention, but Gandhi's career should not be judged solely in those terms. As Thomas Weber notes, the major political events occupied relatively little of Gandhi's time, even though they are the focus of most of the writing on Gandhi.[47] Indeed, as one reads Gandhi's letters and the periodicals he edited, *Young India* and *Navajivan*, one could almost imagine Gandhi as a hands-on, highly motivational executive of a spinning enterprise who happened to have a peripheral interest in Indian independence.

A close examination of Gandhi's spinning program shows Gandhi's own technological development, as he went from a naive romantic to one with a sophisticated appreciation for technical, organizational, and economic considerations in spinning. More than that, Gandhi's program of spinning reveals his vision for creating a different kind of industrial India made up of individual industrious Indians. Rather than the top-down industrialization of India that Parekh's vision of sending a few Indians to MIT implied, Gandhi's vision encompassed every Indian by using spinning as a means to instill in them industrial values. In that way Gandhi's program of handspinning was a remarkably ambitious attempt to restructure a society. He aimed to reengineer Indian society, instilling in its 300-odd million members discipline, time-thrift, organization, and quantitative thinking in a peaceful way that kept most of them in their homes, avoiding the turmoil and disruptions that the Industrial Revolution had caused in the West.[48]

Gandhi's program of cloth production went through several phases between his return to India in 1915 and his 1930 imprisonment following the Salt March. During an incubation period lasting roughly five years, Gandhi and the ashram mastered the technologies. In 1920 the Congress Party's approval of Gandhi's program of noncooperation brought his program of handspinning before the whole nation. After Gandhi was released from jail in 1924, he continued his handspinning under the umbrella of his "constructive program," which focused on providing the Indian people with concrete economic benefits.

When Gandhi returned to India from South Africa in 1915, he returned with an intuitive and romanticized view of the importance of handcloth production, but absolutely no technical knowledge in that regard whatsoever. In his 1909 *Hind Swaraj*, Gandhi had inveighed against Western civilization in general and machinery in particular, calling for the establishment "in thousands of households" of "the ancient and sacred handlooms." But Gandhi later admitted that at the time he was calling passionately for handlooms to be established, he had never even seen one, and in fact had written handloom when he had meant charkha, not even knowing the difference between the two.[49]

As Gandhi set up his Satyagraha Ashram in Ahmedabad in 1915, he made the hand production of cloth one of its top priorities. But how did a barrister with no technical knowledge, even one believing in the value of manual labor, develop cloth production capabilities in his ashram? The question calls attention to Gandhi's partnership with his cousin, Maganlal Gandhi. Writers who focus on Gandhi the politician draw attention to his associations with Vallabhbhai Patel or Jawaharlal Nehru. Without question the most important association for Gandhi's technical work was his second cousin Maganlal, whom Gandhi called "my best comrade."[50]

Maganlal and the Mahatma were a technical team, with the Mahatma providing the grand vision and Maganlal providing the technical skills and the sustained work that brought something like Gandhi's vision into focus. Anachronistically using today's terms, Maganlal might be described as a combination of chief technical officer and head of operations. Maganlal had first played this role in South Africa, when Gandhi had established a commune outside Durban. Here Maganlal took over the job of mastering all the technical skills to run the printing press in the absence of urban tradesmen. Gandhi called Maganlal a "born mechanic" and noted that in civil disobedience campaigns, Maganlal was the one who could not go to jail—his managerial and technical skills made him too valuable.[51]

Gandhi turned to Maganlal to establish cloth production. Gandhi later wrote that "although the conception was mine, his were the hands to reduce it to execution."[52] Gandhi sent Maganlal to Madras in South India, where he and his family spent six months living and working with weavers there before returning to the ashram. By mid-1917 Gandhi could describe the state of handloom weaving in a prospectus for the ashram: "No one in the ashram knew this two years ago. Today practically everyone knows something of it. Some of them may be actually regarded as experts." By October 1919 Maganlal described the inmates of the ashram as weaving for eight hours a day, with Gandhi himself weaving for four hours.[53]

Inexorably the ashram moved into handspinning. Maganlal had wanted to learn spinning during his time in Madras, but was recalled to the ashram before he could. In his autobiography, Gandhi recounted how the ashram's initial handweaving work was dependent on mill-spun yarn, while those in the ashram sought to be free of the mills. However, the desire to bring handspinning into the ashram was frustrated by the inability to find someone who was a teacher. Finally, after a prolonged search, Gandhi found a group of women who had spun before and were willing to begin again, and through their work and Maganlal's technical virtuosity, spinning came to the ashram.[54]

Gandhi's early advocacy of handspinning contained a naive technological enthusiasm. In early 1920 Gandhi announced a prize of

5,000 rupees for an improved charkha capable of producing five times more yarn than previous charkhas. One of the entries was from a Mr. Kale. In several letters to Maganlal, Gandhi expressed his captivation with Kale's design, writing in one letter that "the more I think about the spinning wheel, the more I fall in love with Kale's handiwork."[55] Gandhi urged Maganlal to patent the wheel and then to publicize it. In *Young India*, Gandhi confidently stated that India would soon have a "renovated spinning wheel" that would produce more yarn than a traditional wheel and could be operated by a five year old.[56] However, by 1922 Gandhi was warning the readers of *Young India* against "waiting for revolutionary inventions," because the existing charkha was "perfect of its type."[57]

On the pages of *Young India* and *Navajivan*, Gandhi made a number of arguments for handspinning on a charkha. But fundamentally Gandhi made an economic argument. The peasants of India could not survive solely on their agricultural earnings, but by spinning, peasants could supplement their income and earn enough to live on. Gandhi coupled this positive argument about spinning with a middle-class critique of Indian agriculturists: he argued that they were idle far too much of the time, and spinning would give them something productive to do.[58]

Gandhi was a controversialist who thrived on debate. In the newspapers he edited, *Young India* and *Navajivan*, he relished answering critics of his khadi program. In doing so, his chief arguments were economic ones, backed up by data and calculations. He used the studies of Harold Mann, the principal of the College of Agriculture in Poona, to support his claims that those engaged in agriculture could not provide enough to support themselves—they needed a supplemental industry.[59] Gandhi used the basic facts of India's cotton production and consumption to argue there was a need for more cotton cloth production. His articles were full of calculations: "A spinning-wheel must be worked for twelve hours per day. A practised spinner can spin two tolas and a half per hour. The price that is being paid at present is on an average four annas per forty tolas or one pound of yarn, i.e., one pice per hour. Each wheel therefore should give three annas per day. A strong one costs seven rupees. Working, therefore, at the rate of twelve hours per day it can pay for itself in less than 38 days."[60] Gandhi's frustration grew when his arguments were met by ridicule rather than logic, as they often were.

Handspinning and khadi cloth became overtly political as Gandhi integrated them into his program of noncooperation against the British Raj. But khadi was never political to Gandhi to the exclusion of the technical. Just at the time he was promoting handspinning as a political act against the Raj, Gandhi was also promoting standardization, a culture of expertise, time-discipline, organization, a work ethic, and reporting and performance monitoring, now not just to himself or to his ashram, but to the nation as a whole.

With the expansion of the spinning program to the entire Indian nation came collecting data from the entire nation. Previously Gandhi had collected data to document injustices against peasants or factory workers; now he was turning the tools of surveillance on his own people. Reports, no less than yarn and cloth, became the products of Gandhi's program.

In May 1922 *Young India* described the collection of data in Bardoli. In May 1923, after Mahatma Gandhi had been jailed, the report from Karnatac was held up by *Young India* as a "model of brevity and methodical presentation of detail in striking contrast with some of the haphazard and arbitrarily drawn up reports which we had the occasion to review sometime back."[61] Maganlal Gandhi later announced the khadi department was sending out forms to provincial khadi departments to survey the progress of spinning in the country. Collecting data took on almost a moral imperative; the inability to do so would require a confession that "we are very backward indeed"—reports were the measure of man.[62]

In 1922 the British imprisoned Gandhi. On his release early in 1924 following an appendectomy, he moved away from overt political activity within the Congress party. Instead he concentrated his energies on the constructive work, particularly spinning. Previously the khadi movement had been based on an organization that was part of the Congress Party, the All-India Khaddar Board. Gandhi in effect spun the khadi operation out of Congress, forming the All-India Spinners Association (AISA). In doing so Gandhi wanted a "purely business concern" that was immune to the "fluctuations in the Congress politics." Although Gandhi still maintained an extraordinarily busy schedule, he now had more time to devote to the technical aspects of spinning.[63]

Gandhi's program of handspinning following his 1924 release from prison had two goals: to produce yarn of such a quality that it could be used in handlooms, thus alleviating the poverty of India's farmers, and to produce what can only be called industrial habits in the Indian people. Gandhi pursued both goals relentlessly and with equal vigor. Today one might imagine Gandhi's program as so quixotic that he would gladly welcome any potential ally. However, the pages of *Young India* and *Navajivan* are filled with letters from correspondents who no doubt thought that their well-meaning efforts on behalf of Gandhi's cause would merit praise, only to be subject to withering public criticism for their failings, usually poor work habits.

In 1926 Gandhi reported receiving a package of yarn that was "untidy, badly spun, badly rolled" from a man who expressed his desire to spin to aid the cause. Gandhi defined for him and his readers spinning:

> Spinning does not mean drawing out bits of yarn of any sort as if we were merely playing at spinning. Spinning in fact, means learning all the preliminary processes—sitting down properly, with a mind completely at rest, and spinning daily for a fixed number of hours good, uniform and well-twisted yarn, spraying it, measuring its length and taking its weight, rolling it neatly, and, if it is to be sent out to some other place, packing it carefully and sticking a label on it with details of the variety of cotton used, the count, the length and weight of the yarn, and tying a tag on it with particulars of the contributor's name and address in clear handwriting; when all this is done, one will have completed the spinning-yajna [offering] for the day.[64]

Gandhi's definition of spinning meant time-discipline, work-discipline, numeracy, and reporting.

In all his undertakings, Gandhi wanted precision. In 1926 Gandhi reproduced a report on spinning done by a small society in Tuticorin, in Tamil Nadu, holding up the correspondent as an example of bad reporting practice. "The inaccuracy of the language of the latter is disturbing. Why should there be an 'about' in giving information about a small society? Instead of saying 'most of them spin,' the correspondent could have given the exact number of spinners, the time given by each daily to spinning and the count and the quality spun. Why 'there are about 20 Charkhas' and why not exactly how many? Why 'some paid spinners'? Why not quite how many? Why no mention of the wage given? Are they spinners in need? What is the meaning of 'about 60 towels'? 60 is a round number. A business-like organization should give business-like information. And those who wish to do Khaddar work, i.e. Serve the poorest and the neediest, must be business-like."[65] For Gandhi, a man from the bania (merchant) caste, being business-like was a constant imperative.

In the 1920s Gandhi made a number of tours throughout India to promote handspinning. These tours were not bland pep talks, but had a sense of a production manager going out onto the shop floor to assess the practice of his workers. One outstanding example is Gandhi's tour of Bengal from May to October 1925. As Gandhi prepared to depart for the tour in April, he announced a new arrangement to accommodate his own spinning. Previously, he had brought his own spinning wheel, but on this tour he would ask his local hosts to provide him with a charkha. Gandhi explained his reasoning: "I find that the new arrangement enables me to examine the local wheels and as generally my host tries to provide me with the best working wheel, it enables me to gauge the capacity of the place visited for yarn production. For, when I find the best available wheel to be an indifferent piece of furniture, I know that the production is poor."[66]

Gandhi gave a striking account of his visit to Rajshahi, a town in east Bengal. The women there flooded him with a "never-ending flow" of donations of rupees and jewelry in support of his khadi effort. They then conducted a spinning competition in which over 200 women took part. Gandhi noted that one of the women was "an accomplished spinner of fine yarn." But even with all this devotion to the cause of spinning, Gandhi was unsparing in his technical evaluation: "Almost everyone of the wheels including this lady's was a useless noise-making toy yielding poor results in quantity." He blamed the problem on a local maker of charkhas, whom Gandhi identified by name. He accused him of being a "blind enthusiast" and urged him to remove his charkhas from service until he had studied "the science of the wheel" and become "acquainted with the best wheel."[67]

Gandhi argued for teaching and practicing spinning in the schools, asserting that by doing so, schools could become partially self-sufficient. The effect of spinning in the schools could be judged by the reports of school spinning sent to Gandhi and published in *Young India*. Historians of the Industrial Revolution in England and America have written of institutions such as schools and churches being sites for the inculcation of industrial values in children, but none did so as directly and seemingly intensely as Gandhi's schools.[68]

In 1925 Gandhi published a letter from a principal he had upbraided during a visit for the students' poor spinning performance. The principal wrote back to report on the great increases in production that had happened since Gandhi's visit. Furthermore he reported that in order to prepare for regular spinning competitions, the students had taken four hours in their daily schedule that had been free time and used them to prepare cotton slivers.[69]

The next year Gandhi published a report by the principal of a national school in Dondaicha (in present-day Maharashtra). The principal asserted that the school's twenty-eight boys spun for half an hour every morning and that a "regular record of the daily progress is kept." While initially the average production per student was twenty yards in a half hour, after four weeks, it had increased to thirty yards, with the fastest student spinning fifty yards during that time.[70]

That same year Gandhi proposed the creation of an elite cadre of men and women to help bring his ideal of khadi to fruition. Ironically, the khadi service was modeled to a degree on a British creation, the Indian Civil Service. Gandhi envisioned that applicants to the Khadi Service would go to the Satyagraha Ashram or other sites for two years of training, which would include "all the processes that cotton has to go through up to weaving," as well as bookkeeping. Prospective members of the Khadi Service would then have eight months of practical experience. Members of the Khadi Service (and he hoped to get 700,000) would then disburse to India's villages to promote and oversee khadi work.[71]

Gandhi and his ashram developed a distinctive way to mark the development of the National Week—the week commemorating the satyagraha over the Rowlatt Act and the massacre at Jallianwalla Bagh in April 1919—in a way that both promoted khadi and demonstrated their enormous capacity for work. In the years immediately after 1919, Indians observed the week in a number of ways, including meetings held to pass various resolutions, prayer, fasting, and hartals (strikes).[72]

Mahadev Desai reported on how the Satyagraha Ashram celebrated the National Week in 1926. The ashram decided to do continuous spinning from 4 am on April 6 until 7 pm on April 13, while also working a loom continuously over this period. But in keeping with Gandhi's numeracy, it was not enough just to spin over this period—records were kept. Desai recounted how a sixteen-year-old inmate of the ashram spun fourteen hours in a day and recorded a production of 5,925 yards. Fittingly, Gandhi's production was measured also, and while he had one of the lowest outputs, Desai noted that he was the oldest member of the ashram. Similar celebrations were held in 1927 and 1928. In 1928 three boys spun for twenty-four hours consecutively on the last day. *Young India* published not only the number of yards each boy produced, but also quantified measures of their yarns' quality. An added feature of the 1927 and 1928 celebrations was that it was now possible to make year-to-year comparisons. The 1928 spinning effort had yielded over 200,000 more yards of yarn than in 1926, and the highest individual rate of production had increased from just under 23,000 yards of yarn in the week to over 40,000 yards.[73]

Desai noted that the week was a holiday from school for the boys, but "it was no holiday from the point of view of work."[74] This celebration of a holiday, not by making a hartal or having a parade or other demonstration, but by demonstrating one's capacity for work was extraordinary. A mark of the modernization of European society had been the reduction in holy days, increasing the number of work days and making them more regular. By celebrating a holiday by heroic work, the Satyagraha Ashram showed a capacity for, and embrace of, work that went beyond European or American practice.[75]

In the late 1930s Gandhi would support using crafts such as spinning as the base of the entire Indian educational system. He proposed a system of compulsory education for children between seven and fourteen years old, whose premise was that the fundamental flaws in both Indian and Western education systems required recasting a distinctively new Indian system. Crafts would form the base of the system, taught not as ends in themselves but as a means of imparting education in all other subjects, such as mathematics, history, and geography. Gandhi envisioned that the sales of student-made goods would make the system self-supporting.[76]

Gandhi's ambition was breathtaking. He proposed nothing less than to unravel the Industrial Revolution and then to reweave it in a distinctly Indian fashion. Gandhi's handspinning program obviously did not achieve everything he had hoped it would, and the analysis presented here offers several ways to understand that failure. The production of khadi was a hybrid social-technical system combining preindustrial hardware (the charkha) with industrial values. By putting most of the emphasis on the human rather than the machine, Gandhi's system required an almost robotic discipline. If those most attached to Gandhi, the members of his ashram, could muster that discipline for some time, it can hardly be surprising that the masses did not.

The seeming primitiveness of the charkha masks the complexity of Gandhi's entrepreneurial venture and the technological system he was attempting to build. Gandhi's program of handspinning assumed that he could constitute a country of 300 million people into a labor force and then achieve a level of control over it unprecedented in human history. His system required teaching a nation the demanding skill of spinning yarn and then getting them to continue spinning for months and years. And assuming that was achieved (which it never was), his system then required the coordination of millions of skeins of yarn, keeping track of production while somehow matching yarns of appropriate count and quality with weavers.

Gandhi was captivated by the holism of khadi and the diversity of things it would do for the Indian nation: clothe the poor, teach industrial discipline, unite the classes, and demonstrate Indian defiance to the British. However, Indians appear to have wanted to live their lives more à la carte. Gandhi is hardly the first technological entrepreneur captivated by a dream that could not stand up to the harsh reality of a world populated with real people.

Historian Benjamin Zachariah has suggested that one of Gandhi's strengths was his use of language that could mean different things to different groups, and that alternately used a variety of idioms, ranging from religious to scientific. The Indian middle class doubtless heard language that was familiar to them. But could Gandhi use this language that accepted some of the key principles of industrial society to pull Indians into his alternative society?[77] An impressionistic answer to that question can be had by looking at the paths of a group of his young followers. They showed that however much India seemed to be caught up in a Gandhian moment, being a disciple of Gandhi could be one step on the road to MIT. Devchand's dream was not dead.

4

From Gujarat to Cambridge

IN 1925 Mahatma Gandhi traveled to Jetpur to attend the wedding of Devchand Parekh's daughter, Champa, to T. M. (Trikamlal) Shah, the registrar of the Gujarat Vidyapith, a Gandhian school in Ahmedabad. Gandhi himself wrote a commentary on the wedding:

> Shri Devchandbhai insisted upon my attending it so that I could see how it was celebrated with the utmost simplicity and that nothing but khadi would be found there, and the bride and bridegroom could receive my blessings. I gave in to his sincere and pressing invitation and attended the wedding. There were many men and women present there who had been invited by Devchandbhai's family. On the bridegroom's side, however, there was no one except the bridegroom himself. Shri Trikamlal was determined to marry if he could find a worthy bride, with no more than a *tulsi* leaf as dowry. He carried out that decision of his. The wedding ceremony ended with the bride giving away khadi clothes to the children of the *Antyajas* [untouchables] in their locality. In this marriage, too, music, songs, etc., were completely left out. My request to the *mahajans* of Kathiawar is not to be enraged at such simplicity, but rather regard it as praiseworthy and propagate it. The era of large dinner parties

should be regarded as having ended. Some practices should indeed change with every age.[1]

The wedding was a vivid image of Gandhi's idea of a reformed India: the wealthy Parekh family clothed only in handspun khadi, dowries abjured, untouchables cared for, simplicity valued over ostentation.

But things were not exactly as they seemed. When Gandhi arrived in Jetpur, he needed a charkha for his own daily spinning. Parekh offered Gandhi the use of his own. When it arrived, Gandhi found a spinning wheel in very poor condition—obviously not being used regularly by Parekh, who was perhaps still too aristocratic to spin himself. Gandhi reported that his arm started aching with just a half-hour's worth of spinning using Parekh's charkha. Gandhi launched a semi-humorous public attack on Parekh, saying that Parekh was mocking the spinning wheel.[2]

Parekh had resisted subordinating himself to Gandhi in other ways. He had not given up his idea of sending Indians to MIT. In fact, just a year and a half after T. M. Shah was praised by Gandhi for the way he conducted his wedding, Shah, with the help of Parekh, had left India to study engineering at MIT. And others would follow. When compared to the millions of Indians who stood with Gandhi, either by participating in civil disobedience or wearing khadi, it might seem absurd to dignify the actions of a handful of young men with the term movement. But these men who went to MIT suggest that rather than Gandhi's movement co-opting Parekh's, Parekh's may have actually co-opted Gandhi's—a fact that would only become manifest decades later.

In the 1930s a number of young men went to MIT who had been brought up in Gujarat and had close ties to Gandhi. While at least one later broke with Gandhi, on the whole, most of these young men betrayed no evidence of a disjunction between their Gandhian world and their MIT world. At the same time, the arcs of their careers were ones that moved them away from the kinds of institutions Gandhi had been advocating (small-scale, vernacular, populist, rural) and toward the types of institutions that were compatible with a global, large-scale industrial society.

The lives of these Gandhians who went to MIT and their subsequent careers can be detailed between 1925 and 1945, following the arc they took through Gujarat, then Cambridge in the Roaring Twenties and the days of the Great Depression, and back to World War II India. The effort to combine Gandhian principles and an MIT education was just one challenge that these young men faced. Could they master the demanding MIT curriculum? Could they get along in an alien culture whose values were so different from theirs? Could they reconcile a technologically sophisticated education with opposition to a colonial state, which held most of the technologically sophisticated jobs in country? This chapter details the paths these young men took to MIT as well as their time there, while Chapter 5 will examine their return to India. As this first significant group of men to have been trained at MIT dealt with these tensions, they foreshadowed those of the many more who would come later.

T. M. Shah's Letters Home

When he married Devchand Parekh's daughter in 1925, T. M. Shah, born in 1897, had perfect Gandhian credentials. He had worked for five years teaching at the Gujarat Vidyapith, an institution in Ahmedabad that reflected Gandhi's ideas about education. A 1920 article in Gandhi's *Young India* laid down the fundamental principles of the Vidyapith. Its medium of instruction was Gujarati, but more than that it held "that a systematic study of Asiatic cultures is no less essential than the study of western sciences for a complete education for life." Furthermore, the Vidyapith held that "the vast treasures of Sanskrit and Arabic, Persian, and Pali, and Magadhi have to be ransacked in order to discover wherein lies the source of strength for the nation." The Vidyapith aimed to create a synthesis of cultures "of the swadeshi type, where each culture is assured its legitimate place, and not of the American pattern, where one dominant culture absorbs the rest." Finally the Vidyapith aimed to "broad-cast" its education to the masses so that "the suicidal cleavage between the educated and the non-educated will be bridged."[3]

That was how the Vidyaptih was described in the heady days of 1920, when anything seemed possible. But by the mid-1920s, the

Vidyapith was in trouble. The source of trouble was not colonial repression, but rather apathy on the part of the Indian people. By 1925 Gandhi, acknowledging the Vidyapith's dwindling numbers and its failure to offer "alluring careers," was driven to argue that the Vidyapith should not be closed.[4]

While the Vidyapith struggled on, it would do so without Shah. In arranging his daughter's marriage to Shah (they had been engaged three years earlier), Parekh obviously saw Shah's potential to become one of his dreamed of MIT-trained engineers. Shah's education, a first-class honors degree from Fergusson College in Poona, suggested he could make the transition. In fact, Shah would get a letter of recommendation from the legendary R. P. Paranjape, professor of mathematics at Fergusson and later principal. Paranjape had become a celebrity among middle-class Indians in 1899, when he won the title Senior Wrangler, given to the highest-ranking graduate in mathematics at Cambridge University. At Parekh's urging, Shah applied to and was accepted to MIT.[5]

Shah left Gujarat for the United States in the summer of 1926, leaving his pregnant wife behind. Family testimony is that Shah discussed his path with Gandhi at length, and while Gandhi was not keen on Shah's going to MIT, he ultimately blessed it.[6] A little more than a year after Gandhi had praised his wedding for its exclusive use of khadi, Shah was in a foreign country, studying subjects that would inevitably link India into a global web.

As one of the faculty and a member of the Vidyapith's governing body, Shah had worked alongside and known many of the central figures of the freedom movement in Gujarat. These leaders in Gujarati politics and intellectual life included Mahatma Gandhi, of course, but also figures like Vallabhbhai Patel, Kishore Mashruwala, and Indulal Yagnik. In coming to MIT, Shah was entering a world where these men and their knowledge, respected and revered in Gujarat, meant nothing. Shah was giving up these connections and starting over. After five years of working at the Vidyapith, Shah would be studying at an institution that stood directly opposed to almost everything that the Vidyapith stood for.[7]

Shah's enrollment at MIT in 1926 represented the fulfillment of a long-standing dream for Devchand Parekh. But in the intervening thirty-three years since he had been to America and MIT, so much had changed that one might wonder if an MIT education was an appropriate one for India. Shah did not return to the place Parekh had seen. MIT was no longer a handful of buildings in Copley Square in Boston, not even meriting the term campus. In 1916 it had moved to a spacious location in Cambridge, its main building modeled after the Roman Pantheon, properly reflecting the institute's aspirations: to be a global center of learning.[8]

The 1893 MIT annual report spoke of the "poverty of the Institute," complaining bitterly about the parsimony of the Commonwealth of Massachusetts and its contribution to MIT's annual budget of $300,000. By 1926 pleas of poverty would no longer be credible for an Institute with an annual budget of $2.8 million and an endowment of $28 million. More important than the numbers were the patrons that had made those numbers possible. MIT had forged partnerships with technology-based businesses that provided funds in a variety of ways. The new Cambridge campus was made possible by donations from the du Ponts of the eponymous chemical company and George Eastman, the founder of the photography company, Eastman Kodak. MIT had made an educational alliance with General Electric. Leaders of General Electric, General Motors, and other large corporations were MIT alumni.[9]

One example of the newer MIT was the work of Vannevar Bush, an electrical engineering professor T. M. Shah would study with. In 1925, responding to the needs of an increasingly sophisticated American electrical industry, Bush had begun a program of research aimed at the machine calculation of complex mathematical equations. The result was a series of highly specialized electromechanical machines, which would be harbingers of the modern computer.[10]

In 1893 the Duryea brothers were completing the first American-built automobile. By 1926 Henry Ford had put Americans on wheels, and as Model T production was approaching 15 million, MIT graduate Alfred P. Sloan of General Motors was eating into Ford's market share with "a car for every purse and purpose." In 1893 electric lighting was still a fairly novel technology, and electrical engineering found its place at MIT as part of the physics program. By 1926 not only was electric lighting common, with 58 percent of American homes having electrical service, but American retailers were producing a range of other electric appliances such as washing machines, vacuums, and small cookers. In 1893 no person had yet flown a heavier-than-air vehicle. By 1926 MIT had an aeronautical engineering department, and two years previously MIT alumni Donald Douglas's planes became the first to circumnavigate the earth.[11]

More than any specific technology in the intervening thirty-three years, engineering in the United States was increasingly about extending and sustaining large-scale systems, whether electrical, mechanical, or chemical, and an MIT education trained students to work within those systems. It was an act of great faith, or naïveté, on the part of Shah and Parekh to think that Shah, working as an individual, could appropriate a technical education developed for those systems in American society and use in the very different Indian society.[12]

Devchand Parekh and T. M. Shah exchanged letters on a regular basis while Shah was at MIT. Seventy surviving letters, written between 1926 and 1929, document Shah's experience at MIT as well as his relations with those back home.[13] In the 1960s and afterward, Indians would often describe their time at MIT as the best years of their lives. Shah's letters show this was not the case for him. He faced difficulties that few Indians who came after him would have to endure. Shah was in his late twenties and early thirties, redoing an undergraduate curriculum, having left behind his pregnant wife as well as his friends and other family, living in a strange place with virtually no friends. He was studying an entirely new discipline, piling up substantial debt, without having any prospects of a job when he finished, all the while being subjected to continual postal hectoring from his father-in-law. At his most pessimistic moments, Shah seemed to believe that going to MIT was a big mistake. But Shah persevered, working through his inexperience and embracing the challenges to become a competent engineer.

Parekh had asked Shah to write to him in English, making the letters unreadable to most of the rest of those back in India and suggesting that they had a special partnership. But it was more than a partnership. Parekh had maintained the dream of sending Indians to MIT for over thirty years. To do so had required a great deal of faith. Parekh's drive to achieve this dream was so great that at times he seemed to see Shah as an actor in a drama in which he was the scriptwriter and director. Parekh's attempt to direct Shah from 7,000 miles away based on thirty-year-old knowledge was a formula for frustration.[14]

From Shah's responses, Parekh's early letters made a number of demands upon Shah, often having to do with Shah's health. In the early twentieth century some Indians believed that their constitutions were not suited for the cold climate of the northeastern part of America and furthermore that a diet for a cold climate required meat, which Shah and Parekh, as Jains, could not eat. Parekh, who is unlikely to have known another Indian who lived in the United States for an extended period, seems to have been concerned that by sending Shah to MIT, he was jeopardizing his health.[15]

Shah and Parekh expressed these worries in a form appropriate both for a follower of Gandhi and for a student at MIT: by quantification. Parekh demanded weekly reports on Shah's weight, down to the nearest tenth of a pound. Shah who had weighed only 98 pounds when he arrived at MIT, dutifully complied, each week offering with his weight some words that he must have hoped would assuage Parekh's anxieties. But Parekh always found something to worry about. With the onset of winter, Parekh worried that Shah's weight was fluctuating weekly. Shah was flummoxed. In February 1927 he wrote that he couldn't imagine reaching his current weight of 109.1 pounds if he had remained in India. His "underweight" had not kept him from doing anything. Eventually Shah exerted the control that 7,000 miles distance made possible. In April he offhandedly wrote that he hardly took "the trouble to know my weight." By July he was positively refusing to tell Devchand Parekh his weight anymore.[16]

Shah used the distance to break free from Parekh in other ways. In March 1927 Shah announced to Parekh that contrary to his requests, he would no longer write to him in English. Although he wrote to his wife separately, Shah said she complained bitterly about her inability to read his letters to Parekh. Furthermore Shah wanted other members of the family to be able to read the letters.[17]

Parekh had seen Shah's weight as a proxy for his diet, and while Shah made the transition to American vegetarian food effortlessly,

Parekh saw the process as full of peril. Parekh doubtless saw his family's chemical business as giving him some authority in this matter. (In 1928 Gandhi was asking Parekh for dietary advice.) Parekh made many suggestions to Shah, asking for regular reports on his diet. Again Shah provided quantitative data: five pounds of honey eaten in the month, a half pound of butter eaten a week. Initially Shah accepted Parekh's suggestions with good grace, thanking him for them while hoping Parekh "will leave something to my judgment." Here again Parekh's efforts to direct Shah grated over time.[18]

Shah wrote to Parekh that "America has much to teach us regarding food."[19] Both men developed a fascination with ready-to-eat cereals. Shah associated them with a more scientific and industrial approach to eating, which was perhaps what he meant when he said America had much to teach India regarding food. But at the same time, Shah may have been taking marketing material at face value. In one letter Shah forwarded to Parekh a description of how shredded wheat was made, asserting it was "produced without any touch of human hand, shredded and packed all by machines."[20] In response to Parekh's suggestion that he try porridge, Shah was emphatic: "Oatmeal porridge is an impossibility? Who will care to spend some 4 or 5 hours after it when the far more nourishing and delicious food is shredded wheat?"[21] Shah further noted that there were other ready-to-eat cereals "prepared scientifically and they are quite nutritious." In November 1926 Parekh asked for samples of shredded wheat, grape nuts, cornflakes, and all bran to be sent to him. Shah grumbled, noting the difficulties of sending parcels, but promised to do something after Christmas break.[22]

Shah seems to have insisted on paying his own way to MIT. A 1921 article for Indians who were considering studying in America listed MIT as among the "universities for the well to do," with MIT's tuition being at the time $300 a year, with land-grant colleges being as low as $20 a year. Parekh may have been well-to-do, but Shah was not, and he had to borrow money from relatives. In an April 1927 letter to Parekh discussing the money he had borrowed, Shah wrote that he had "often felt uneasy" about coming to the United States. The debts he was piling up gave him a feeling of "bondage" that he said "remains constantly with me." This fighter for India's freedom had put himself in bondage for its technological development. He wrote of plans to apply for a "Tata Studentship" and to approach a "Sheth Giranchand" for support.[23]

Even though Shah was paying his own way, Parekh seems to have required regular reports on his expenses and often wrote back with comments on them. Parekh worried at times that Shah was not spending enough, but Shah claimed that with his desire to minimize his debt, he lived "frugally, but not stingily."[24] Parekh seems to have wanted these reports in part out of desire to understand life in America, perhaps in part to help those who would come later, but in part from genuine personal curiosity. Parekh demanded to see restaurant bills and even Shah's personal expense book. Shah told Parekh that his perceptions of things were based on nineteenth-century England, and since that time there had been "huge changes," which happened in America with "tremendous force." Shah promised to send his expense book, but a few months later thought better of it. Sending it would have merely extended the inquisition.[25]

In 1928 Parekh, with his usual insensitivity to Shah's concerns, suggested that Shah stay on at MIT to earn a doctorate. Shah had been at MIT two years and had two years to go to earn the master's and bachelor's degrees they had planned—a doctorate would be years more. Shah, who had neither the time nor the money to return to India and his wife and daughter since he had started at MIT, wrote that he felt "like rushing to India this very moment," but only stayed because he knew he needed to complete his education to be able to pay off his debts. Such was his bondage.[26]

Devchand Parekh expected Shah to be his agent in America. He was particularly interested in chemistry materials. Parekh requested a book on lead, a list of chemistry periodicals, a catalog of GE products, and materials on soapmaking. Parekh acted as an armchair entrepreneur, thinking up schemes that often involved work on Shah's part. Parekh had the idea of hiring an MIT professor as a consultant to begin some business in India (possibly with the financial assistance of a princely state). Whether part of the same scheme or an entirely different one, Parekh also had the idea of developing small gadgets for sale in India and of developing a machine to roll out Indian flat breads.[27]

Shah threw cold water on these plans, which would often require a lot of running around on his part. While he told Parekh that MIT professors could be hired on a consulting basis, it would be expensive and he would not begin inquiries until Parekh had a specific plan. He was cynical about the princely states. He thought India was too poor for a gadget business to be successful and he was sure that rolling machines would be the ruination of India, denying women one of their primary means of exercise.[28]

Shah also served as a vanguard for Devchand Parekh's plan to send more family members and acquaintances to MIT. He reported back a variety of information and helped ease the entire application process. Because of the family's involvement in the chemical business, Devchand appears to have slotted most of his sons and nephews to study chemical engineering at MIT. Shah met with Maneck Kanga, a Parsi from Bombay who had graduated from MIT, to get details on the chemical engineering practice school. (Kanga suggested that six months' expenses at the practice school would amount to $720.) Shah first made arrangements to help a family friend, Nandlal Shah, come to study electrical engineering in the summer of 1927. The next year he assisted Parekh's nephew Maganbhai, in securing admission to the master's program in chemical engineering. He also helped another family friend, Mansukhlal Mehta, obtain admission to the bachelor's program in electrical engineering.[29]

Shah served as the intermediary between the Indians and MIT, sending out application blanks to India, answering questions from applicants, and making sure they knew the requirements for admission. The issues came from incompletely or incorrectly filled-out forms or missing documents more than anything else, for Shah asserted that "no student faces difficulties in getting admitted here."[30] In the case of Nandlal Shah and Mehta, Shah was able to submit the application forms and find out immediately that they had been accepted. Shah then sped the process by working with MIT officials to send telegraphic notification to American consular officials in India to enable the students to get visas.[31]

Shah and Parekh's most productive and friction-free exchanges came over securing Shah's admission to MIT Course VI-A, the Cooperative Course in Electrical Engineering. This course, fully implemented in 1920 combined training at MIT with work assignments at General Electric sites. (Other companies were added later.) For a student from India, which had no comparable industrial sites, getting such training would be invaluable. Shah would come back to India not with mere book knowledge but with practical experience that he could hope to put to immediate use. Initially Shah seemed to think that he could combine a standard electrical engineering program with several years of work, but he came to believe "for acquiring practical knowledge there is no other way but to enter the industrial stream [the cooperative course]."[32] This was because he claimed that immigration restrictions made it impossible for him to work for any length of time in the United States after he got his degree.

Admission into the program required the approval of both MIT and General Electric. General Electric paid particular attention when foreign students were involved. While students from Turkey, China, and Soviet Russia graduated from the program in the 1920s, at least one Indian student was denied admission to the program. MIT professors discouraged Shah about his possibilities of getting accepted into the program, but he applied anyhow.[33]

Throughout his letters home, the slight, deferential Shah gave the impression that he would have largely been anonymous at MIT, always slipping under people's notice. But when it came to pursuing admission into the cooperative program, Shah did so with a fierce tenacity. Shah had discussions with MIT faculty running the program, who gave him a "sporting chance" to gain admission. Shah then approached Dugald Jackson, the head of MIT's electrical engineering department and the driving force behind the program, and he agreed to support Shah.[34]

In February 1927 Shah traveled to Lynn, Massachusetts, for an interview with the GE official running the program at Lynn. Shah came away from the interview with a sense of how difficult admission to the program would be, but he was not at all deterred. The official told Shah that the program was designed to serve GE's interests by serving the interests of its customers. If GE had some inkling that a foreign student had the support of a GE customer, GE would be inclined to support the applicant. Otherwise they would not.[35]

The most prominent GE customer in India was the House of Tata, which had a large hydroelectric power plant near Bombay. Shah took vigorous action. Usually Parekh made demands of him; now it was the other way around. He forcefully asked his father-in-law to secure an interview with Dorab Tata, the leader of the House of Tata, on his behalf. He wrote to his former principal at Fergusson College in Poona, R. P. Paranjape, asking him to intervene for him with Tata. Shah wrote personally to Tata, India's leading industrialist. Both Shah and Parekh scrambled trying to find someone in India who could move the Tatas on Shah's behalf.[36]

In March 1927 Shah received a letter of rejection from GE, saying they had too many foreign students. Shah was not deterred, still seeking a letter of recommendation from the Tatas. Then in late April he got a telegram stating that the Tatas would not nominate him. He immediately sought a meeting with an MIT professor to see if he could help him gain admission. Shortly thereafter, Shah received notice that he had been accepted into the program. The whole process was opaque, with Shah not ultimately sure why he got admitted to the program, but the decisive factor in Shah's acceptance seems to have been the recommendation of the MIT professor.[37]

After Shah was admitted to the GE cooperative program he had an interview with GE personnel. In his description of the interview, Shah made it clear that they saw him more as an exotic, from a land they knew very little about, than as a member of a global fraternity of engineers, recounting that they talked "mainly about India and about caste." Shah reported that one GE official had read in a newspaper that when an Indian woman was widowed, she was driven out of the house. The GE officials described having one Sikh in the program, but they apparently did not know anything about Sikhism. They asked Shah, "Who is a Sikh? Are you a Sikh? Do you wear a turban?"[38]

Other than the fact that he had a degree in mathematics (and the fact that he had done spinning as part of Gandhi's movement), Shah had no special qualifications that would have prepared him to study engineering. By the time T. M. Shah arrived at MIT, most of his fellow students would have had experience with a wide variety of technologies that were common in the United States. Given the ubiquity of the Model T in America, most of the young men would have doubtless spent time working on automobiles. Many likely had built or used radios. Many would have had experience with electrical systems. Many had some done work that had exposed them to modern technologies. Shah had none of this.

In his first year, Shah did well in his physics, chemistry, and French courses, but drawing proved to be his nemesis. He reported having to redo several assignments and feared that his struggles in drawing would hold him back in other courses. However, even as he struggled in drawing, he expressed his commitment to the principle of education that required it: "You will realize how necessary an all-round training is to a man. It is not alone the brain, but all the parts of the body that make a man efficient."[39] This sentiment, here applied to drawing, was one that Gandhi would have agreed with.

During his second term Shah's big challenge came from electrical lab. In addition to the fact that he was the only one lacking experience with electricity, a further challenge came from MIT's efforts to instill what Shah called a "habit of self-reliance" in its students.[40] In electrical lab, this meant that students were given books and references, but no help from instructors, and yet were expected to find their way. Self-reliance was a key term of Gandhi's, and indeed the entire Indian independence movement, and while its exact meaning was different in MIT and the Satyagraha Ashram, the term had similar resonances. Shah thought of it as a positive thing. But it imposed practical difficulties. Shah noted that an electrical lab scheduled for three hours, and that took some Americans six hours, took him eight hours to complete. And while Shah's drawing had improved, it was still causing him difficulties. He wrote that the drawings this term "were understood with great difficulty and I spend hours and hours without drawing a single line." However, Shah was confident in the MIT approach to engineering education and in his ability to surmount his difficulties through "time and patience alone."[41]

The cooperative program Shah had fought so hard to enter involved students alternating time between MIT and GE factories. Over his time in the United States, Shah spent 70 weeks working in General Electric factories in Lynn and West Lynn, Massachusetts,

Pittsfield, Massachusetts, and Philadelphia, Pennsylvania. In July 1927 as Shah finally started working in a General Electric factory, alongside both MIT students and young men not going to college but undergoing apprentice training at General Electric, he simultaneously enjoyed the experience of working with young people and felt lost. He was a thirty-year-old working with teenagers who knew more than he did. He wrote to his father-in-law, "I am astonished by the way the students in my class talk. All of them have acquired considerable experience with electricity and electrical machines by working at one or the other place. I regret my ignorance."[42]

In February 1928 Shah entered the machine shop at West Lynn; the man who had spun on a charkha in India was now working with lathes, drilling presses, and milling machines. Machine shop was not central to electrical engineering and some students sought to avoid it, but Shah specifically asked for it. While he initially despaired of his abilities, writing that "it appears that I cannot become a skilled craftsman," he kept at it. Within a week he wrote that he was gaining confidence that with time "I can certainly become a craftsman." Perhaps his charkha spinning paid off, for he later wrote that he found machine shop "an enjoyable experience." He was becoming an engineer, writing, "I have learned to think in terms of one thousandth of an inch."[43]

Becoming an engineer required other skills as well. In a July 1928 letter, Shah wrote to Parekh that "There is a better way" was the slogan that was responsible for American business success, with workers presenting suggestions for improvements and companies evaluating and implementing them. However, Shah lacked confidence, saying, "I do not have the ability to make spontaneous suggestions."[44]

In working at General Electric, Shah was entering the quintessential early twentieth-century American company of high technology and complex systems. General Electric had its origin in Thomas Edison's work in the 1870s and 1880s developing a system of electric lighting, but in the following decades, it had moved far beyond the work of the world's most famous inventor. General Electric made the products that made it possible for electricity to enter into every aspect of American life, both in the home and in business.

In 1926 General Electric's sales were a third of a billion dollars, with over $44 million in profits, while employing over 77,000 workers. Shah started working in West Lynn, Massachusetts, and he gave a sense of that plant's scope to his father-in-law: "G. E. Company makes all kinds of machines. Machines needed for big power houses, railway, telephone, and radio are made here."[45]

General Electric had pioneered the industrial research lab in America, where it brought together teams of highly trained scientists and engineers to extend its existing technologies. Its large sales base made it possible to sustain groups whose work might take years or even decades before paying tangible dividends. By the time Shah had arrived, General Electric chemist Irving Langmuir had done work that would be recognized with the Nobel Prize in Chemistry for 1932.[46]

In the 1920s an MIT education, and especially an education within the cooperative program, was designed to enable someone to work within complex American technological systems. As Shah began to understand the state of American technology and the American technological system, he became more pessimistic about the technological possibilities of India along those same lines. He wrote to his father-in-law: "I do not think India is in a condition to make Electrical machines today. There is no scope for small gadgets either. Investment of half a million to a million will not suffice in such a business. Moreover new inventions are done here every day and machines are constantly improved. It will take India a long time to get into this competition."[47]

Shah also came to see that all his training at General Electric would not necessarily translate into a job in India. The most likely employer was Tata Electric, but it was a user of General Electric equipment. The specialized skills Shah was learning to make this equipment were not necessarily ones that Tata needed. Dorab Tata, the head of the House of Tata, brusquely refused to promise Shah a job on his return. In fact, Shah noted that an Indian who had worked at Tata had essentially trained himself out of a job by coming to General Electric for advanced training. This man was now looking for work in the United States because "there is no place for him in India."[48]

As Shah learned more and more about advanced American technological systems, he became more and more pessimistic and cynical about the possibilities of India competing on those grounds. In November 1927 he reported on a presentation he had heard from Bell Telephone, stating that he "learned about the huge sums of money spent by industries for research and development." He concluded, "I doubt if India can ever compete with people here. Our investors only care for profit, not for improvement in business."[49]

Parekh continued to see America as the model that India should use in every way for technological development, while Shah had serious doubts. In response to a query to gather books that would teach young people about machines, Shah said, "These countries are for the wealthy. If one wants to take advantage of everything here, one has to have ample money in one's pocket." Shah suggested that instead of buying books on machines, Parekh expose young people to spinning wheels, bicycles, and watches.[50]

Shah's letters provide a sense of his social interactions in America. While Shah was not unhappy per se, he faced a challenging social situation. He had few true peers. He had a small group of Indian contacts. While he enjoyed Americans and felt accepted and loved by them, it was difficult for an undergraduate student his age, living on a tight budget, hard-pressed to keep up with his schoolwork, to fully enter the Americans' world.

Upon his arrival Shah lived with two Bengalis who had a third-story apartment in a house near campus. Just how important they were to Shah can be seen by a suggestion the ever-insensitive Parekh made that he could eat more nutritiously and cheaply by striking out on his own. Shah replied, "I can get their love and affection only by living with them. So I simply do not have the inclination to separate from them."[51]

Before Shah had left India, he and Parekh had set up contact with D. D. Kosambi, a scholar of Buddhism who had come to Cambridge to work with a Harvard professor translating and editing works on Buddhist philosophy. In Shah's early months in Cambridge, he would go out regularly for a meal and entertainment with Kosambi and his son. Finding acceptable food was a challenge. They ate at a cafeteria where Shah said, "Of course the meal was not to my taste." Better was a Greek restaurant, where he found the preparation "almost Indian." They then went to a theater, for a night mixed with films, dancing, and music, where Kosambi apparently paid.[52] Over Christmas break Shah and his roommates found time to cook for Kosambi, with Shah noting that it was "really a pleasure to have Indian dishes."[53]

Occasionally Shah could meet with larger groups of Indians. He attended meetings of the Hindustan Association. In February 1928 he reported a dinner of the Hindustan Association where over 200 people were present. In September 1928 he went to New York to meet two Indians coming to study at MIT. Shah wrote that because of New York's centrality, "many Indians pass through it," and when he and his friends met Indians "we feel happy as if we are in our own country." New York even had an Indian restaurant.[54]

But most of the time Shah was with Americans. And he poignantly described his situation in his second month in America: "The American students are quite sociable and you can make friends with them if you have money and time. I lack both and so I have no friends to boast of, but there are many acquaintances found. Intimacy does not come easy with me."[55] While most students at MIT were financially well-off, Shah had to carefully husband his funds. His work took him longer, and he probably took it more seriously than most students, so he had less time.

Shah's uncertain relations with his classmates and his challenges in making his way in American student culture were clearly evidenced in an act of uncivil disobedience, the 1926 Field Day riot. Field Day was an interclass competition between the MIT freshman and sophomore classes, but in 1926 MIT's tradition of pranks got out of hand. MIT freshmen, a classification that ostensibly included Shah, invaded the sophomore's competition eve banquet, breaking into the hall with battering rams and then throwing tear gas bombs. A full-scale riot ensued that spread from Cambridge into Boston, resulting in five arrests, significant injuries, and property damage.[56]

Shah described the occasion as a "signal for the outlet of the animal in man." Shah went on to tell his father-in-law that the "occasion can best rival our Holi minus its lewdness." Shah, writing of the students in the third person, noted that they had "visited the cities of Boston and Cambridge with their mischief" and had "inconvenienced a great deal the public," ultimately causing $15,000 worth of damage. Although Shah could write home about the incident with the detached eye of an anthropologist, he was a nominal participant in the high jinks because MIT intended to assess a token fine on all the first- and second-year students to partially pay for the damage they had caused.[57]

Shah wrote home that he could avoid the fine by making a declaration before a notary that he had not been present at the events. However, he also noted that "an American dislikes to make any such declaration as he considers it a shame not to be with his class in all its doings." (It is unclear where Shah picked up this idea. It seems mostly likely to have come from his roommates or his landlady.) Although Shah wrote that he appreciated the strong loyalty of Americans, he considered his own position precarious. He was willing to pay a fine of a few dollars, but he did not feel that he could afford to pay a fine of $20 or $25.[58]

Shah finally took up the problem with an MIT dean, who told him he was justified in taking an oath because "fellows who have to care for money should do so." In spite of this, Shah felt it better if he would "stand by the American sense of loyalty to one's associates." Ultimately the assessment against the students was a pittance—two dollars—which Shah paid.[59]

The Hindustan Association had a "league of friends and neighbors," apparently non-Indians who were interested in Indian culture. A member of the league was a Miss Gulberson, a college graduate who had been born in India to missionary parents. Shah reported home that she had invited him out for tea.[60]

In November Shah wrote of a reception where he was invited to meet MIT's president, Samuel Stratton. He wrote also of Dean Talbot, who had counseled Shah about not paying the riot fine, as being "almost fatherly in his cares and worries about the 1,500 or 1,600 students of the Tech."[61] Because Shah was not following a typical course (he had come to MIT with a bachelor's degree), he had to make individual arrangements with professors for every course. In doing so, Shah later wrote that he experienced "all kindness from them."[62]

Another example of how Shah felt accepted but still not part of the Americans' world came on the occasion of a tea, possibly a Christmas tea, held by an MIT professor and attended by twenty students. Because the students and faculty were "very loving," he told Parekh, "you do not feel that you are a foreigner," even though he was the only foreigner in the group. However, it was "not easy to be a part of their lives." He couldn't participate in their music. When they talked about "warships, cannons and fighter planes," he lacked the knowledge to join in on the discussion. Henry Ford's newly introduced Model A was all the rage, but Shah, who spent his days focused on his schoolwork and paid limited attention to American culture, seemed to have had nothing to say.[63]

In his first term at MIT a significant part of Shah's interactions with Americans outside the university came through formal outreach programs to international students. For the Christmas season, the Rotary Club had invited MIT's foreign students to a luncheon meeting. Here again, Shah played the anthropologist, noting that the clubs "rituals" were quite interesting and conducive to the development of intimacy, particularly their habit of calling members by their first name, singing songs, and shaking hands. (That Shah thought calling members by their first names was a club-specific ritual suggests how little informal life he had experienced in America.) Shah met a Mr. Stockwell, the vice president of the manufacturing firm Barbour-Stockwell, and hoped to develop a closer acquaintance with him. Shah was particularly struck by how all the Rotarians, whom he called "great industrialists," mixed with students so freely claiming that "it is all so different from India or even England, where the rich will not condescend to look upon the poor."[64]

One of the most frequent and contentious subjects of Shah's correspondence with his father-in-law had to do with Champa, Shah's wife, and their daughter, Saroj, born in August 1926. Shah and Champa had been engaged when Champa was only fourteen and then married when she was seventeen. Shah wanted to give Champa freedom so that she could be an independent thinker while he was in America, but his young wife could not maintain that freedom in the face of her father's designs for her. Shah's inability to intervene from 7,000 miles away led from frustration to anger.

Traditionally Indian wives would stay at the home of their husband's family when the husband was away, but instead of staying at Ranpur with Shah's family, Champa stayed in Jetpur with her family. Parekh was wealthy, Shah's family was not. Shah saw that Parekh used the superiority of his living situation to encourage her to stay in Jetpur. Shah believed that the aristocratic style the Parekhs lived in made Champa soft.[65]

When Shah was engaged, Parekh had made promises about the education Champa would receive. Shah believed those promises were broken, and he saw Parekh as determined to thwart his efforts to give Champa a formal education. Shah saw learning English as "the key to inexhaustible treasures of knowledge," but doubted Parekh would allow her that key. Shah wanted her to go to a university; Parekh wanted her to receive English lessons at home from family members.[66]

Shah's biggest source of resentment toward Parekh was that he stoked fear in an anxious and insecure Champa, making her believe that their daughter, Saroj, could be kept healthy, indeed be kept alive, only by staying in Jetpur. This so enraged Shah that he wrote to Parekh that if that were truly the case, he wished Saroj would "die tomorrow."[67]

Shah's ultimate priority was for his young wife's freedom and education. He was willing to be satisfied with a variety of possibilities for her so long as it was Champa's choice. But Shah believed that she was paralyzed by her father. Finally in frustration, in July 1928, Shah, who rarely sent telegrams because of their expense, sent one ordering Champa to go to the Satyagraha Ashram. He noted that even if the ashram did not offer a literary education, it was preferable to staying somewhere that offered no educational opportunities.[68]

Champa would stay at the ashram less than six months. In January 1929 Saroj developed a cough. Champa was worried about it, and Parekh amplified those worries, calling her back to Jetpur. As Champa left, Gandhi wrote to Parekh about her departure, showing no awareness of how Parekh had constantly manipulated her.[69]

Shortly thereafter, in May 1929, Shah could take no more. He wrote a letter to Parekh stating that "our relationship ends today." He accused Parekh of "trampling on the ideals" of his and Champa's marital obligations. On the envelope as received in India was scrawled "Trikubhai [Shah] is angry, I do not know why." If these words were written by Parekh, they suggest his complete insensitivity to everything that was going on. The scriptwriter paid no attention to the feelings of his characters.[70]

While both Devchand Parekh and T. M. Shah were both in a sense Gandhians, in practice this meant something different to each man. Even though Parekh was objectively closer to Gandhi, having by this point worked with Gandhi over a quarter of a century and being a regular correspondent of his, Shah regularly challenged Parekh's Gandhian credentials. At the heart of Shah's critique of Parekh was that he was an aristocrat, used to a life of luxury, who was only super-ficially Gandhian. Parekh did not lead the austere, disciplined life that Gandhi espoused. When Gandhi returned to India from South Africa, Parekh was in his forties and had already established a pat-tern of living. As much as Parekh tried to support his longtime friend, he could not change who he himself was. Gandhi's exposure of Parekh's unused spinning wheel had shown this, but Shah saw it more deeply.[71]

In March 1927 Parekh asked Shah for advice about the possibility of marrying a daughter in the family to Gandhi's son, Devdas. Shah, who seemed to know Devdas fairly well, said it would be great fortune for the Parekh family to have him as a son-in-law, but he also gave Parekh a warning. Devdas was a "great devotee of Bapu" [Gandhi] who would "try his best to incorporate his principles into his own life." This would make it a challenge for any woman to be married to him, and Shah urged Parekh to let the woman make the choice herself. But Shah, who had at one point implied that Champa had had "pam-pering," pointedly doubted whether any woman in the Parekh family "would be ready to follow such disciplined and industrious life."[72]

In August 1927, as he told Parekh that he had ordered some khadi to be shipped to him, Shah complained that if he had been in India, "foreign cloth would not have touched Saroj's body." While he blamed his wife, given the family dynamic, one can imagine Devchand or other family members suggesting that the child would be more com-fortable in some clothing other than the rough khadi. But what really bothered Shah, both in Parekh and in his wife, was that they seemed to accept a principle but then not live by it. Shah tellingly said to Parekh, "Perhaps you might not understand my idealism," saying he wanted Parekh to know "we walk on different paths."[73]

Parekh served as the secretary of the Kathiawar Political Conference, and Gandhi would occasionally speak at its annual meetings. Shah regularly ridiculed the Conference to Parekh for its inactivity and its insensitivity to the plight of the people of Kathiawar. In 1927, after floods had struck Kathiawar, Shah wondered why the Conference wasn't involved in flood relief. Shah several times said that if he were in Kathiawar, he would try to "topple" the office bearers of the Conference, which would have of course included Parekh himself.[74]

Between 1928 and 1929 Shah and his wife Champa, 7,000 miles apart, were each separately connected to two of the twentieth century's most extraordinary institutions and two of the century's key figures. And the seeming diametric opposition between these two institutions and these two people might be seen in their almost literally being diametrically apart on the globe.

For Shah it was MIT, which used the power of science, capitalism, and organization to promote technological development. No one better exemplified MIT than his professor, Vannevar Bush, who would later become MIT's dean of engineering. He would go on to invent for himself the position of science adviser to President Franklin Roosevelt, where he would apply scientific technology to the practice of war, giving it a destructive power it had never before had in human history. Later he would lay out a vision for new ways of accessing and handling information that would lead his biographer to call him "the Godfather" of the information age.[75]

For Champa, it was the Satyagraha Ashram and Mahatma Gandhi. Here too was an institution straining to create a new world, but in this case based on nonviolence and renunciation. Here the entrance requirements were not courses in math or science, but a series of vows, including not stealing, truth telling, celibacy, acceptance of untouchables, and not wearing foreign cloth. It was in some ways an organization more concerned with people's souls than their possessions.[76]

But there were similarities between the two institutions. In spite of its overt anticapitalism, the ashram was sustained by its links to Indian capitalists. Just as MIT had George Eastman, Pierre du Pont, and Alfred P. Sloan, the Satyagraha Ashram had among its supporters leading Indian businessmen such as Ambalal Sarabhai, Jamnalal Bajaj, and G. D. Birla.

But more than that, both institutions were built on disciplined labor and self-reliance. Just as Shah would be struggling to do his electrical laboratory experiments or his drawing assignments, Champa would be spinning or doing her own dishes or laundry. Both were slaves to a rigid time-discipline, as Shah ran off to his courses and Champa responded to the bells ringing in the ashram to call inmates to their assigned tasks.

One indication of the similarities between the two institutions comes through a little-recognized name change in the Satyagraha Ashram. In November 1928 Gandhi wrote that the name Satyagraha Ashram represented an ideal that its members aspired to, but had yet to achieve. For the sake of honesty, Gandhi was thus changing the ashram's name to better reflect what went on there. The new name would be Udyoga Mandir, "temple of industry." Both the ashram and MIT could have in fact shared that name for at both places one might see industry in a variety of forms.[77]

Family testimony is that Champa stayed in the ashram in the room of Gandhi's wife, Kasturba. Her work partner was Prabhavati Devi, the young wife of Jayprakash Narayan, who was himself studying in the United States. At times Champa's young daughter was left near Gandhi himself, in order that Champa could complete her chores.[78]

T. M. Shah was the first Indian with direct connections to Devchand Parekh to attend MIT; others would follow. There was Devchand's nephew Maganlal, followed by nephew Kantilal, followed by son Vasantlal, followed by son Mansukhlal, followed by nephew Rasiklal, followed by grandson Suresh Nanavati. Almost all studied some type of chemical engineering, presumably encouraged in that area by their family's work in the Bhavnagar Chemical Works.

Bhavnagar and MIT

T. M. Shah seems to have experienced a good deal of loneliness during his early years at MIT. His early correspondence reveals no contact with other Gujaratis, with whom he could speak his native tongue, and only a very small circle of Indian friends. By the time he left MIT in 1930 that had changed, as three other Gujaratis, two from the Kathiawar Peninsula, had arrived. In the 1930s, a small wave of Gujaratis would attend MIT. In the years 1930 to 1940, MIT awarded thirty-two degrees to people from India: nineteen of these degrees went to people from Gujarat. More striking still is how these Gujaratis were localized. Almost all were from the Kathiawar Peninsula and, more specifically, the princely state of Bhavnagar, where the Parekhs had set up their chemical factory. Fifteen of the thirty-two degrees earned by Indians at MIT during this period came from people associated with Bhavnagar, making this small princely state, which composed less than 2 percent of India's population, responsible for almost 50 percent of the degrees won by Indians.

The explanation for this remarkable overrepresentation of Bhavnagar at MIT is an informal partnership between Devchand Parekh and the dewan of Bhavnagar, Sir Prabhashankar Pattani. Pattani's interests in industrial development had led him to grant the Parekhs concessions to locate their chemical plant in his state. Devchand Parekh's passion for MIT, joined with the men's mutual interests in industrial development and Pattani's resources as the chief minister of a princely state, led students to MIT.[79]

During the 1920s and 1930s, admission to MIT was not a matter of standing out in some fierce competition, where many applicants vied for a few positions. Rather, getting into MIT was a matter of three factors: wanting to go to MIT, having an adequate academic preparation, and having the financial support. For both domestic and foreign students, particularly in the era of the Great Depression, these three factors served as a sufficient filter such that students meeting those conditions were likely to gain admission to MIT. What made Bhavnagar such a rich source of students to MIT was that it was a place where a disproportionate number of Indians met all three conditions.

Bhavnagar's status as a princely state provided an environment that was more favorable to funding education in the United States than was possible in the parts of India directly controlled by the British. Up until the mid-1950s, with the exception of doctoral students who might get a research or teaching assistantship, the money to fund an Indian student at MIT had to come from India. This money could come from several sources such as families, private voluntary organizations, or philanthropies, such as the Tata Endowment. Indian princely states were another potential source. The British often claimed that India did not need more Indian engineers, particularly engineers with advanced training, because these engineers would not be able to find jobs. The British funded some Indians to come to the United Kingdom for technical training, but the funds were meager and the numbers were small—sixty-six in a nine-year period between 1904 and 1912. Here, however, a princely state had a certain amount of freedom to spend money as it wished.[80]

Most of the MIT students from Bhavnagar, with the exception of the Parekhs, appear to have been funded by the princely state. The grandson of the dewan of Bhavnagar entered MIT in 1936. When the dewan himself came, he treated all the Indian students at MIT to a luncheon party at Boston's Ritz-Carlton Hotel. Later, the maharaja of Bhavnagar visited Boston and also treated the MIT Indian students.[81]

Bhavnagar also provided the first Indian to earn a doctorate in engineering from MIT and one of the most important Indian MIT graduates of the colonial period, Anant Pandya. Pandya's life and career demonstrate the interactions between an Indian middle class, Gandhian nationalism, and a more Western technological nationalism. Pandya was born in Bhavnagar in 1909, and when he was three his father left to study agriculture at Cornell while Pandya stayed in Bhavnagar under the care of his grandfather, a railroad station master. Pandya's father, in addition to graduating from Cornell, studied at Berkeley before serving as an agriculturalist for several Indian princely states, spending large amounts of time away from the family home. Pandya's cousin, Upendra Bhatt, five months older than Pandya, was raised with him as a virtual twin and wrote a memoir of their years together. He recalled their early introduction to engineering when they were four: "We played with wooden blocks and pictures received from America. We asked questions about the skyscrapers, Niagara Falls, mammoth bridges, and engineering marvels of Panama Canal."[82]

Pandya and Bhatt attended the Dakshinamurti Bhavan, a nationalist school in Bhavnagar where some of Devchand Parekh's children were also educated. The school emphasized "ancient culture and character building," but also had influences from the Dalton and Montessori movements. In 1925, when Gandhi came to Bhavnagar to chair the Kathiawar Political Conference, Pandya and Bhatt worked in his camp as volunteers.[83]

When Pandya and Bhatt completed their education at the Dakshinamurti Bhavan, they faced a decision: they could continue their education on Gandhian lines by going to the Gujarat Vidyapith, or if they wanted to seek a scientific education, they would have to enter the government-controlled Gujarat College in Ahmedabad. After long discussions with their teachers at the Dakshinamurti Bhavan, the two young men decided to go to Gujarat College to study science. Even though they studied at the Gujarat College, their hearts were with Gandhi and the Vidyapith. At times they slept at Gandhi's ashram so that they could participate in morning prayers. They spent much time with the faculty and students at the Vidyapith, frequently spending the night there.[84]

In 1927 Pandya entered the NED Engineering College in Karachi, with Bhatt following the next year. Pandya, Bhatt, and a friend, Ramesh Mehta, maintained the following schedule, which in its discipline could have been appreciated by a member of the Satyagraha Ashram: "We got up early morning at 3:15, concentrated on studies between 3:30 and 7:30 in the quiet of the morning. Physical exercise from 7:30 to 8:00. Bath and personal laundry between 8:00 and 9:00." Bhatt's conclusion was again almost Gandhian: "With full control of the body and the mind, the examinations were conquered."[85]

In Karachi, Pandya and Bhatt continued their involvement in social and political activities. The students of the Gujarat Club skipped dinner and contributed the funds for the relief of the Sindh flood victims. Gandhi and Bhatt met weekly in the salon of a former principal of the Gujarat Vidyapith, lately moved to Karachi, to discuss the problems of India's freedom. Pandya and Bhatt organized, and got the principal to support, a one-day-a-month student strike in sympathy with Gandhi's imprisonment.[86]

Pandya finished at the top of all University of Bombay graduates. Upon his graduation he again faced a decision. The two clearest options were taking a job with the Indian Engineering Service or accepting a Prince of Wales Scholarship. However, the position in the Indian Engineering Service meant being a part of the colonial state, a condition unacceptable to him. The Prince of Wales Scholarship came with similar limitations in that it was tenable only in the United Kingdom or other dominion countries. Instead, Pandya applied to and was accepted at MIT, supported by a scholarship from the state of Bhavnagar. Bhatt followed the next year.[87]

On September 5, 1930, Pandya arrived in New York City en route to MIT. He traveled with Vinayak Shah, another young man from Bhavnagar also headed to MIT. On the customs form, Pandya listed the person he would be staying with at MIT as Maganlal Parekh, Devchand's nephew. Pandya would have had many factors pushing him to MIT. It is inconceivable that Pandya and his family would not have known the Parekhs, members of a small middle class in Bhavnagar. Furthermore Pandya's father, with his American education, would have had a good sense of the educational opportunities available in the United States.[88]

At MIT the programs of the Parekhs and the other Bhavnagar students did not reflect particular tailoring for engineering in India. They got the same education that American students did. For example, Devchand's son Mansukhlal did a doctoral thesis in chemical engineering working under Edward Gilliland on techniques for petroleum refining, at a time when India had neither petroleum production nor refining capabilities.[89]

The Parekh young men and the other Indians from Bhavnagar appear to have had a much easier time adjusting to MIT and the United States than T. M. Shah did in the late 1920s. By and large, they had had engineering training in India that gave them some preparation for coming to MIT. They formed a group whose members were much younger and without the cares of a wife or children. Neither did they have financial concerns, with several being from wealthy families. While Devchand Parekh required a detailed accounting of expenses from T. M. Shah, Parekh's son Mansukhlal asserted that he had no such requirements placed on him.[90]

Neither the Parekhs nor any of the other Indians from Bhavnagar appear to have lived at MIT in such a way as to call attention to their connections to Gandhi. Pictures of their time in America always show them wearing Western clothes. The student newspaper shows no sign of their being involved in formal political activities. The best sense of their daily lives comes through the memoirs of L. M. Krishnan, a young South Indian who earned a master's degree at MIT from 1936 to 1939. Krishnan's memoirs show a small group of wealthy Indians who had a healthy social life with one another, maintaining aspects of their Indian culture while largely fitting into American culture.[91]

Not surprisingly, Krishnan's memoir showed that the Parekhs and other students from Bhavnagar dominated social life at MIT. Krishnan had an uncle who lived in Cambridge, but as soon as Krishnan arrived he took them over to the apartment shared by four Parekh cousins so that they could help him get established at MIT. After living alone for one term, one of the Parekh cousins took a cooperative job, opening up a space for Krishnan to move into the apartment.[92]

The four did their own cooking "Indian style," sustained by a shop in Boston that sold twenty-seven different spices. As might be expected, their cooking was Gujarati, not the South Indian fare that would have been more familiar to Krishnan. Occasionally the roommates would invite other Indian friends to the apartment for a special dinner that consisted of deep fried vegetables and "milk flavored with spice and boiled to creamy consistency." When they were not cooking themselves, the students often ate standard American fare at MIT's cafeteria or local restaurants, like Howard Johnson's. When they sought Indian food or something resembling it, they either ate at the Athens restaurant, at the Boston Vedanta Center, run by the Ramakrishna Mission or finally, in 1939, at the first Indian restaurant in Boston, run by a Syrian Christian.[93]

The students showed great creativity in amusing themselves in ways that were distinctive of Indian culture and in ways typical of MIT students. Once the students had an Indian night with a Garba Gujarati folk dance. American women were enlisted to participate and were wrapped in saris by one of the men. Krishnan noted that the man "did a good job and had a good time doing it." However, as might be expected, with little dance practice, Krishnan explained that "most of us got out of rhythm and began facing the wrong direction and we were unable to make hand contact with partners for clapping. It was an utter fiasco."[94]

The Indian students also got into MIT's spirit of pranks and practical jokes. When Devchand's son Mansukhlal completed his doctoral examination, he offered to take a group of four Indian students out to dinner at Hartwell Farms outside of Boston. Parekh and a coconspirator with a car excused themselves to go to the bathroom midway through the dinner and then took off for home, leaving the others to pay the bill and walk home. Parekh then booby-trapped the hapless victims' apartment so that the first one who opened the door was doused with water.[95]

The Mahatma and the Engineer

In the 1930s no young Indian had better Gandhian nationalist credentials than Bal Kalelkar. His father, Kaka Kalelkar, himself the son of a treasury officer for the Raj, had developed nationalist and anti-British leanings in the early part of the twentieth century through reading the work of Tilak. The senior Kalelkar worked primarily in Indian schools, going in 1914 to Rabindranath Tagore's Santiniketan, where he was to meet Gandhi in February 1915. Shortly thereafter Kalelkar joined Gandhi at his newly established Satyagraha Ashram in Ahmedabad. Kalelkar became Gandhi's main educationalist, serving for a time as principal of the ashram school and later professor at the Gujarat Vidyapith. Kalelkar spent time in jail with Gandhi, organized events, and took over some of Gandhi's publications when he was in jail.[96]

Gandhi later wrote that Bal Kalelkar, born in 1912, "was brought up under my hands," and in a memoir Kalelkar wrote of his many experiences growing up in Gandhi's ashram. He wrote of Gandhi's involvement in the early days with the physical aspects of the ashram, "from clearing the ground for open air prayers, to digging ditches for movable latrines, there was nothing that he did not personally supervise and actively participate in." He wrote further of Gandhi's insistence on "everyone learning and meticulously observing rules of hygiene."[97]

While the rest of the world was coming to see Gandhi as a larger-than-life figure, a saint, who by the power of nonviolence had taken on the mighty British Empire, Bal experienced Gandhi as someone who himself could be taken on. He and one of Gandhi's grandsons convinced Gandhi to give them each a five rupee a month allowance so that they could learn photography. When Gandhi dictated that the supply of soap for washing clothes at the ashram be cut as a show of solidarity with villagers, Bal—who had regularly engaged in a competition with the other boys as to who could wash their clothes the whitest—argued against the new dispensation on the grounds that the ashram should not lower itself to village practices. He got 90 percent of the ashram boys to sign a petition asking for the return of the soap, and at that point Gandhi relented. Bal also showed his commitment to the discipline Gandhi espoused by rapidly learning the 700 verses of the Bhagavad Gita and by setting records for handspinning—even, according to a later reminiscence, spinning for twenty-four consecutive hours.[98]

Bal stood with Gandhi in some of the key moments of his movement. At age eighteen, Bal became one of a select group chosen to participate with Gandhi on the Salt March. For Gandhi this was a political, spiritual, and moral exercise, and he required each marcher to perform daily duties: spinning, praying, and keeping a diary, which Gandhi read. Kalelkar was imprisoned three times and on several occasions nursed Gandhi during his fasts.[99]

In an autobiographical sketch written in 1944, Kalelkar seemed to be the model of the young person Gandhi had hoped to train to build up his new India:

[I]n the year 1930, the author found the country seething with political unrest, and though still in his teens, he decided to plunge into the social and political activities carried out by the Indian National Congress under the leadership of Mahatma Gandhi. In the years 1930–35 he devoted his entire time to organizing political activities in the villages of India and was imprisoned for the same. During this period and after his release from prisons, he also did extensive social and constructive work. In collaboration with three other colleagues the author started a social-educational institute in a village of Gujarat and acted as the secretary and teacher at the institute. He also acted as the treasurer in two responsible nationalistic organizations during this period and took an active part in organizing the relief work in the province of Bihar during the Great Earthquake of 1934.[100]

Being part of Gandhi's movement was not all Kalelkar had been doing; he had also been studying engineering, earning his degree from NED Engineering College in Karachi in 1940. And in 1940 he was off to America to pursue graduate studies in engineering at MIT.[101]

How did Kalelkar come to his interest in engineering? Kalelkar had some connections with Devchand Parekh's network. Bal wrote of ministering to Gandhi in 1939 with one of Parekh's daughters who ultimately married Bal's brother. Kalelkar's 1946 memoir of Gandhi points in another direction, however. The memoir, largely an apologia for how a disciple of Gandhi could end up in engineering, a discipline in some ways so antithetical to Gandhi's ideas, creates an image of Gandhi as an engineer of human souls. In other parts of the memoir, Kalelkar is much less metaphorical about Gandhi as an engineer. The Gandhi he experienced was an engineer, pure and simple, trying to bring order out of his physical world. Indeed, the whole Satyagraha Ashram was consistent with engineering with its attitude of discipline and its focus on the physical aspects of life. It can hardly be surprising that someone who spent time each day spinning or washing his clothes to make them as white as possible would then become interested in technology more broadly. In that way the Satyagraha Ashram paralleled medieval monasteries, where the practice of daily manual labor encouraged monks to look more broadly at technology. Kalelkar noted that he was made the "head of the Workshop Dept." in Gandhi's ashram, the manager "having seen the budding engineer" in him.[102]

Gandhi supported Kalelkar's quest to become an engineer in a variety of ways. Going to MIT required money and Bal and his father had none. Kalelkar wrote to G. D. Birla, Indian business magnate and close associate of Gandhi, asking for a scholarship of 9,000 rupees. Whether or not this was Gandhi's idea, Gandhi, like Benjamin Franklin, a newspaper editor, helped Kalelkar revise his letter.[103]

Kalelkar laid out his case to Birla in a straightforward way in both the draft and the final version. He sought funds to earn a doctorate in engineering at MIT, which "is considered the best institute of its kind in the whole world." While Kalelkar had funded his education up to that point with gifts from friends and scholarships based on his examination results, he felt he could not rely on those in America. Furthermore, Kalelkar did not want to be encumbered by debt on his return to India—he wanted a gift not a loan, although he wrote that he would consider any gifts to contain a "moral obligation" that he would seek to discharge upon his return to India. Kalelkar also made clear that he was not undertaking a doctorate for the pursuit of personal glory, but "to serve our motherland through my profession and to see her in a better position."[104]

Gandhi, giving the conflicting signals he would sometimes send, personally forwarded Kalelkar's letter to Birla with his own disingenuously distancing cover note:

BHAI GHANSHYAMDAS, This from Bal. He wishes it sent just as it is. I said if it must be sent then let me do the sending. But no special significance should be attached to fact that I am forwarding it.
   Blessings from
   Bapu[105]

Although Kalelkar assumed that Birla did not know him, Gandhi's letter seems to presume that he did. It can hardly be surprising that Birla agreed to support Kalelkar.[106]

When Kalelkar left for America in July 1940, Gandhi wrote him the following letter of support: "This is to introduce young Kalelkar to all my friends in America. He was brought up under my hands. He is one of the most promising among the boys brought up in Satyagraha Ashram. Any help rendered him will be appreciated."[107]

In the late twentieth and early twenty-first centuries, Gandhi has become an icon rather than a real flesh and blood person, so that the adjective "Gandhian" has come to have unambiguous meaning. However, one would do well to ponder what it says about Gandhi's movement that a young man Gandhi claimed was "brought up under my hands" ended up at MIT, the antithesis of what the stereotypical view of "Gandhian" envisions. And he ended up there not in open rebellion against Gandhi, but as a committed disciple.[108]

On July 13, 1940, Kalelkar set sail from Bombay for New York on the *President Garfield*. Whether or not the Parekhs had anything to do with him going to MIT, by this time he had been integrated into the Bhavnagar/Parekh/MIT network. His traveling companion was Suresh Nanavati, Devchand Parekh's grandson, who was entering MIT as an undergraduate. Both young men listed Mansukhlal Parekh as their contact in America, and he likely met them when they arrived in New York on August 18.[109]

Kalelkar, like the other Indian students of his generation, pursued an engineering education at MIT that was not at all Gandhian in its content, training him in the state of the art of American engineering. After completing a master's thesis on the stress on connecting rods in internal combustion engines, Kalelkar moved to Cornell where he completed his doctorate in mechanical engineering, working on another aspect of the design of internal combustion engines.[110]

While Kalelkar dedicated his doctoral dissertation to Gandhi as "That Grand Old Man of India," there were tensions in the relation between the teacher and his disciple. Kalelkar family testimony is that when Kalelkar left for the United States, Gandhi proposed that the two reserve a set time each week when the two would think of each other. But when Kalelkar got to the United States, he found with the pressure of his schoolwork, he could not maintain this once-a-week focus on Gandhi. Family testimony says that Kalelkar was forced to write to Gandhi confessing his inability to maintain their plan. One kind of discipline had crowded out another.[111]

Another measure of the distance that Kalelkar had moved from Gandhi is suggested by the fact that for a time Kalelkar worked for the American aircraft manufacturer Lycoming in Williamsport, Pennsylvania, as part of the effort to build a 5,000-horsepower aircraft engine, the largest in the world at that time. This engine was intended to power a large bomber that could launch attacks on

Europe from the United States. This disciple of the twentieth century's greatest advocate of nonviolence thus became a part of the most powerful system of organized violence that the world had ever seen.[112]

In November 1944 Gandhi wrote Kalelkar: "I have your beautiful letter. I can understand that western music has claimed you. Does it not mean that you have such a sensitive ear as to appreciate this music? All I wish is that you should have all that is to be gained there and come here when your time is up and be worthy of your country."[113]

Gandhi's words raised fundamental questions for all Indians who went to MIT, not just Kalelkar. What did it mean to have "all that is to be gained there"? And what did it mean to return and "be worthy of your country," particularly a country whose very definition was up for grabs? The Indians who went to MIT, from Shah to Pandya to Kalelkar, would face these questions upon their return to India.

T. M. Shah, Anant Pandya, Bal Kalelkar, and the other Indians who went to MIT in the late 1920s through the early 1940s had demonstrated several things. Although not in serious question, they had demonstrated that Indians could survive in the cold Northeast climate of the United States. They had demonstrated that based on the education they had received in India, they could successfully complete the demanding MIT curriculum, becoming, at least by academic qualifications, engineers trained to the highest level. To a certain extent they had demonstrated that they could take part in American culture while retaining their Indian culture. It remained to be seen what an MIT education would mean in India.

5

Engineering a Colonial State

Hazaribagh, "A Thousand Gardens," is a district in eastern India where two Indian MIT graduates had dramatically different experiences between 1939 and 1944. In 1939 Anant Pandya was the newly appointed principal of the Bengal Engineering College, placing him at the top ranks of Indian engineers. He drove up to Hazaribagh from Calcutta, where he met a group of third- and fourth-year civil engineering students who were there for survey camp. He and his wife had tea with the students; one student later reported that Pandya's "radiant persuasion and homely behavior endeared him to us all within a very short time."[1] T. M. Shah's experience was far less pleasant. Hazaribagh was also the home of an infamous jail, where participants in India's freedom struggle were detained. Shah spent eighteen months there, a punishment he earned for leading a strike at the Tata Iron and Steel Works in support of Mahatma Gandhi's "Quit India" movement.[2]

The group of Indians who had gone to MIT beginning in the 1920s finished their education and returned to India with the best engineering training America could provide. Devchand Parekh had a vision of what these engineers might do in India: start new industries that would benefit the country. As these Indians came back from MIT they faced a country in a struggle for freedom. They then had to face the question of what role they should play in this struggle:

Were they engineers first, or should they be part of the freedom movement? The more specific question was, Should they take up some of the many technical jobs requiring them to accede to the colonial system or should they stay outside that system? Each person came to his own answer, which did much to shape the course of his own career. The different experiences of Pandya and Shah, along with that of Gandhi's disciple Bal Kalelkar, showed the tensions elite engineers faced in the colonial state.

Pandya's Progress

Of all the engineers from Bhavnagar who went to MIT, Anant Pandya had the most prominent career. That prominence and a unique collection of letters he wrote to an American friend allow his life to be followed with greater specificity than any of his contemporaries. After leaving MIT, the most challenging part of his life began as he worked to find a way to use his education and his talents in a colonial state.

Pandya had come to MIT as a Gandhian; at some point during his time at Cambridge, his politics shifted more toward Marx. Between 1932 and 1934 he regularly wrote to one of his close leftist friends, Frances Siegel, a Radcliffe graduate who worked as a secretary in Harvard's Widener Library. While these letters allude to their leftist inclinations, not uncommon at a time when a global economic crisis had given rise to doubts about the viability of capitalism, they also paint a picture of how the first Indian to earn a doctorate in engineering from MIT juggled his identities as an American-educated Indian engineer.[3]

By going to MIT, Indian students had become a part of a community of engineers that spanned the United States and the world. In June 1933, after completing his doctoral dissertation, Anant Pandya traveled throughout the United States with two fellow Indian students, and then in September, he left to return to India by way of Europe. When Pandya landed in England, he did so as a subject of the British Empire, but also as an engineer trained to the highest standards in America and as someone who had experienced American life. By virtue of his time spent in the United States and of the values he had imbibed there, he saw himself, in some way, as an American even if the United States government did not. While Pandya's critiques of Western society could be overly simplistic (as the British critiques of India often were), he brought an expertise and personal experience to his reports that were not found in the *Mahratta* of the nineteenth century.[4]

On arriving in England from the United States, Pandya wrote to his friend that the "queerest thing" he noticed was the number of bicycles on the road, compared to the cars he would have seen in the United States. The bicycles demonstrated to Pandya England's diminished economic and technological position, prompting him to say England is "no longer the leader" and "can never aspire to regain her important position again in the rapidly developing world of the twentieth century." Pandya also noted the smallness of houses, roads, cars, and railroads in Britain, which he said "could not escape the eyes of an American like me!"[5]

Pandya's letters to Siegel occasionally make oblique and cautious reference to leftist activities. After eight days in London, Pandya was off to Leningrad to see a Russian friend and the new society being built in the Soviet Union, where he saw positive signs everywhere. While he commented on the new construction, he was far more impressed with the new people being created whose faces "displayed a remarkable optimism, freedom and energy." Throughout his visit, Russians asked Pandya to stay and be a part of their new society.[6]

Pandya's letters to Siegel from Europe show how his affiliation with MIT had created connections that linked him with people everywhere he went in Europe. Sometimes those connections would be with those he had known in Cambridge; sometimes the connections were with people Pandya had never met before, but whose own connection to MIT linked them to Pandya. At the most basic level, by virtue of his MIT education, Pandya was a member of a scientific and technical elite and he found fellowship within that community throughout Europe.

On the ship to Leningrad, Pandya met a Soviet student who had studied at Cornell and had known the Soviet students Pandya knew at MIT. On the ship back from Leningrad, Pandya encountered the distinguished Soviet physicist Abram Ioffe, noting to Siegel that he had once lectured at MIT. In the course of Pandya's discussions with him, Pandya became convinced that within a decade, "the USSR will be the leader also in science and technology."[7]

Pandya arrived at his next stop, Germany, at a remarkable time in history. In January 1933 Hitler had been appointed chancellor and Pandya saw signs of the Nazification of Germany everywhere. On October 14, while Pandya was in Berlin, Hitler withdrew the country from the League of Nations, and Pandya reported on freshly printed news accounts as well as recently hung posters with appeals from Hitler to the German people. Pandya saw many other ominous signs portending war, concluding, "I would be more surprised if there is no war in five years (to say the least) than if there is one," showing more insight than many contemporary European leaders.[8]

From Berlin Pandya went to Saxony to stay with a German engineer who had been a fellow student at MIT. To Pandya's surprise, the friend was now an ardent Nazi and anti-Semite. Pandya noted that villagers greeted one another with the fascist salute and "Heil Hitler." Later in Heidelberg, the whole city was closed between the hours of ten and two to allow people to hear Hitler speak.[9]

In spite of Pandya's "admiration for German laboratories and German science" (and he had seen much of both during his trip), the Nazis cast a pall over his trip. Although the swastika had been a sign of "bright auspiciousness" to him during his childhood in India, Pandya had come to hate it in Germany. When Pandya finally left Germany for Amsterdam, he wrote that he felt as "free as a bird."[10]

The freedom Pandya felt in Amsterdam did not last long. Pandya had been outside the British Empire for three and a half years since he had left India to come to MIT, but on December 8, 1933, when he boarded the P&O liner *Mantua* in Marseilles for the voyage to Bombay, he wrote, "Here I am already in India—for the atmosphere here is no longer the free European or American one, but rather unpleasant, sort of stifling."[11]

As he wrote about the unpleasantness of the British, he lapsed into general stereotypical characterizations of them, just as they had done of Indians. He called the British "too dull to have any musical sense," continuing, "They seem to have the toughness and moods of a bulldog, their talk sounds like low heavy growl and though not possessing much brains they do have a tenacity and strength which gives them mastery over an empire 'over which the sun never sets.'" Pandya concluded this paragraph by saying that British mannerisms were "too much for an 'American' like me!"[12]

As Pandya implied, he was doubly alienated from his British shipmates on the voyage to Bombay: once because he was Indian and again because of his orientation to the United States. An Oxford-educated member of the British Labour Party complimented Pandya for having "escaped the bad influence of corrupted American language!" Pandya noted that this man, "like most other passengers," has "a strong dislike for the States and for things American—excepting perhaps American movies."[13]

Nevertheless, in an overly serious, moralizing letter to his American friend, Pandya professed to see advantages to being with the British. Pandya wrote that this was "the first time I have come in close contact with the British people—my 'rulers,'" and as such he got an opportunity to observe their manners and social life. Pandya wrote that he imagined that he would have to deal with these people his entire life, "a difficult and bitter struggle" he was not looking forward to.[14]

On December 22, 1933 Pandya landed in Bombay. Although he had not anticipated it, "very many" came to receive him, including his parents, his younger brother, an uncle, and his good friend Ramesh Mehta. (Some had traveled a long distance to be able to welcome him home.) Pandya spent three days in Bombay meeting people before traveling by train to his parents' house in Gwalior, where his father served as an agricultural adviser to the princely state.[15]

In his first letter from India to his American friend, Pandya was introspective about how he had changed and how he now saw India differently. He felt "surprise and strangeness" on returning to India, seeing that it was "poorer and more wretched" than he had thought. Bombay, the richest city in India, was dirtier than the dirtiest city Pandya had visited in the United States.[16]

But even in Gwalior, a city of 127,000 that was by no means cosmopolitan, capitalism periodically brought America to Pandya. His father drove a Canadian-made Ford. The annual fair in Gwalior included as one of its attractions an American variety show run by "a very typical American from Chicago," with whom Pandya had a pleasant chat. One evening in January 1934, Pandya went to a talking picture show in Gwalior. An Indian talkie, *Puran Bhakt*, was the main feature, but it was preceded by a Walt Disney Silly Symphony, *Noah's Ark*. Pandya described the experience to Siegel: "It was very good to hear American tunes and to enjoy the peculiarly American humour after a long time. I almost felt as if I were in a Boston theatre!"[17]

A trip to villages with his father brought home to Pandya how far he was from America. The only traffic they encountered while traveling on a poor Indian road were bullock carts. Besides the "sharp contrast between the ultra-modern automobile and the ancient (if not prehistoric) village cart," there were practical problems. Groups of carts moving in the same direction would span the road, bringing the Ford, going in the opposite direction, to a halt. Sometimes the Ford frightened the bullocks, who would run down the berm of the road carrying their carts behind them.[18]

In a town of 6,000 they visited, Pandya was not able to find a single newspaper of any sort. The town had only one primary and one middle school, but no water other than wells and no sanitation. And this town was incomparably better provisioned than most villages. Pandya gave a bleak assessment of the Indian masses: "People live in ignorance, superstition, fear, and vague hopes day after day and year after year!"[19]

Seeing America doubtlessly made Pandya see the villages of India in a different light, but Pandya did not change the way he saw everything in India. Pandya went to Agra and the Taj Mahal, curious as to whether having been abroad would change his response to it. However, after all his travels and studies, he still saw the Taj as incomparable, writing that "there is a certain nobility, grace, delicacy, charm about Taj which distinguishes it from all structures I have seen so far!" And he still found joy in his old friends, and the rounds of parties made the days go by quickly when he returned to Bhavnagar.[20]

In January 1934, shortly after returning to India, Pandya attended the annual meeting of the major Indian engineering society, the Institution of Engineers, held in New Delhi. Pandya commented that the meeting, which included tours of the new buildings of the new

Indian capital, New Delhi, was "Imperial." The official report of the meeting, published in the *Journal of the Institution of Engineers*, makes clear what Pandya meant and gives a picture of the colonial engineering profession Pandya saw.[21]

The membership of almost 1,300 was largely made up of British engineers who had served the Raj in some official capacity. The year's president was Sir Guthrie Russell, the chief commissioner of railways. The Viceroy of India, the Marquess of Willingdon, as well as most of his executive council attended the annual banquet, held at the exclusive Maiden's Hotel. The Marquess, who had been in office less than two years, had taken an extremely hard line against those in the Indian freedom movement, jailing Gandhi and tens of thousands of his supporters. His remarks suggest that he looked on the Institution as being essentially equivalent to the Indian Engineering Service—a state body. In the only recorded remark that acknowledged any discord in the country, Willingdon thanked members of "this great service" for the "their steadiness and staunchness," for without it, he would have "found it extremely difficult to maintain government control throughout the length and breadth of this country." The global economic crisis and its effect on India also dampened the evenings celebration as Guthrie Russell apologized to the Viceroy because he had wanted to describe "spectacular engineering projects which were in the course of construction," but was unable to do so because of the difficult financial situation, which he hoped would soon be over. The dinner, of course, included drinking a toast to both the Viceroy and the King.[22]

Pandya was not as hostile toward the Institution of Engineers and its proceedings as might have been expected, writing, "I have got to know Indian engineers and the sooner I start the better." His first impression of the meeting and of his "brother engineers" was "very assuring," and he announced plans to join and present a paper at the next meeting, confidently asserting to Frances, "I shouldn't find much difficulty in pushing myself forward!"[23]

But finding a job proved to be difficult for Pandya. His attendance at the Institution of Engineers meeting did not lead to a job. While the global depression was one problem, his good friend Upendra Bhatt felt the basic issue was that no one in a responsible position in

India could properly assess Pandya's abilities. Bhatt noted a prejudice in India against those with American degrees. Pandya's MIT
education seemed to confer no advantage in India. One firm offered
Pandya a junior position at the humiliatingly low salary of 150 rupees
a month.[24]

Pandya's inability to find a job took a personal toll on this gregar-
ious, high-energy young man. In May 1934 Pandya wrote to Siegel
after an interval of five weeks, apologizing for not answering her
many letters written in the interim and confessing not knowing why
he hadn't written. He continued: "It is almost three months I have
been here and I haven't accomplished anything—I am simply idling
away my time. And that I find very distressing—unbearable. I must
have plenty of work and activity but that is not to be found in this
corner of India."[25]

Finally, after six months of looking, Pandya took a job with Mc-
Kenzies, a construction firm, working with British engineers in
Bombay. Pandya started out on an optimistic note saying the work
was "interesting" and likely to prove useful when he started work
independently. But the job turned out to be the routine work of a
typical engineering firm. He supervised small jobs in small towns,
writing to Siegel from Bulsar, where he was overseeing the construction of a foundation for an electric installation for a textile mill as
well as a water filtration plant. Pandya's job was not one that would
have required an MIT doctorate.[26]

In America, Pandya's political inclinations moved from Gandhi to
Marx, and on his return to India, his correspondence shows him in
the role of a spectator and commentator on the Indian struggle for
freedom. In June 1935 Pandya fully laid out his attitude toward
Gandhi: "Mr. Gandhi as usual has always some new stunt—the latest
being the Village Industries Reconstruction. The man has not the
faintest idea about present-day economics and here he is out to
revive dead hand-industries and all primitive methods of production.
In the face of his magnetic personality and almost perfect selflessness
there is no voice of protest—nor even a question. I cannot say what
terrible national economic waste he is expounding and fostering!
He has left up all real political work and is side-tracking the whole
country to obscure by-lanes."[27]

Pandya's life challenged aspects of traditional Indian society in a variety of ways. In September 1934, after being back in India for less than nine months, Pandya announced his engagement to a young woman living in Bombay, Lily Shah. Their meeting was almost inevitable. Both of their families were from Bhavnagar. Lily's father was Hiralal Shah, a Bombay cloth merchant who had corresponded with Gandhi. At the time the engagement was announced, two of Shah's sons were studying in America, one at MIT. The previous summer Pandya had traveled across the United States with Lily's two brothers.[28]

Pandya and Lily developed their relationship on their own even though their families had much in common, each being a "typical middle class Gujarati family—vegetarian, admirers of Gandhi," with (in Pandya's Marxist language) a "petit bourgeois outlook on life."[29] Even with so much in common, some in Pandya's family saw an insurmountable problem to the marriage. Shah, a Jain Vaishya, was from a different religion and caste than Pandya, a Hindu Brahmin. In May 1935, after months of difficulties, Pandya finally broke off relations with his family over their objections to the impending marriage. While Pandya acknowledged the split was difficult and painful, he expressed happiness with the thought "that as an individual I can undertake any risk or act without hindrance or setback from any side from now on."[30]

That same kind of personal confidence led Pandya to take action regarding his job. After a year with McKenzies, Pandya came to see that with most big engineering jobs done by British engineers working in the colonial government, there was "no scope for private initiative and enterprise," with little "original and intelligent" work left for those outside the state.[31] After spending an entire night with his friend Upendra Bhatt pondering the alternatives and plotting a course, Pandya decided to quit McKenzies and head off to seek work in London. After six weeks in London, Pandya considered taking a position without pay as a way to make contacts and was at least willing to entertain the idea of going to the Soviet Union to look for work, but he finally landed a paying job with Trussed Concrete Structures.[32]

Here Pandya finally had a job that took advantage of his abilities, as he worked on new materials and techniques for buildings. Specifically, he was part of a group developing a system of reinforced concrete and welded diagonal frames to be used in the construction of large industrial structures, such as aircraft hangers and warehouses. A paper describing his work won him and his coauthor $11,000 from the James Lincoln Arc Welding Foundation as the second-prize winner in a global competition. In 1937 Pandya received a promotion to chief engineer at Diagrid Structures, Ltd.[33]

Pandya's intuition that he could advance his career more successfully in London, the colonial metropole, than he could back in India proved correct. In 1939 an advertisement appeared in England for the position of principal of Bengal Engineering College near Calcutta, one of India's four leading engineering colleges. Pandya applied, and the selection committee judged him the most qualified applicant. In 1939, at the remarkably young age of thirty and with only one year of work in India, Pandya took the position at Sibpur, thereby becoming one of the highest placed Indian engineers in India.[34]

It would be impossible to overestimate what Pandya's appointment would have meant to students at Sibpur. With Pandya's ascension to the head of the Bengal Engineering College, a group of British engineering faculty worked under the authority of an Indian. The *Bengal Engineering College Annual* for 1940 gave a sense of the student feeling toward Pandya. It asserted that Pandya had won the "hearts of the students" through his "sound judgement and amiable character" and had removed the "feeling of awe and suspense that usually clouds the Bengal Engineering College sky."[35] The report of the school's annual sports day noted a new event, the "staff-race," in which Pandya came in second place.[36]

A student's poem described Pandya:

> The image
> Of a tall but slim stature, I was at a loss
> To behold that a jug full of water behind
> And attaché of leather inscribed P&O
> Like procession did follow the broad-eyed savant
> A. H. Pandya: the Pandit who bosses all show.
> Got impressed with the gait as he walked on the way

With the left hand in pocket and neck as a swan
And his hat made of pith was inclined to the spine
At an angle ϕ so that tan ϕ is one.[37]

Here the students see a technological Indian effortlessly spanning several worlds that might have been thought unbridgeable. Pandya unself-consciously bears on himself several of the prime emblems of the colonial regime, with the pith helmet—covering his prematurely bald head—and the P&O attaché—the P&O being the steamship line connecting Britain and India. Even so, Pandya is described with a Hindu term for scholar or teacher—pandit. Across from the poem was a cartoon of Pandya, depicting him as a Roman lictor, an imperial guard. Wearing Western clothes, his weapon was fashioned out of a slide rule, and on the blade the words "love" and "discipline" were inscribed. The student's overall picture of Pandya was of love and respect.[38]

In his welcome address given to the faculty, students, and staff at Sibpur on November 7, 1939, Pandya gave his bold vision of engineering—a vision diametrically opposed to the one he accepted when he had listened to Gandhi speak at the Satyagraha Ashram in Ahmedabad, a little more than a decade previously. Pandya gave a brief history of humanity based on changes of a material nature, such as the invention of fire or the bow and arrow. Pandya asserted that the material changes made moral and mental advancement possible—a claim that Gandhi would have vigorously protested. Pandya claimed that the world was entering a new epoch, one in which humans had learned to manufacture power. He provided a definition of engineering by Karl Compton, his president at MIT, who would play a major role in harnessing science and technology for use in the war that had just started. In this new epoch, Pandya, a member of the priestly Brahmin caste, asserted that engineers were "the priests of material development." And society would be increasingly technocratic, with rulers being selected from engineers and the importance of planning in society increasingly coming to the fore. Pandya stated that India's "crying need" was an increase in its people's standard of living, which could only come through "engineering advance and progress." As he gave this broad vision of the future, he implicitly showed one of its problems, as he made no mention of India's status as a colonial power or any efforts to win independence for India—those were not issues that could be accommodated by technocrats.[39]

By becoming principal at Bengal Engineering College, Pandya had, to a large extent, cast his lot with the colonial government. He was now part of the colonial system and had apparently made the decision that he could do more for India within the system than fighting it from without. World War II presented two opportunities for Indians: one was to use the demands for production as a way to advance the Indian economy; the other was to use the period of instability to give the British the final shove out. Pandya focused on the former. Like Vannevar Bush and Karl Compton did at MIT, Pandya used the war to show that his institution could be used to put technical knowledge to work for the war effort. Pandya took the lead in converting the college over to an institute for training technicians who would then work in munitions factories or the technical wing of the air force. Admission to the entering class of 1940 was suspended to devote the college's facilities to the war effort, and temporary barracks and classrooms were built to accommodate the new students.[40]

The exigencies of the war brought a dramatic change in Pandya's status over what he had experienced when he had first returned to India from the United States slightly over five years previously. Then no one understood his abilities, and he languished for months at the family's house in Bhavnagar. With the outbreak of the war, Indians with Pandya's talents were rare, and he rapidly rose through a series of increasingly responsible positions. The Bengal government seconded him to the Government of India, where he served as director of metals, then he was promoted to deputy director of munitions production, the first Indian to hold that position. Just how far Pandya had risen was seen by the fact that his immediate superior had become Sir Thomas Guthrie Russell, who had given the Presidential Address of the Institution of Engineers when Pandya had just returned to India, largely unknown and looking for his first job.[41]

The Travails of T. M. Shah

In June 1930, after four years in the United States, T. M. Shah returned to India with both bachelor's and master's degrees in electrical engineering from MIT. He returned at an auspicious time in the history of the Indian freedom movement, but at an inauspicious time for his engineering career. In March of that year, Gandhi had set off on his march from Ahmedabad to the sea to make his own salt in defiance of the colonial government's salt tax. Gandhi's march had electrified millions of people throughout the world and sparked civil disobedience throughout India.[42] By December, the colonial government had imprisoned 60,000 for acts of civil disobedience. Shah later said that on his return he "was immediately engaged with the freedom movement in India that had begun with the historic Dandi March of Mahatma Gandhi in 1930." Shah had been four years outside the freedom movement at MIT, but whatever it meant for his engineering career, Shah would from here on be both an engineer and freedom fighter. Perhaps Shah's priorities can be judged by his statement for his MIT twenty-fifth reunion brochure: "I had some professional work in between the prison goings."[43]

A common theme in the lives of those leaving their culture for education and then returning is a sense of double-mindedness: alienation or anomie. They were present to some degree in Shah's career, but not for the reasons usually given. The alienation did not come inevitably from cultural differences, but was due to the tension between pursuing the kind of career an MIT education presumed or offered and working for the cause of Indian freedom, which could relegate career considerations to a secondary role.

Shah came back to India with a combined theoretical and practical education qualifying him for some of the most responsible positions in electrical engineering in India. He had worked for almost a year and a half at GE plants in Lynn, Massachusetts, Pittsfield, Massachusetts, and Philadelphia, Pennsylvania. Just as he had tenaciously pursued admission into the cooperative program, he sought high-level employment in India. By mid-1930 he had letters of recommendation sent to Tata Hydro-Electric Power and to Dorab Tata himself for a position. For some reason, Shah was either not offered a job or did not accept it.[44]

In 1931 the Indian Institute of Science advertised an opening for a lecturer in electrical technology. It would be hard to imagine anyone in India better qualified than Shah. He orchestrated a truly impressive set of recommendations in support of his application, although those at the Indian Institute of Science may not have completely appreciated them. Five MIT professors, including Vannevar Bush and Dugald Jackson, chair of the electrical engineering department at MIT, wrote on Shah's behalf attesting to his practical knowledge, keen mind, and pleasing personality.[45]

Again Shah was either not offered the position or did not accept it. While one can only speculate about the reasons, Shah's commitment to the freedom movement may have played a part, either on Shah's side or that of the Institute. The Indian Institute of Science depended on support from the colonial government and may not have wanted to cause controversy by having a potential troublemaker like Shah on the staff. On the other hand, Shah may have seen joining the Institute as requiring him to compromise his commitment to the freedom movement, since the Institute was in the princely state of Mysore, where the ostensible direct ruler was an Indian who was generally considered to follow enlightened policies. Here civil disobedience would be against a native Indian government, making it much less effective than civil disobedience in Ahmedabad.

With his MIT classmate Nandlal Shah, Shah set up a small contracting company, Shah and Shah, which electrified textile mills. Shah would have been greatly overqualified for such routine work. In May 1932 a visiting Indian, Ramesh Mehta, told a group at MIT that both Shah and Shah had been imprisoned for anti-British activities. Shah and Shah was a business that could be run (or not run) according to the needs of the Indian freedom movement. Shah's oldest daughter asserted that when as a child she was asked about her parents whereabouts, she would reply that they were "in the movement."[46]

That the intersection of Shah's professional career as an electrical engineer and his political career as a supporter of Mahatma Gandhi came at the Tata Iron and Steel plant in Jamshedpur had a certain tragic appropriateness to it. The technological vision that Shah had pursued in going to MIT had its epitome in India in ventures established by the House of Tata, whether it was the Tata Iron and Steel

Company, the Tata Hydro-Electric Company, or the Indian Institute of Science. However, Shah seems never to have received the support from the Tatas that he had hoped for, either at MIT or when he sought employment in India. Now Shah, finally part of the Tata organization, would play a central role in halting work at Jamshedpur, putting the entire plant, the industrial jewel of India, in jeopardy.

The Jamshedpur plant had been the ultimate industrial dream of J. N. Tata, realized only after his death. Like Shah himself, the plant was an example of Indians using American know-how to develop Indian technological capacity. The Tatas had enlisted American help to design and initially operate and manage the plant, but by 1938 the plant had its first Indian general manager. By 1939 Jamshedpur employed 21,000 workers and had a production capacity of 800,000 tons of steel a year, making it responsible for three-quarters of the steel used in India.[47]

At some point in 1939 or 1940, Shah took a position as assistant to the power engineer at the Jamshedpur plant, a position he was overqualified for. Family members have spoken of practical considerations behind Shah accepting the job. Because of the Jamshedpur plant's size, there was a large enough Gujarati population to make it possible for the children to receive a Gujarati education. While the Tatas had supported Gandhi's work in South Africa, as the interests of each grew, it inevitably led to tension. The Tata's development of large-scale businesses, such as iron and steel or hydroelectric power, inherently required the cooperation of the British Indian government. As these businesses succeeded, they became more important to the economy of India, and of more interest to the Indian government. Gandhi was willing to defy the British and call for noncooperation in ways that the Tatas could never support. The Tatas maintained the general Parsi position of being accommodationists toward the colonial government, subordinating political matters to their business interests.

The Tatas and the Jamshedpur plant wholeheartedly supported the British war effort. The August 1942 *TISCO Review*, a magazine clearly written for senior-level English-speaking employees, asserted proudly that the company was "rendering the maximum help possible to the Government" in the war effort. A notice encouraging employee donations to pay for a fighter plane included the slogans "Now is the time to Play Your Part in the Battle for Freedom," and "No Sacrifice is Too Great in the Cause of Freedom."[48] Some at Jamshedpur, T. M. Shah among them, would have different ideas about what constituted the "cause of freedom."

World War II dramatically raised tensions between the colonial government and the Indian freedom movement. When the colonial government declared India at war against Germany alongside Great Britain without consulting the nationalists who were its ostensible partners, members of the Congress Party resigned en masse from their elected positions. With the outbreak of the war, Britain needed India's resources more, but it was weaker and less able to give the nationalists what they wanted. In 1942, with the British even further hard-pressed by Japan's military successes, which had put them right at India's doorstep, the Congress Party passed a resolution urging the British to "Quit India."[49]

The Congress resolution was lengthy and complex, but for most Indians the call to action would have been given not by the resolution itself, but by a speech by Gandhi made shortly after the resolution passed. While Gandhi mentioned his expectation that he would talk to the Viceroy, it would no longer be in the form of negotiations over fine points. Gandhi asserted that he was no longer going to be "satisfied with anything short of complete freedom." He then clearly let his followers throughout India know what he expected of them, saying, "Here is a *mantra*, a short one, that I give you. You may imprint it on your hearts and let every breath of yours give expression to it. The *mantra is:* 'Do or Die.' We shall either free India or die in the attempt; we shall not live to see the perpetuation of our slavery. Every true Congressman or [Congress] woman will join the struggle with an inflexible determination not to remain alive to see the country in bondage and slavery. Let that be your pledge."[50] The Indian freedom movement as of August 9, 1942 had come down to three Hindi words, "karenge ya marenge"—"do or die."

The colonial government responded quickly to the "Quit India" resolution, arresting Gandhi and hundreds of leaders of the Congress Party throughout India within hours of the passage of the resolution. Indians responded more slowly but with great effect. The Congress resolution and the continued repression by the colonial government prompted waves of spontaneous demonstrations against the British throughout India. Some areas declared their independence. Indians opposed to the colonial government cut down telegraph lines and burned down police and telegraph stations. The Viceroy of India was to call it "by far the most serious rebellion since that of 1857."[51]

On August 20, 1942, workers and supervisory personnel at Jamshedpur went on strike, forcing the management that stayed on the job to hurriedly shut down the plant. Company records provide its perspective on the strike. In a huge plant such as Jamshedpur, employing 23,000 people of widely varied social and linguistic backgrounds, there were doubtless different groups acting out of various motives. Tata management had loyalists throughout the plant reporting back to them on conditions and activities throughout the workforce, and Tata management came to see supervisory personnel's support for the strike as the key factor beginning and sustaining the walkout. And they saw the key leader in support of the strike among supervisory personnel as T. M. Shah.[52]

The precise nature of Shah's role in precipitating and sustaining the strike cannot be determined with precision from the available documents. One Tata document identified him as the president of the "Jamshedpur Strike Committee."[53] In several of the company's reports of meetings of supervisory personnel, Shah was said to be serving as the chair. How Shah came to this position is unclear. Perhaps it was his association with Gandhi or his experience in civil disobedience. He was a relatively recent employee, and as he was Gujarati, he was part of a minority within the plant.

Although rumors of a strike or hartal had been circulating for days, late on the night of August 19, 1942, senior management at Jamshedpur received word that the supervisory staff of the plant had instructed workers under them to strike, assuring them of their support. The strike began on the afternoon of August 20, and by the next day was total throughout the plant.[54]

In spite of the rumors, the strike took Tata management by surprise. A large steel mill is a complex system that operates based on the control of an enormous amount of energy; a loss of control puts the entire system in jeopardy. Normally shutting down a mill would take place in an orderly fashion over an extended period of time. This was not possible at Jamshedpur, and senior management believed that only through the extraordinary efforts of the few nonstriking managerial personnel (some clearly English or American) was the plant spared serious damage.[55]

If the fate of India hinged on the outcome of the "Quit India," movement, the fate of industrial India hinged on the outcome of the Jamshedpur strike. Jamshedpur was a critical component of the colonial government's war effort, and the government carefully monitored all activity there. The Tatas and the Indian government had a plan to bring in government personnel to at least run the power plant, and possibly the entire mill, if the strike continued. Indian workers had threatened sabotage against any attempts to forcibly resume production. Were violence and sabotage to erupt, India's premier industrial plant could have been destroyed. A major incident with significant casualties would have severely compromised the position of the House of Tata, India's leading industrial concern, in the eyes of the Indian people.[56]

Tata management was in a particularly delicate situation. They wanted to get the plant back up and running, but they wanted to do so with the minimum of government involvement. As such, in meetings with strikers, they frequently warned of the possibilities of government intervention, while at the same time they urged Indian government officials to let them manage the affair on their own.

On Saturday August 22, 1942, the third day of the strike, senior management of the Tata Iron and Steel Company held a boardroom meeting with superintendents and foremen at the plant, a group which included the leading instigators of the strike. The central figures at the meeting were two American-educated engineers, Shah and J. J. Ghandy, the general manager of Tata Iron and Steel. Ghandy, who had joined Tata Iron and Steel in 1919, had earned a master's degree in metallurgy from Carnegie Institute of Technology and then studied business at Columbia University. His career and the growth of Jamshedpur during this same time showed the possibilities of technological development in India. One price of that growth had been a willingness to accept British political control of India, a price Shah was no longer willing to pay.[57]

The Tata argument for going back to work had many points, some mutually contradictory. The one consistency was their realism in seeing the absolute necessity of accommodating the colonial state. A Tata official, S. K. Sinha, spoke first, stating that the strike had achieved its objectives and its continuance risked bloodshed and damage to the mill. Ghandy then spoke, the voice of reason and conciliation. He made what the secretary called a "sporting offer," that if everyone returned to work, all would be forgiven and there would be no repercussions. He portrayed the company as the one force that stood between the workers and the strong hand of the government. He warned, however, that although he had been able to keep the colonial government from getting involved, that might not be possible if the strike continued; the consequences of government involvement were impossible to predict.[58]

Ghandy then asked for the views of the supervisory workers, and Shah spoke. Tata Iron and Steel's records of the meeting give his statement:

> He said that he had no intention of manufacturing steel which would be made into bullets and then used later against our own countrymen, and possibly against the workers of Jamshedpur. He said they were only following the Congress plan, and Ghandhiji's [sic] last message was "to do or die," [which] called for a complete stoppage of all plants engaged in the war effort. When asked how long they would continue to strike, he said that the strike would continue as long as our leaders were in jail, or until they directed the struggle to stop. When asked if they had a definite plan or local leaders to direct the strike, and whether his opinions also voiced the views of others, Mr. Shaha [sic] replied that they had no leaders, no organization, no definite plan, and that every man spoke for himself.[59]

The report further noted that others expressed similar views. Shah's position showed that he had heard and taken in Gandhi's message, and it did not allow for the kinds of reasonable considerations that Ghandy and Sinha appealed to. Throughout Tata management's discussion of the strike, their ideal worker was "moderate," implicitly meaning one who was willing to repress personal political views in light of the evident futility of challenging the colonial state. Shah was not moderate.

Whether Shah was being disingenuous about the organization of the movement is impossible to say. When asked about the leadership, it was wise to claim the movement was leaderless when any leaders would likely be marked for detention by the government. However, throughout India the response to Gandhi's call to "Quit India" was to a large degree spontaneous and not rigorously planned.[60]

On Wednesday, August 26 the supervisory staff met. A report of the meeting made by an attendee loyal to management reported 300 people in attendance inside (or outside) the bungalow of one of the strike leaders. Shah chaired the meeting. The report of the meeting gave the appearance of solidarity behind the strike and unhappiness with the way the Tatas had dealt with it. Ardeshir Dalal, the former Indian Civil Service officer now the director in charge of TISCO, was called a "mouthpiece" of L. S. Amery, the British Secretary of State for India. Dalal's claim that in the face of a continued strike the government would take over the plant and bring in foreigners to run it was seen as bluster. The workers saw through the contradictions in Dalal's claim that the Jamshedpur plant was too small on a global scale to make a difference in the war effort, wondering why, then, ending the strike was so critical. Even the threat of a British official to break the strike within thirty minutes by bringing in the military was taken as "another instance of the high-handed attitude of the Government and was a challenge which the strikers were determined to meet." Finally the men settled on two possible ways of resolving the strike: either the company intervene to have Gandhi, Nehru, and the other leaders released, or Gandhi or Nehru explicitly ask them to call off their strike. After some discussion of the advantages of having a formal working committee to organize the strike, V. G. Gopal, representing the worker's union, stated that the Congress Working Committee in Bombay had instructed them to strike without a formal organization to avoid government arrests.[61]

Both sides conducted furious propaganda campaigns. Tata senior management instructed department superintendents who had stayed loyal to meet with their supervisory personnel, both to listen to their grievances and to persuade them to return to work. Ardeshir Dalal had come over to Jamshedpur from Tata headquarters in Bombay to oversee the situation. On August 28, a week after the strike began, he addressed the workers by radio, rehearsing the company's arguments. He repeated a claim he had made earlier that "India had achieved its independence," a status that would only be manifest at the end of the war. He claimed that loyalty to the country did not conflict with "loyalty to the Company." Winning the war was in the best interests of both. He promised that no disciplinary action would be taken against whose who returned to work by August 31, and he further promised that if the workers returned, he and J. R. D. Tata, the leader of the House of Tata, would go to Delhi to "personally communicate your feelings and sentiments to the Government of India."[62]

The strikers conducted a house-to-house campaign urging people not to return to work. Tata management collected a wide variety of propaganda material produced in support of the strike. The pamphlets, written in English, Hindi, Bengali, and Gujarati, ranged from fairly straightforward advocacy of Gandhi's program to wild fantasies, including a Gujarati publication from Bombay claiming that thousands of workers had been slain in Jamshedpur.[63]

Part of the Tata strategy was based on dividing the supervisory workforce. Superintendents met with the supervisory staff of each of their departments, listening to the workers' grievances against the government and repeating the company's line of threats of government intervention along with claims that the strike had already achieved its purpose. Tata management believed that the power house, where Shah had worked, was the key to the whole situation. On the first of September, Ardeshir Dalal met with representatives of Power House No. 3, with Shah not present, telling them bluntly that arrangements had been made to bring 150 men in from Calcutta to occupy the positions necessary to run the plant. Dalal further asserted that the government of Bihar had invoked the Essential Services Ordinance, requiring workers to return to their job on the pain of a year's imprisonment. The notification of the invocation of the ordinance had been flown down to Jamshedpur, seventy miles from the state capital in Ranchi, the air dispatch likely prudent because other methods of communication, such as rail, road, and telegraph, had become unreliable due to "Quit India" violence.[64]

After a brief period discussing the matter among themselves, those present returned to Dalal professing their good feeling toward the company, promising to return to work, and begging to avoid punishment. The workers further pleaded with Dalal to make efforts to get the Indian political leaders released, but when Dalal refused, the supervisory workers agreed to return unconditionally. On Thursday, September 3, two weeks after the strike had started, Tata management reported that the majority of the supervisory staff had returned to work and the total attendance at work was 14,000 (out of a normal workforce of 23,000). One sign of how effective the Tata methods had been during the period of labor unrest was that violence was avoided and only eighty-five people were arrested during the strike.[65]

The effects of the strike lingered on. Many of the striking workers had left Jamshedpur to return to their villages. The effort to get these workers back and to return the plant to full production led Tata Iron and Steel into conflict with the colonial government. Tata Iron and Steel wanted to broadcast widely that the strike was over and that the mill was returning to normal operation so that workers who had left would know that they should return. The colonial government, whose priority was to suppress any news of the strike at all, opposed Tata's proposal. While Ardeshir Dalal bragged about their approach to the strike leading to a settlement without even "a pane of glass being broken," there were costs. Tata managers doubted the plant would be back at full operation before the beginning of October and estimated the loss of production at 60,000 to 70,000 tons of steel—almost a full month's production.[66]

On September 14 Shah was arrested by the Government of India and confined for eighteen months in the infamous Hazaribagh jail, which detained many nationalist leaders. Remarkably, on his release he was reemployed at Jamshedpur, even though he had told Tata management that "he could not say when he would again feel the urge to take an active part in politics." Shah felt oppressed at Jamshedpur, complaining that the company was behaving vindictively toward him, with men watching his movements. He resigned after working only six more months at Jamshedpur to return to Ahmedabad.[67]

One could imagine that Shah's involvement in the Tata Iron strike and his reaffirmation of his commitment to Gandhianism might have made him rethink his association with MIT and its relevance to his aspirations. MIT no less than Jamshedpur was part of a large technological system, where individuals were subject to its logic. But Shah did not see things this way. Just as going to MIT did not mean Shah was repudiating Gandhi, so Shah's actions on behalf of the freedom movement did not mean he was repudiating MIT. In the first half of 1945 he wrote to his former professor, Karl Wildes, asking for MIT application forms for a friend and recommendations for American schools with good programs in chemical engineering. Shah had left money with Wildes so that the MIT professor could buy and forward him engineering books. Wildes noted to Shah that based on his reading of the news, "You may not have to spend so much time in jail in future years as you have spent in recent years." In 1947 Shah made a $10 contribution to a scholarship in honor of a professor who had led the cooperative course. Wildes responded with gratitude, noting that if all the graduates of the program had been as generous as he had been, the fund would have been three times as large as it was. That same year Wildes's secretary, Bertha Goodrich, sent Shah a short note telling him that she remembered him as "one of our best students" and apologizing for what a poor correspondent Wildes was. After independence, Shah worked as the chief electrical engineer for the state of Saurashtra in Gujarat. Family testimony asserts that in the later years of his life, Shah spun daily with a charka while listening to the news on the radio.[68]

Pandya and Shah represented the two extremes of engineering career paths in India in the late days of the Raj. Pandya completely incorporated himself into the colonial system, while Shah resisted it to the point of imprisonment. Pandya's alignment with power allowed him to have, by conventional standards, a more successful engineering career. Other Indian MIT graduates from this period took middle courses: avoiding direct involvement with the colonial government, but not courting imprisonment. Private businesses established by Indians were a common work choice.

Bal Kalelkar's case was particularly poignant. His middle path after his return from the United States helped him avoid both jail and the colonial state. Kalelkar returned to India in the fall of 1945 after having earned his master's and doctorate degrees in mechanical engineering from MIT and Cornell, respectively. He later reported he was feeling "diffident" about his reception by Gandhi, but when he was reunited with him in Poona on the Hindu holiday Divali, he felt the "same depth of love and affection" from Gandhi as he had before. But in spite of Kalelkar's positive portrayal of his reunion, all was not the same to Gandhi. Shortly thereafter Gandhi wrote to his grandson Kantilal, Bal's great friend growing up in the ashram, devoting a paragraph to a description of his reunion with Bal: "Bal has obtained the highest degree in engineering and has become a PhD. He was here for four or five days. He has gone with Kakasaheb to Kashi. He does not seem to have given up everything that he had learnt in the Ashram. He is still unaffected in his speech. He participated fully in the prayers here. He sang bhajans for our benefit with great enthusiasm."[69]

Gandhi's words emphasized loss and looked to the past. He was clearly examining Bal closely and expecting or fearing that his time studying engineering in America had totally transformed him. Although some things seemed the same, Gandhi was unwilling to make a definite assessment of Bal. He did not seem confident that he knew him anymore. It must have been a particularly poignant moment for the seventy-six-year-old Gandhi, being physically reunited with the young man of whom he had previously said, "He was brought up under my hands." As he had worked with Kalelkar when he was a boy, Gandhi had doubtlessly hoped that Bal would carry on the work that he had started, developing an Indian alternative to Western industrial society. If Bal's speech was the same, and the bhajans were the same, Gandhi clearly saw much that was different.

For his part, Bal seems to have had second thoughts about the path he was taking. In December 1945 he wrote to Gandhi apparently asking if he could work either serving him or his own father, who was still in Gandhi's movement. Gandhi turned down the offer, writing to Bal that "your dharma is to keep up what you are doing." Gandhi's letter implied several reasons for denying Kalelkar's request. G. D. Birla had supported Bal's education, and now Bal had a duty to serve him. More basically Gandhi, with the sense of a practical businessman, implied that Bal had a duty to take the expensive specialized education he had received in the United States and put it to use in India. In offering this counsel to one he had once called "one of the most promising among the boys brought up in Satyagraha Ashram," there was a sense of acceptance on Gandhi's part of the path that Kalelkar had taken, but there must have also been a sense of loss: if even Bal had taken this path, what chance did Gandhi have of enlisting millions of Indians in his cause? Gandhi gave Bal a final exhortation that, while accepting Kalelkar's path, also contained a hint of Gandhi's vision for India: "In the end you have to let the masses utilize your knowledge without any thought of fame or fortune."[70]

In 1946 Kalelkar took a job with Texmaco, Birla's textile machinery company, as an assistant works manager in the company's Calcutta plant. Although Kalelkar was working for the man who had supported his education in America, it failed to satisfy Kalelkar on a number of levels. Birla's industries were, by and large, not technologically sophisticated. Texmaco was producing textile equipment that had largely been copied from other designs. To make matters worse, Kalelkar was working under a British general manager. The job at Texmaco was not one that required the skills of a Cornell doctorate in mechanical engineering.[71]

These three young men who had grown up in Gujarat had, from the perspective of the Indian nation, gone to MIT idiosyncratically. When they returned, they had no well-defined career path to follow. Indeed, they faced the challenge of reconciling their MIT education with the colonial state and with their earlier commitment to Gandhi. Each came to a different point of balance, but the post-MIT careers of Pandya, Shah, and Kalelkar demonstrate the difficulties of fully using an MIT education while in a Gandhian opposition to the colonial state. After renouncing his Gandhian ideals and serving a period in the wilderness, Anant Pandya's talent and charisma enabled him to break into a position of responsibility with the colonial system. And the needs for technical talent during World War II offered

Pandya further opportunities. The paths that Shah and Kalelkar took were more complicated. Shah's commitment to the Indian freedom movement led him to jail and seems to have kept him from the positions he was technically qualified for. If Kalelkar did not go to jail after his return from the United States, he felt ambivalence about the distance that his education had taken him from his earlier Gandhian self.

But if World War II did not bring the independence that Gandhi had sought, it did bring fundamental changes in India. It would bring the United States and India together as never before. It led to a changed environment for technical education in India. In its last years the colonial state would pay for hundreds of young Indians to study in the United States, and it would accept MIT as a model for technical education.

6

Tryst with America, Tryst with MIT

IN THE CONTEXT of a war where 60 million people were killed and unprecedented destructive powers were unleashed on the world, two meetings held in India in early 1945 could easily be seen as having a vanishing degree of significance. But those inaugural meetings of the MIT Club of India, one in Bombay and one in Calcutta, told something important about India's past and foreshadowed India's technological future.

These meetings of Indian graduates of MIT were the first time that this group had a corporate identity in India. By and large they had gone to the United States driven by individual initiative, and even now, most of them had kept themselves aloof from the Indian state. But what brought them together in 1945 was a plan of the Indian government, announced the year before, to establish an Indian MIT. As welcome as the Indian government's proposal would have been to them, it was not a new idea. There was the *Kesari*'s proposal sixty years before, and Mokshagundam Visvesvaraya, the eighty-five-year-old doyen of the old dispensation of engineering, reminded the Bombay audience that a committee (that he had chaired) had made a similar recommendation about an Indian institute modeled on MIT in 1922. That proposal had failed when the committee's British majority refused to support it. Now Visvesvaraya heartily endorsed the government's plan, with the *Times of India* reporting his claim that "no better model could be thought of than MIT."[1]

Six guests attended the Calcutta meeting: Brigadier General Stuart Godfrey and five other U.S. Army officers, all alumni of MIT. Godfrey and the others were informal representatives of the roughly 200,000 GIs stationed in India, an unprecedented American presence in India. Godfrey later wrote that he and the MIT men on his staff were doing what they could to assist the efforts at technical development in India. While many American soldiers in India were anxious for the war to end so that they could return home, Godfrey pointed to levels of American society that sought to establish a strong and lasting connection with India.[2]

World War II in general, and these meetings in particular, marked the beginning of the transformation of the role that Indian graduates of MIT, as well as MIT itself, would have on Indian technological development. Previously a few Indians had gone to MIT through a highly individualized and idiosyncratic process, and when they returned to India their careers had been equally individualized and idiosyncratic. That would now change, with MIT being the standard to which Indian technical education would aspire. While only a few had sought to go to MIT previously, at the Calcutta meeting Anant Pandya told Godfrey that far more Indians wished to attend MIT than could be accepted, a situation that has continued to the present day.

More than this, World War II marked a critical period of rewiring India's technological connections, loosening those with Britain while strengthening those with the United States. Like the processes that had previously brought Indians to MIT, earlier connections between the United States and India had been idiosyncratic, often based on individual initiative and if not operating against the colonial state, then at least outside it. This changed with World War II. A number of factors and actors were responsible for this transformation. At its most basic level the World War II alliance between the United States, Great Britain, and its Indian colony brought American power and resources to bear on India as never before, while simultaneously making clear the relative weakness of Great Britain. In this new environment, even Britons were involved in the process of connecting

India and the United States. A group of Indian elites wanted closer connections to the United States, while some Americans sought to expand their country's presence in India. World War II provided both groups an opportunity.

At the same time there was ambivalence on both sides. Any expectations in India that connections with the United States, whose president had uttered words about the right of self-determination, would be backed by concrete American actions regarding India were met with disappointment. The formal discriminatory laws that the United States had in place against Indians posed fundamental limitations on how close relations could be between the two countries. Whether an overarching American attitude toward India existed or not, specific episodes demonstrated a well of American antipathy for India, where Indians could be portrayed not as allies or friends or strategic partners, but as a people who were different.

Perhaps only during World War II would it be possible to describe 200,000 American soldiers in India as a second order effect. The China-Burma-India theater of war got far less attention than the European or Pacific theaters. The war in India included a famine that killed millions, a movement to ally with the Japanese, and a popular uprising to expel the British: the Indo-American aspect of it could easily be lost.

Big Plans, Small Steps

The entry of the United States into the war brought both Americans into India and America into Indian affairs in ways that were unprecedented. The British maintained a fiction that India was a sovereign government, and as the United States entered the war, it became a formal ally of India, with a much more direct relation with India than it had ever had before. India had two sources of strategic importance to the Allies. First of all, it was the main link to China. A China actively fighting the Japanese kept large numbers of Japanese forces occupied and prevented them from being employed elsewhere. The only way to provision Chinese forces and keep China in the war was through northeast India. Secondly, after the Japanese victories of early 1942, India was Japan's most attractive target. The colonial government's worst nightmare was of a Japanese invasion accompanied by a popular uprising in which Indians welcomed their fellow Asians' help in evicting the British.[3]

The first American soldiers arrived in Karachi in March 1942. Their primary mission would be logistical, getting supplies from Indian ports to the northeastern Indian region of Assam and then airlifting them to China. The work of the American soldiers in India would involve more engineering than fighting, as they built or reengineered pipelines, ports, railroads, airfields, and roads.[4]

World War II challenged the ideals of the United States and Franklin Delano Roosevelt. Roosevelt was no friend of colonialism. As Franklin Roosevelt drew the United States closer to formal entry into the war, in August 1941 he met with Winston Churchill to lay out a set of idealistic aims for the war, which themselves had the practical aim of winning the support of the American public. These aims included the restoration of self-government to all peoples who had been forcibly denied it. While many Americans might have immediately thought of German-occupied Poland or France, many Indians would have thought of themselves. The United States entry into the war in December 1941 gave Roosevelt the license to discuss India's status with Churchill as never before. It was no longer merely a matter of American ideals, but also a question of what would make the Indian people the most effective partners in the war effort and also what would make the American people the most supportive of the war effort.[5]

In March 1942, faced with Japanese advances into neighboring Burma and American senators who were increasingly demanding that India be given a "status of autonomy," FDR wrote a series of telegrams to London inquiring about the possibilities of what was delicately put as "new relationships between Britain and India." Roosevelt's pressure was one source of the impetus to a British mission to India led by Stafford Cripps. While offering concessions, including the promise of independence after the war, Cripps's proposals were rejected by the Congress Party in India. In any case, the "failure" of the Cripps Mission offered Churchill the best of both worlds: he could tell FDR he had made a good faith offer to the Indians, while in fact nothing had changed. During World War II,

India was important not only in and of itself, but also for its effect on the Anglo-American alliance.[6]

Ironically, one of the key figures in connecting India and the United States was the highly Anglicized, Oxford-educated member of the Indian Civil Service, Girja Shankar Bajpai, who in 1941 was appointed Agent General to the United States. Although India as a British colony had no formal direct ambassador in the United States, Bajpai served the equivalent function. Historian Kenton Clymer has argued that Bajpai, who had the implicit trust of the British, used his position in Washington to surreptitiously argue for American actions that were compatible with the interests of Indian nationalists and opposed to British interests.[7]

In late December 1941 and early 1942, Bajpai made a series of visits to U.S. State Department officials. It would have been a massive understatement to say that the war was going poorly for the allies. In the Pacific theater, the attack on Pearl Harbor was but one of the Allies' problems. The Japanese had launched a simultaneous attack on the Philippines and the Malaysian Peninsula. The fall of Singapore in February 1942 brought Japan to India's doorstep. The front was thousands of miles from both Britain and the United States.[8]

Bajpai proposed that the Americans help India develop its production capability so that it could equip itself militarily. In contrast to the lend-lease program as conducted in Europe, where the United States sent Great Britain and other nations materials, a self-sufficient India would not require huge volumes of materials from the United States. Bajpai presented data that showed that of 60,000 items required in modern warfare, India was capable of manufacturing 85 percent of them herself, with the major problem being such heavy equipment as tanks and airplanes. Bajpai suggested that the United States send a mission to India to examine the possibilities of U.S. assistance to increase India's military production. Bajpai presented the proposal as, at least in part, his own initiative, which lacked formal approval from both the governments of Britain and India. President Roosevelt considered Bajpai's idea "worthy of pursuing."[9]

In March 1942 the United States announced the formal creation of a four-man technical mission that would go to India and offer advice on questions of production. The mission was headed by Louis

Johnson, who was appointed Roosevelt's personal representative to India—the rough equivalent of ambassador. One effect of the mission, perhaps intended by Bajpai, was to get the United States involved in Indian politics. Johnson was present during Cripps's mission to India and informally participated in the negotiations. He later wrote several memos back to Washington proposing an Indian War Production Board, analogous to one established in America, suggesting that such a measure could increase India's industrial capacity by two and a half times. However, in October 1942, with the Allied position better than it had been six months previously, but with India astir from the "Quit India" movement, the United States government shelved the report's recommendations based on claims of insufficient American resources to devote to the project.[10]

If Bajpai did not get what he had hoped for here, the United States and India were connected in smaller ways. Among the millions of people put in motion by World War II, an extremely subtle effect was the movement of a handful of Indian elites, in particular Indian students who came to study at MIT. More Indians studied at MIT than had at any previous time, with twenty-four enrolled in 1944, as compared to an average enrollment of seven in the 1930s.

Far more important than the numbers were who they were: children of those at the highest levels of Indian society. A large portion of them were children of Indian government officials, themselves newly on the move, posted to the United States as part of the closer relations between the United States and India that was developing during the war. In other circumstances, these families may have well sent their children to Oxford or Cambridge. Oxford and Cambridge were under a war footing and both travel to and living in England were hazardous. Whether the war forced these families' hands or whether it merely gave them the opportunity to do something that they would have liked to have done but would not have, because sending a child to America would have been too unusual, cannot be disentangled at this point, but both were likely factors.

Bajpai himself was a part of this smaller-scale linking of the United States and India, for when he came to America, his son Durga entered MIT to study architecture. A 1942 letter that followed on the heels of Bajpai's acceptance showed that his going to MIT in fact had some large-scale forces behind it. The letter, to the MIT dean of admissions with a copy to MIT President Karl Compton, came from Gordon Rentschler, an MIT trustee and the president of National City Bank, one of the country's largest banks, with branches in India. Five years earlier, a National City Bank official in India had given a talk to potential exporters in New York telling them that India was a "natural market" for mass-produced items and that its citizens' purchasing power would increase as the country industrialized. Rentschler wrote to let MIT know of the younger Bajpai's pending arrival. Rentschler said that the senior Bajpai, whom he knew well, "like many of the leading people in his country believes that there is a great future for their young men who have a good American technical training." Rentschler asked the admissions dean to "keep a helpful eye" on the younger Bajpai. He concluded by saying, "I think we can do something that will be of great value to ourselves as well as the Indians if we give them the benefit of the best of American education," doubtlessly thinking about the possibilities of business between the two countries.[11]

Indian elites continued to come. In 1942 Cambridge-educated K. C. Mahindra, the former editor of the *Hindustan Review*, came to Washington to head the Indian Supply Mission. With him came his son Keshub, who entered MIT. Shortly after the war, the elder Mahindra started the company that would be called Mahindra & Mahindra, which began with an agreement with the American firm Willys to assemble jeeps. In November 1944 two sons of Kasturbhai Lalbhai, one of India's leading industrialists, came to the United States to study at MIT (to be discussed in further detail in Chapter 8).[12]

S. K. Kirpalani was an Oxford-educated member of the Indian Civil Service who in 1944 won an appointment as the Indian Trade Commissioner in New York. One reason that Kirpalani sought the appointment was so that his eldest son could receive an education in the United States. During Kirpalani's posting in New Delhi in 1943, he had made friends with some American officers. One of them, a Colonel Smith in the statistical branch, gave Kirpalani an introduction to "influential friends" in the United States who helped Kirpalani gain his son's admission to MIT.[13]

In 1944 Munshi Iswar Saran, a prominent attorney and Congress leader from Allahabad, wrote a letter to MIT President Karl

Compton. Saran told Compton that while he had a brother and two sons who had studied at either Oxford or Cambridge, his grandson was now studying at MIT. The elder Saran confessed that although he had been to Europe several times, he had never been to the United States and was anxious about his grandson, asking Compton to look out for him. The elite family backgrounds of these handful of Indians who went to MIT during World War II put them in a position to have both influence among other Indians and success in their careers when they returned to India.[14]

As a physicist, Karl Compton was a joiner of worlds. During his tenure as MIT's president he joined the worlds of modern physics with that of engineering. During World War II, working with Vannevar Bush, he joined the world of modern science to the military, leading to unprecedented military technology. In a smaller and more idiosyncratic way, Compton also served to join MIT and India. Compton, the son of a Presbyterian minister, had a sister and brother-in-law who were Presbyterian missionaries at the Forman Christian College in Lahore. Forman itself had a particular scientific focus, with a program in industrial chemistry begun by Peter Speers, a missionary who was a Princeton graduate. His goal was to develop a chemistry program, giving training "equal to that in the best American colleges," which would simultaneously promote Christianity and the industrial development of India.[15] Forman's most prominent graduate was S. S. Bhatnagar, who would become one of India's leading chemists, with a career that exemplified the practical chemistry ideal of Speers. In 1926 and 1927 Karl Compton's brother, Arthur Holly Compton, who won the Nobel Prize in Physics in 1927, came to India to lecture at Forman, working with Bhatnagar and other Indian scientists on cosmic ray experiments.[16]

Although he never went to India, Karl had personal relationships with Indian scientists. In 1944, when he heard about Bhatnagar's impending visit to the United States with a group of prominent Indian scientists, Compton personally invited him to visit MIT and to be his houseguest. Indian physicist M. N. Saha sent a number of his students to MIT, and in 1942 his journal *Science and Culture* wrote that Compton and his brother Arthur were "amongst the best friends of the Indian students in the States, who find their doors always open to them."[17] In 1944 Saha sent several telegrams to Karl Compton

seeking MIT admission for specific Indian students. The telegrams
suggested that Saha and Compton had some agreement where MIT
would admit a few students based on Saha's support.[18]

A. V. Hill and the Idea of an Indian MIT (Again)

During World War II, there was a great deal of ambiguity about
India's future. The most basic unanswered question was whether
India would be given its independence after the war. Looking backward, one might consider India's postwar independence foreordained,
particularly given that the Cripps Mission had made such an offer.
But following the Cripps Mission, the question of India's future
was not officially discussed openly, allowing a wide variety of views
to exist at the same time. British imperialists may well have imagined that after the end of the war, they would be able return India
to the status quo ante, while nationalists may have been certain that
India's independence merely awaited the cessation of hostilities.

Historians Benjamin Zachariah and Sanjoy Bhattacharya have
argued that during World War II, the Government of British India
saw the key role that publicizing its intentions for the development
of India could have in convincing both Indians and Americans of the
beneficence of the Indian government.[19] A frequent theme of these
promises was the development of technical education in a way that
had never been done before. Just as there was ambiguity about India's
future, so was there ambiguity about these promises. Some members of the Government of India seem to have thought of them
cynically, as a cheap way to tamp down protests; others saw them as
sincere.

An example of the colonial government's prospective developmental program came with the visit of A. V. Hill to India. In 1943
Hill, the biological secretary of the Royal Society as well as the
winner of the Nobel Prize in Physiology or Medicine in 1922, went
to India to advise the government on scientific research. His mission seems to have had its origins in London and came, at least in
part, as a response to the Government of India's slowness in inte-
grating Indian scientists into the large Allied community contributing to the war effort. Hill had been a leading figure in the effort

to mobilize Britain's scientific resources for battle. While Hill had been involved with Britain's radar work since 1935, in 1940 he traveled to the United States, meeting with Vannevar Bush, the former MIT professor who had taken the initiative in getting Franklin Roosevelt's approval for the organization of scientists for the war. Hill had catalyzed the British transfer of its radar work to a team of Americans based at MIT. Hill had made the case for this transfer in dramatic terms, arguing that a British "impudent assumption of superiority" "may help to lose us the war." Hill understood the new dispensation, where science was used to develop new technology and where Britain accepted American leadership.[20]

Hill was in India for four months, visiting the major centers for scientific, medical, and technological research. His report, *Scientific Research in India*, presented informally to the Government of India in April 1944, then reprinted in Britain in April 1945, is too rich and wide-ranging to be fully discussed here, covering science, technology, medicine, and agriculture. The report began by essentially apologizing to the Indian scientific community for its being left out of the great scientific cooperative efforts of World War II, and implying that the Government of India was responsible for this slight.[21]

Hill's report set up Anglo-American norms for Indian science, medicine, and technology to aspire to. This is striking in two ways. First, since at least the 1880s, Indian colonial officials had claimed that Western norms were inappropriate for India. Indian hopes for technical institutes in the 1880s and 1920s had been dashed by Indian officials saying that type of sophisticated education was not needed for India. Hill made no argument about whether the Western models were appropriate—he assumed they were.

Second, Hill set up Anglo-American norms for India to aspire to, not just Anglo ones. If the editors of the *Mahratta*, looking hard and wishfully in the late nineteenth century, had thought they had seen evidence of British relative decline, by now it was obvious that Britain no longer led in many areas of science, technology, and medicine. Hill called for the development of an Indian "Johns Hopkins," in terms of medicine. In his discussion of industrial research, Hill began by writing about Bell Labs, Eastman Kodak, and General Electric, and then writing more briefly about British labs, which were "not on so magnificent a scale."[22]

The most striking example of Hill's recognition of Britain's position vis-à-vis America came in his discussion of technical education. Most of Hill's report consisted of looking at Indian deficiencies, with the implicit or explicit view that British and American practice represented the norm that India should aspire to. Hill dropped any pretense of Britain serving as the norm in his discussion of technical education. Instead, he spent a full paragraph lamenting Britain's lagging state in technical education, specifically its failure to develop an institution like MIT. He then went on to argue that if Britain needed an MIT, then obviously India did as well, writing that "there ought to be founded in India a few Colleges of Technology on a really great scale, like the MIT at Cambridge, Mass."[23]

What did Hill mean by invoking MIT? There is no evidence that Hill, a physiologist, had any specific knowledge of MIT's educational program. By and large his discussion of MIT was secondhand, citing the views of "responsible technical people" in Britain. He referred to what MIT was only in the vaguest terms, pointing to the "quality of equipment" and "excellence in teaching and research." In this section Hill wrote about Indians "trained to the highest level" and Indian industry developed "to the highest level," suggesting a unitary scale existing both within India and globally.[24]

In calling for an Indian MIT, Hill was echoing a call that was first made in the *Kesari* in 1884 and implicitly made regularly by Indian nationalists who argued for the importance of advanced technical education. But Hill's support seems to have been crucial in moving an Indian government newly sensitized to at least appearing to be supportive of Indian aspirations. In June 1944 the Government of India appointed Ardeshir Dalal to a newly created position on the Viceroy's Council for Planning and Development. Dalal had served as a member of the Indian Civil Service and then, as we have seen, a senior executive for the House of Tata. He had also been one of a group of Indian industrialists who had developed what became known as the Bombay Plan for the long-term development of India. The appointment of Dalal to this position was at least a tacit endorsement by the Government of India of the industrialists' work.[25]

In September 1944 the *Times of India* quoted Dalal as saying that "the question of personnel" was the most serious one India faced in its efforts at economic development, and the paper described plans to implement several of Hill's suggestions. India was planning on sending large numbers of people to Great Britain and the United States for training. The paper also noted that "the establishment of a very high grade technological institute on the lines of the M.I.T. in America" was being considered.[26]

In January 1945, as a follow-up to Hill's visit to India, a group of seven leading Indian scientists visited the United Kingdom and the United States. Ardeshir Dalal had asked the scientists to gather information about MIT that would be useful in their effort to develop an Indian MIT. When they visited MIT, they met with President Compton, leading administrators, and a group of Indian students. They also visited major American industrial research labs, such as those of Bell Labs, RCA, and Gulf Oil. The scientists' subsequent report affirmed the centrality of scientific research to industry, stating that "no industry which is based on scientific knowledge can prosper in a competitive world unless continually improved upon by new inventions based on scientific research." The Indian scientists made the "competitive world" more concrete by reporting that a hundred Chinese students were studying at MIT, while a group of Chinese engineers at Niagara Falls had told them of planned Chinese hydroelectric projects.[27]

In January 1945 Mansukhlal Parekh, who had gone to MIT in response to the dream of his father Devchand, earning a doctorate in chemical engineering, wrote to MIT Director of Admissions B. Alden Thresher. Parekh announced that the Government of India had decided to establish a technological institute "on the same basis as MIT," and he had been appointed to a committee "to work out the details." Parekh then asked Thresher to send him a wide variety of materials about MIT, including catalogs, drawings of the Institute's layout, as well as any personal suggestions for establishing such an institution.[28]

In the first half of 1945, Dalal announced the appointment of a committee headed by N. R. Sarkar to consider the development of an institution of higher technical education, "possibly on the lines of the Massachusetts Institute of Technology." Although given Hill's endorsement and Dalal's public statements, as Parekh had said, it was a foregone conclusion that the committee would recommend an Indian MIT. Sarkar's leadership of the committee signaled a strong commitment to this new institution, for Sarkar was one of India's leading businessmen, who had long been a voice calling for the industrialization of India. He was the head of the largest life insurance company in India and had served on the Viceroy's Council, the highest position held by an Indian in the colonial government.[29]

The seriousness with which the proposed institution was conceived of as an Indian MIT can be seen in several ways. Between late 1944 and mid 1945, Dalal and three members of the committee traveled to MIT, including the well-known scientists J. G. Ghosh and S. S. Bhatnagar. Also on the committee were two young Indian engineers with doctorates from MIT. One of them, Anant Pandya, was the head of the Bengal Engineering College. The other, Mansukhlal Parekh, worked at Delhi Cloth Mills. The fact that S. R. Sen Gupta, a professor at Bengal Engineering College and a protégé of Pandya, was the committee's secretary suggested that Pandya would have a central role on the committee.[30]

Previously in India, government committees had been charged to look at technical education with the tacit premise that not much would be done. Implicit to this committee was that it would quickly recommend a maximal program. It made no survey of the state of technical education in India, but began on the premise that India's needs in this area were obvious and urgent.[31]

The committee's report was approved at its second meeting in December 1945. It called for the creation "as speedily as possible" of four institutes of higher technical education. The first was to be established in or near Calcutta, the second in the Bombay area. The third would be located in the South of India, and the fourth, focused on hydraulic engineering, would be in the north. The committee estimated that the cost of setting up the Calcutta institute would be 30 million rupees, and that it would have a recurring budget of 6.7 million rupees. The committee estimated that the cost of educating a student would be less than one-half the cost of educating a student at a comparable university in the West.[32]

While an "Indian MIT" had great power as a slogan or an aspiration, in the end, after all the committee's exposure to MIT and after all the MIT materials it had gathered, MIT seems to have been less compelling as a model for a group of Indians, experienced in technical education, to build a technical institute around. One reason may have been that the Sarkar committee may have seen that MIT was itself sui generis, not something that other institutes in other countries could reasonably hope to aspire to. In the report approved by the whole committee, MIT itself was mentioned only once outside the statement of the committee's original charge. That reference stated merely that the new technical institution's standards for graduation should not be lower than a "first class institution abroad," with MIT being one of the examples.[33]

The MIT that the pilgrims from India had seen was very different from the MIT that Anant Pandya had experienced in the early 1930s. The school's civilian student population had declined by half over its normal number, partially compensated for by students in Army and Navy programs. The war had begun a further transformation of MIT, which while giving it access to tremendous resources, would make it problematic as a model for India. Former MIT professor Vannevar Bush's creation, the Office of Scientific Research and Development, a government agency whose purpose was speeding the development of technologies with military relevance, had funneled huge amounts of money through MIT. MIT's Radiation Laboratory brought together nearly 4,000 engineers, scientists, and technicians and $100 million to make radar into a technology that would be useful for the war effort. While the money would decline (for a while) after the war, what would remain was a system where the United States government would fund research at MIT and other American universities on an unprecedented scale.[34]

The Sarkar committee report was an implicit vote of no confidence in India's existing technological institutions, bypassing them in the reconstruction of India. The strongest and most poignant response of an established institute came from the College of Engineering and Technology at Jadavpur, a suburb of Calcutta. At a time when the colonial state controlled a large portion of the discipline of engineering in India, this college had been established in the early twentieth century as an expression of technological nationalism coming out of the swadeshi movement. It had relied on Indian donors, rather than the government, for its creation and sustenance. By 1939 Jadavpur had a diverse engineering curriculum, ranging from civil engineering to chemical engineering to electrical engineering, graduating one-seventh of all the bachelor's degrees in engineering produced in India.[35]

In 1946 Benoy Kumar Sarkar, a history professor at Jadavpur, wrote a history of the university. Sarkar's work was largely devoted to demonstrating the role that Jadavpur had played in India's technological development. Sarkar's claim for Jadavpur became most clear near its end, when he devoted a section to "The Glib Talk of an 'Indian MIT.'" Sarkar ridiculed the idea, saying that India lacked the financial and industrial base for such an institution. He accused its advocates of imagining that through paying high salaries to professors and through building "imposing brick and mortar edifices" the "atmosphere of an MIT can be automatically engendered."[36]

Sarkar claimed that India already had an MIT in Jadavpur. It had proved it by the fact that its students were admitted at such prestigious institutions as Imperial College, Carnegie Tech, and MIT itself. Sarkar recommended upgrading Jadavpur by providing for more (Indian) foreign-trained faculty, setting a goal of two-thirds of the faculty having foreign training and one-third having doctorates. He concluded: "Jadavpur College—the existing MIT of today—can become a more efficient MIT in case a few million rupees can be spent within five years. Let the MIT-Wallahs, both official and non-official ponder over this reality."[37] Perhaps most stinging to Sarkar was that N. R. Sarkar, who was on the executive council of Jadavpur, had become an MIT-Wallah. MIT's power as a symbol was nonetheless strong, because even in arguing against the establishment of an Indian MIT, Benoy Sarkar had to invoke the title of MIT for Jadavpur.

America's Ambivalence toward India

Even though the moving figures in the plan to develop an Indian MIT were Indian and British, at least a few Americans at the highest levels of business and government shared an interest it. In July 1945, just prior to N. R. Sarkar's visit to MIT, Gordon Rentschler again contacted MIT officials trying to use MIT to build a connection between India and the United States. This time he called James Killian, MIT's executive vice president who was responsible for the day-to-day running of the Institute, to lay the groundwork for the visit of Sarkar and his group to MIT. A follow-up letter by one of Rentschler's assistants provided further details, suggesting an almost seamless interface between MIT, National City Bank, and the U.S. government, stating, "We want to make sure he [Sarkar] sees everything that is open to him" (without ever specifying who the "we" was). The letter went on to assert that "the Secretary of State is particularly anxious that members of the mission be afforded every facility,"[38] providing names of people in the Secretary of War's office who could provide the necessary security clearances for Sarkar, if necessary.

The United States government showed an interest in India that went beyond the war. Its propaganda arm, the Office of War Information (OWI), had a substantial operation in India, seeking to win Indian support for the Allied cause. Given American actions, convincing Indians of the purity of American democratic ideals was an impossible task. But in some areas, the OWI could be more effective, particularly as it sought a basis for long-lasting connections between the countries. One such area was the establishment of libraries, first in Calcutta in 1943, then in Bombay in 1944, and finally in New Delhi in 1946. Although Calcutta was where the most American soldiers were, and New Delhi was the seat of government, the OWI put priority on Bombay, India's leading business city.[39]

The Bombay library was opened in an elaborate ceremony that gave a sense of the American purpose and commitment behind it. Acting U.S. Secretary of State Edward Stettinius sent a congratulatory cable, while Vice Chancellor B. J. Wadia of Bombay University was there in person. The American consul in Bombay, Howard Donovan, explained the library as an outgrowth of a war that had caused both a shortage of information about the United States in India as well as a greater interest in the United States on the part of Indians. Donovan said the purpose of the library was to make available "accurate and authoritative information about America," so Indians could interpret America and its people as they really were.[40]

If the proximate cause of the library was the war, the speeches at the opening clearly implied that the library was not going away when the war ended. Wadia called it "a permanent cultural mission," hoping for the day when there could be an analogous Indian institution in the United States. The American official representing the OWI hoped that the library would lead to a "more intimate understanding by India of the institutions, character, and life of the American people," asserting that a product of the war must be "a closer understanding between friendly people." Stettinius's message expressed the hope that the library would foster understanding between the two people and so "tighten the bonds of friendship that both people value so highly."[41]

The *Times of India*'s account of the event clearly showed the resources that the United States had put into the library: this was not a slapdash effort. The Library of Congress had provided assistance in its establishment. The library consisted of 1,500 books, which would have been given space inside U.S. cargo ships alongside material in more direct support of the war efforts. Flora Belle Ludington, the librarian of Mt. Holyoke College, was in charge, working with a staff of three others.[42]

Whether intended or not, one of the main missions of the Bombay library and its staff became disseminating information to Indians about coming to American universities and colleges. An April 1945 report asserted that the library handled 75 to 100 educational inquiries a month (141 in March), with half of them being made in person and the other half by mail. The library had developed a collection of college catalogs and the staff prepared detailed personal replies to mail queries. Washington responded to a report on the Indian libraries with the assertion that "time is well spent when it is devoted to young students who are interested in, but uninformed about American universities and colleges." By late 1946 the New Delhi library was also building up its college catalog collection.[43]

The U.S. diplomatic staff in India took an active role in encouraging Indians to come to the United States. In a 1944 letter to the Bombay consulate, MIT admissions officer Paul Chalmers, who would later be named MIT's first full-time adviser to foreign students, noted that the State Department had developed a standard application form, which it was placing with Chinese consulates, to be used by Chinese students applying to American universities. MIT officials asked that this standard application form be sent to India and used there. It was, and through the Bombay library, copies of the standard application form to American universities were sent to the other American consulates as well as a number of Indian universities.[44]

After noting the contributions of MIT personnel to technological development for the war, MIT President Karl Compton in his 1944 annual report wrote that MIT's prestige had never been at so high a level. Although Compton was writing of MIT's prestige in the United States, in India, his statement was true because of the recommendation of A. V. Hill for creating an Indian MIT and Ardeshir Dalal's acceptance of it. In the educational correspondence directed to the American consulates in Bombay and Calcutta and with the American mission in New Delhi, no university figured anywhere near as prominently as MIT. In December 1944 the editor of the *Journal of Scientific and Industrial Research*, published by the Government of India's leading scientific research organization, wrote the New Delhi mission asking for the name of someone who might contribute an article on MIT. The mission obliged by providing one written by the OWI. The All-India Manufacturers Organization wrote asking for an MIT prospectus.[45]

The consulate worked with a variety of students who wanted to go to MIT. In March 1945 George Merrell, the secretary in charge of the U.S. mission in New Delhi, acting at the request of Girja Shankar Bajpai's brother, sent to the MIT admissions dean the application of a V. G. Rajadhyaksha for a graduate program in chemical engineering, noting that he was the son of a distinguished judge on the Bombay High Court and asking that his application receive "benevolent treatment." (Rajadhyaksha, who ended up going to the University of Michigan, went on to be the chairman of Hindustan Lever, one of India's leading companies.) In 1944 the New Delhi mission inquired if the Calcutta consulate had an MIT catalog that could be sent over and examined by the son of a government official. In May 1945 an American diplomat wrote on behalf of a principal of an engineering college in Madras, hoping to get an honorary appointment at MIT to enable him to learn about American engineering education.[46]

Less prominent people wrote in to the consulate asking for information about MIT as well. In May 1944 the University of Calcutta wrote saying that two of four scholarship winners seeking training in the United States hoped to go to MIT. In early 1945 N. K. Choudhuri, an assistant examiner in an opium factory, wrote to the consulate in Calcutta to arrange for training positions in the United States. He had had previous education and training in both Germany and the United Kingdom, but he desired to have the "most up-to-date methods" of chemical technology and was especially interested in the "world famous" Massachusetts Institute of Technology. In December 1946 Debabreata Dutt from Assam sent a handwritten note to the Calcutta consulate asking for an MIT application form. Although the United States had many first-rate engineering colleges, Indians wrote to the U.S. consulates about none of them in the same way as they did MIT.[47]

An early patron of one of the American libraries was Jamshed Patel, a Parsi in his late teens, who spent time in both Bombay and Calcutta. He later remembered seeing brochures of American planes in the library and being particularly impressed with a twin fuselage plane (perhaps the Northrop P-61 Black Widow fighter). He also saw an MIT catalog and later recalled that as he looked through it, MIT "somehow fixed itself in my mind."[48]

Patel's further experiences suggest how the American presence in India during World War II shaped a young Indian man. Patel had come from a middle-class Indian family and read *National Geographic* as a child. In recounting his time as a young man in India, Patel emphasized how the presence of the United States in India had affected him. He recalled a building under construction near his home in Calcutta. The work has proceeded for years without the building being completed, but when American soldiers came and took over the construction, the building was completed and occupied in six months. However, Patel also remembered segregated American bases.[49]

Even as the American diplomats worked to encourage Indian students to come to the United States, they were aware of potential problems. In October 1945 W. F. Dickson, an agricultural adviser in India, wrote expressing concern that Indian students would experience race prejudice upon coming to America. He noted that the schools that had good programs covering important crops in India such as rice, cotton, tobacco, and sugar were in the South. Although Dickson had met many Indians who looked forward to coming to the United States and were very "pro-American," he feared what would happen to them if they experienced racial prejudice. A State Department memo asserted that an examination of the issue was ongoing, but also expressed a broader concern about how Indians generally should be prepared for "certain unfortunate experiences" they might meet in America. The memo further stated that an unspecified "many" believed that "every Indian" coming to America should wear a turban during the first weeks of "his stay in any community," whether he had worn a turban in India or not, apparently as a way of establishing that he was Indian and not African American. (This memo obviously made the incorrect assumption that only men would be coming.)[50]

While American diplomats worked to establish a path from India to American colleges and universities, one of the suggestions in A. V. Hill's report had the same effect. Hill had proposed sending Indians abroad for advanced scientific and technical training. In his report Hill noted the changed financial calculus that would support such a program. While previously India had been a debtor to Britain, its war contributions had made it a creditor. A state-supported scholarship program would provide a way for India to spend that balance in the United Kingdom.[51]

In 1944 the Indian government announced a plan for scholarships that would send 500 Indians to Britain or the United States for advanced technical training. The government distributed 30,000 application brochures and ultimately received roughly 9,000 applications. The scheme sought applicants in forty-two areas, most of which were defined in terms of a specific technology such as plastics, radio engineering, or automobile engineering. Although biochemistry, agriculture, and economics were included, the vast majority of the areas were engineering related. The selection committee noted that this process brought out a talented range of candidates, with a "surprising" number of students who had stood first in their subjects in the university. Many of the students had master's degrees in their subject and some of them had doctorates. The committee ultimately selected roughly 600 students.[52]

Hill's recommendation on Indians studying abroad had focused on them coming to Great Britain at least in part to strengthen their connection to the metropole. But just as Hill's report had been of an Anglo-American scientific and technical world, here too America would enter the picture. In April 1945 the *New York Times* described the plan, stating that India would be sending up to 300 students to the United States in the fall. (Ultimately 380 came to the United States, over half of those selected for scholarships.) Furthermore, the Government of India had appointed M. S. Sundaram to serve as the educational liaison to the United States, working to place Indian students in American universities. The *Times* further stated that India planned ultimately to send 1,500 students to the United States. Two months later the *Times* reported that Sundaram was making a tour of leading American universities with an eye toward placing Indian students. Perhaps trying to win American support for the program, Sundaram made a statement that accepted American claims of idealism at face value, saying, "Our students are bound to be impressed by the ideals of democracy and freedom for which America has stood throughout the war."[53]

In May 1945 Sundaram visited MIT, doubtless to see what kind of arrangements could be made for Indians to study there. But Sundaram would be disappointed because just as the Government of India was making money available for Indians to come and study in the United States, MIT was removing the welcome mat for foreign students. The period of easy entry for foreign students, where T. M. Shah could present a friend's credentials and get an acceptance that day, was gone. MIT President Karl Compton commissioned a 1944 study on the position of foreign students at MIT that was nothing less than a definition of MIT's place in the technological training of the world, and implicitly a definition of the world's nations with respect to MIT. This study presciently recognized that for the first time an MIT technological education would be a globally scarce commodity whose distribution MIT would actively control. In that process the study implied that people from some places, Europe in particular, would be favored over people from other places.[54]

The stimulus for the report was the large numbers of foreign students who had entered MIT during the war and the recognition that those numbers were not sustainable after the war had ended. Before the war, foreign students had made up 10 percent of the graduate student population. The war greatly reduced the numbers of American students at MIT, but MIT's main foreign hinterlands, China, India, the Near East, and Latin America, were still able to send students, and in fact they sent more. By 1944 foreign students made up half the population of the graduate school. The role of non-Westerners was unprecedented.[55]

The MIT study anticipated that large numbers of Americans whose studies had been postponed or interrupted because of the war would be looking to return to the classroom. Quite understandably, MIT considered that it had a particular obligation to ensure that there was room for these students. But oddly the MIT study claimed a special privilege for another group whose claims on MIT were not as clear: European nations that had been affected by the war. American veterans had a claim on MIT because of their military service; European countries had a claim because of the destruction their countries had faced. The MIT report seemed to assert that countries such as India had a less valid claim on MIT because they had "suffered little devastation." In the committee's eyes World War II was the overriding factor in MIT's new admissions policies. The fact that India was on the verge of independence and was in need of building technical infrastructure seemed to be irrelevant. The fact that China, occupied by Japan since 1937, was so easily put on the side of the countries that had "suffered little devastation" suggests the committee saw Asians differently.[56]

The report itself provides evidence that other considerations were important in the committee's deliberations. One example of this comes in an aside about Canadians: "Canadians are excluded from the foreign group on the grounds that both by language and institutional background, they are entirely assimilable with our own students, and raise no problem of control."[57]

The ability of Canadian students to assimilate with "our students" (presumably meaning American students) meant the fact that they came from another nation was irrelevant. Later, as the report discussed the arguments that could be made for restrictions, it made the point more directly, but in a negative way: "The Institute in its origin and character is characteristically American and this character would be altered if foreign groups became too large a part of the total enrollment."[58]

The committee saw the very identity of MIT as an American institution at risk. The differential treatment accorded to Canadians and Europeans as compared to the areas where most of MIT's foreign students had come from—China, India, the Middle East, and Latin America—suggests that the MIT faculty saw students from the latter areas as representing problems of assimilation, and a threat to MIT's identity.

The report asserted that after the war, the United States would be for a time, "perhaps a very long time, the intellectual center of the world."[59] The unspoken corollary was that MIT would be the technological center. Previously, in a period when an MIT education had not been in such demand globally, MIT had educated students from China, India, and Latin America, giving it a "worldwide reputation."[60] But if MIT were now to be the world's center of scientific and technological education, it would have to make sure that students from those countries did not crowd out students from European countries, whose presence would be the definitive sign of MIT's position in the world.

The committee anticipated that the demand for an MIT education would dramatically increase in the postwar world. The committee admitted that MIT had previously accepted "any foreign applicant" whose credentials showed them to be "reasonably well prepared."[61] The committee anticipated that this situation would no longer hold after the war ended, with many more highly qualified candidates applying than could be admitted, requiring much greater selectivity and scrutiny of foreign credentials. The MIT committee recognized that this situation would cause public relations problems for MIT, with many not being able to understand how their highly qualified candidate was denied admission.

The committee recommended the establishment of a quota of 300 foreign students subject to a "reasonable" geographical balance. The committee further recommended that renewed efforts be made to ensure the qualifications of foreign students. MIT maintained this limit until 1948, and further limited foreign students to 10 percent of the graduate student body until 1952.[62]

The statistical figures for India gave a particularly powerful picture of the situation that MIT faced. In 1945, with the Indian government offering unprecedented funding for graduate training abroad and the plan for the "Indian MIT" giving MIT enormous publicity, MIT had almost 500 applications on hand from Indian students, more than double the number it had received from China, long the source of the largest non-American contingent at MIT. Although MIT admitted twenty-four Indians for the fall semester, it had placed 180 Indian students on its waiting list, implicitly stating the students were qualified, but that there was no room for them. With this huge increase in applications from India, MIT decided it had to take action to keep Indian students from dominating its foreign student population. Anant Pandya had assuredly been correct when he had said that far more Indians wished to attend MIT than could be accepted.[63]

The ambivalence that the MIT faculty report showed about an increasingly global role for the United States could be seen elsewhere—and not just among Americans. In November 1945, with the war over, the *Washington Post* ran an on-the-ground account of what the American presence in India had meant for Indo-American relations, which emphasized disillusionment and ambivalence on both sides. Indians had initially looked at America as "the Galahad among nations" that might rescue the Indians from British rule. Not only were American soldiers under orders not to get involved in Indian political matters, but they had no interest in them. American soldiers had a far greater affinity for the British in India than the Indians. The article claimed Americans had "not succeeded in reaching the heart of the people nor effected any substantial change in their way of life," with the greatest effect being a million tons of American military equipment left behind.[64]

The tension that Americans held toward Indians came out most markedly during what should have been a period of unalloyed joy in America. The war was over and the American soldiers were coming home. On December 7, 1945, the troopship *Torrens* docked at Staten Island, New York, bringing 1,700 American soldiers from Karachi back to American soil. But the *Torrens* brought with it controversy that appeared in newspapers throughout the country.

Along with the 1,700 American soldiers were fifty Indian students, bound for American universities as part of the Government of India's overseas scholarship program. A group of American soldiers, frustrated by years of war away from their home, in a foreign land, subject to death and the more mundane insults of military life, made the presence of these students on the ship a subject of controversy. And while the GIs would have the universal support of the American public, the Indian students would become convenient scapegoats, put into an impossible situation.

The controversy began with a lengthy article in the *Daily Gateway*, the American Army base newspaper, published in Karachi at the time the *Torrens* departed India. The article loudly announced that Indian students were being carried on the ship, their rumored presence confirmed by an examination of the manifest. The authors claimed that by accepting fare-paying passengers on the ship ahead of American soldiers, the U.S. government was valuing money over the soldiers' sacrifice. They described American GIs as citizen-soldiers who had stifled their natural inclination to question while in the service, but who could do so no longer. At the root of it, though, the article shows American soldiers tired of having their lives made subject to strategic considerations determined in Washington. The authors said that the Indians on the *Torrens* represented "mixing the cement of foreign relations with the tears of wives and sweethearts."[65]

It was obviously true that in some sense, every Indian student carried on board the ship meant that an American soldier was not carried. And seeming to put Indian students over American soldiers would be impossible to justify in the pages of American newspapers. But in fact, American troopships had long been carrying American civilians, as well as civilians of other nationalities. State Department officials clearly wanted to accommodate these Indian students and seem to have been unaware of the controversy they were fomenting. George Merrell, the chief American diplomat in India who had been working to develop ties between the United States and India and was doubtless tone-deaf from not having to justify policies publicly, tried to make the argument that the Indian students had not in fact bumped the American soldiers, because the space had been allocated through Washington for Chinese, Indian, and Polish civilians.[66]

The article in the *Daily Gateway* of Karachi was angry but measured, refusing to blame the Indian students; however, the story that appeared in American newspapers had metastasized to one with a much stronger anti-Indian tone. From this distance it is impossible to know how the story got transformed. American journalists appear to have taken the original *Daily Gateway* article and supplemented it with reports they got from GIs on the ship. Perhaps there were specific events on board ship, but more likely the story became driven by anti-India stereotypes held by Americans, combined with the fact that the Indians made easy scapegoats.

Among the most anti-Indian articles was a story carried by the *Chicago Tribune*, reported by a correspondent from Karachi. The article, "GIs Beef as U.S. Shares British India Headache," accused the Indians of bad behavior that had "added flame to the fire of hatred."[67] The *Tribune* claimed that at Karachi the Indians had staged a three-hour protest, sitting down on the docks and refusing to carry their bags onto the ship as the soldiers had, with the protest ending only when coolies were brought in to carry the students' bags. The American-based *Voice of India* subsequently published an article giving the Indian view of the troopship experience, claiming that the Americans had failed to provide Indian students with "proper food" and had exhibited a "shocking colour hatred."[68]

Seventeen hundred American soldiers, bound together by a common national and military culture, shared space with fifty Indians from whom they were alienated by an American racism, which was compounded by the fact that these Indians were the proximate cause of American comrades being left in India. It was clearly a situation ripe for enmity. The American soldiers were used to military discipline, which carried the expectation that they would carry their own bags and clean their own quarters. The Indian students may have expected the conditions prevailing on commercial carriers. Whether the American ships were equipped to provide vegetarian meals is unclear.

On the *Torrens* were two Indians who would be among independent India's most prominent engineers: Satish Dhawan and Brahm Prakash. Dhawan would receive a doctorate from Caltech and then go on to be the director of the Indian Institute of Science and the head of the Indian Space Research Organization. Prakash, who is discussed in the Chapter 7, would go on to earn a doctorate from MIT and hold key positions in India's atomic and space programs. Their experience with America did not get off to an auspicious start.[69]

Indian students would continue to come, but the U.S. State and War Departments rushed to correct the mistake they had made in the eyes of the American public. On December 4 the State Department announced that no more Indian students would be carried on troopships until "material progress" had been made in repatriating American troops from India. As reported by the *Times of India*, the communiqué tried to justify why the Indian students were on the ships in the first place, saying that they were coming as a part of "programmes of industrial and commercial expansion" in India that would have "a direct bearing on American foreign trade and on the long-range American relations with the 400,000,000 people of India."[70]

A final level of American ambivalence toward India can be seen in what was, at the time, viewed as a great victory for Indo-American relations: the passage by Congress and the signing by President Truman of the Luce-Celler Act of 1946. This bill made it possible for Indians to become American citizens, something previously denied them. As this bill was being considered in Congress, the State Department made clear its strong support for the change. A memo by Acting Secretary of State Joseph Grew to Truman quoted an Indian paper saying that the removal of the "humiliating immigration barriers" was a prerequisite for Indians to view America "with anything like the enthusiasm its propaganda seeks to inspire." Grew went on to suggest the possibility of a "color war" if Asian people believed that they could not receive fair treatment from the white races. Grew called India a "great potential market for American goods."[71] The bill did not put India on anything like an equal footing with European nations, allowing only 100 Indians a year to permanently immigrate to the United States, and was a victory only in that it removed some of the worst measures of racial discrimination that Indians had faced in the United States.[72]

World War II was a fundamental event in history, leaving imprints on our world visible even to this day. The ascendancy of the United States became an undeniable reality. In World War II and after,

American diplomats increasingly saw their remits as intervention throughout the world to maintain an order consistent with American interests.

The basis for American power to a substantial degree was technological, whether in the assembly lines that produced planes, tanks, and guns in unprecedented numbers or the research laboratories that developed weapons never seen before, such as the atomic bomb or radar.[73] In both India and America, the American system of technological education was becoming an instrument of foreign policy.

Great Britain, although nominally a victor, can be seen as one of the war's big losers. The war had drained Britain, leaving it without the power or the will to maintain its grasp on India, the most important colony in the British Empire. The Indian people won their independence against a weakened Britain. Even before the British left, American involvement in India was increasing. And there were Indians who welcomed that involvement.

In the years before World War II, the technological connections between the United States and India, seen most vividly in Indian students at MIT, had largely been ones that operated outside the governments of the two countries, sustained largely by individuals. Beginning in World War II and continuing thereafter, the connections would be increasingly shaped by the governments of the two countries. But the effects of those connections would not always be what either government desired.

7

High Priests of Nehru's India

I N OCTOBER 1949 Jawaharlal Nehru, now prime minister of independent India, made his first visit to the United States. Both
India and the United States had undergone dramatic changes since
the end of World War II. After long years of struggle, India had
won its freedom from the British in 1947, a freedom that entailed a
violent partition of the country along religious lines; in the ensuing
communal violence, millions were displaced and hundreds of thousands killed. In the United States, the confidence that had followed
the end of the World War II had given way to an increasing obsession with the spread of global communism. Earlier in 1949 China, a
reliable American ally during World War II, had "fallen" to com-
munists, while the Soviets' detonation of an atomic bomb broke the
American monopoly on that weapon and refuted any ideas of an
absolute American technological supremacy.

In spite of the obligatory pageantry and speeches, Nehru's visit had
to have been a disappointment to both sides. While the United States
sought reliable allies against communism and made India's acceptance of that role a condition of economic assistance, Nehru made
clear his intention to maintain a nonaligned foreign policy. The
disjuncture between the two governments became abundantly clear
in the months after Nehru's visit. In November India put on hold
consideration of a trade agreement with the United States, while the

next year India became one of the first countries to recognize Communist China. Shortly after Nehru's visit, the U.S. State Department rejected a proposed $500 million program of economic aid to India, while it also turned down India's request for a million tons of wheat.[1]

During his visit, Nehru traveled widely throughout the United States, and on October 21 he made a pilgrimage to MIT. There he was met by MIT President James Killian and a crowd of several hundred, including eighty to ninety Indian students. While his audience was more friendly than the one he faced in Washington, as he addressed a group of his fellow countrymen, he faced questions of identity and alignment of a different sort than the ones raised by America's Cold Warriors. Here (and indeed during most of his trip to the United States), Nehru dispensed with his eponymous jacket and hat in favor of a Western suit and tie. He began addressing the students in English, but midway through his talk, he asked in Hindustani, "How many of you know this language?" When the response was primarily laughter and the raising of hands in affirmation, Nehru continued on in Hindustani, in spite of the fact that he himself was more fluent in English.[2]

Perhaps unsurprisingly, given his location and audience, Nehru's remarks were technocentric, asserting that the history of a nation "must be looked at from a technological viewpoint." He went on to claim that India's technological lag had led to its colonization. As he considered India's future he said: "It is most important not only that our country advance along known technological lines, but that our technicians should show initiative and add to the existing fund of knowledge." He expressed his happiness that so many Indians were studying engineering, stating that "India has too many lawyers and too few engineers." He urged the students to "work hard to make India once again a first class nation."[3]

After independence, India immediately faced the need to define itself as a nation. What was India, religiously, linguistically, diplomatically, economically, but also technologically? Nehru's speech at MIT suggested that technology was fundamental not just to India's identity, but to its very existence. His visit to MIT implied a close relationship between Western and Indian technology.

Mahatma Gandhi was no more by the time Nehru visited MIT, felled by an assassin's bullet in January 1948. Just before his assassination, Gandhi had proposed that with political independence achieved, the Congress Party should disband. He believed that India's remaining work, the task of achieving "social, moral, and economic independence," should be done through a "Lok Sevak Sangh," a small-scale, village-level, voluntary organization. Gandhi was skeptical of centralized power.[4]

When Gandhi was assassinated, there could be little doubt that the weight of history was against his vision. The twentieth century up to that point could make one think that technology, when backed by science and given support by a powerful state or the corporation, really was a Kalpavriksha tree that granted every wish, as M. M. Kunte had said in 1884. Atomic energy, developed by scientists and engineers in the United States during World War II, unleashed unprecedented destructive powers on the world, but also promised a coming era of limitless electric power. The development of products such as nylon implied the possibility of designing a range of new materials whose precisely engineered characteristics were superior to those that nature provided. A group of researchers at Bell Labs in the United States had just invented the transistor, a device that in the coming decades would make almost infinite computing power available.

If technology was a Kalpavriksha tree, Nehru surely needed one when he became prime minister. He faced daunting economic problems in addition to the immediate political and social ones associated with India's independence and partition. Nearly 200 years of British rule had left India a poor agricultural nation, when country after country around the world had used industry as a path to wealth. By most standards of economics or human development, India was not "first class." It had a literacy rate of only 16 percent. Eighty-five percent of its population was employed in agriculture. India's per capita income at the time of independence was roughly 180 rupees, a figure that had increased scarcely at all in the course of the century. The average male life expectancy at birth was less than thirty-two years, while in the United States it was more than twice that. While Gandhi had frequently said that the real India was its 700,000 villages, less than 4,000 had access to electricity.[5]

Nehru had decisively rejected Gandhi's technological dreams in favor of a more Western model. In 1951, after long delays, the institute that A. V. Hill and the Sarkar committee had proposed finally opened its doors as the Indian Institute of Technology at Kharagpur, eighty miles west of Calcutta. In spite of all the pilgrimages to MIT, as the institute was built, the connection to MIT was attenuated both in rhetoric and conception. The term "Indian MIT" might still be used, but primarily as a way of asserting its position as India's apex technological institute.[6]

MIT was to play no direct role in IIT Kharagpur, nor did any Indian MIT alumni. Instead, Kharagpur received help where it was offered, resulting in a very ecumenical school. Kharagpur got assistance, such as visiting professors, from the Soviet Union, Great Britain, Germany, and the United States, with the American contribution coming from the University of Illinois. Kharagpur's goal was to produce engineers who would help lay the foundation of an industrial India, while as Chapter 10 will show, MIT was moving in a different direction. At Kharagpur at that time, civil engineering had the place of preeminence among the disciplines, well suited to an India doing lots of building, but far from the leading edge of engineering in the West. Within electrical engineering at Kharagpur the focus was on electrical power, again appropriate given the work still to be done to electrify India, but electronics was becoming predominant in the West. The first batch of Kharagpur graduates primarily got jobs in Indian industries or in government service.[7]

Kharagpur, receiving contributions from a number of nations, was a visible demonstration of Nehru's policy of refusing to choose sides in the Cold War. In establishing the remaining three (later four) Indian Institutes of Technology, India restructured them so as to add a level of foreign competition that had not been present before. India reached an agreement with the Soviet Union to sponsor IIT Bombay and with West Germany to sponsor IIT Madras. The expectation was that the United States would sponsor IIT Kanpur. Later the British sponsored IIT Delhi. Each institute was to showcase the technological education system and the technology of its patron.[8]

At the same time, the process, begun during World War II, of building connections between the United States and India continued.

In October 1946 Piroja J. Vesugar, the director of the J. N. Tata Endowment, arrived in the United States. The endowment was the organization established by J. N. Tata in 1892 to enable Indians to study abroad. A friend of Vesugar's wrote to an American friend explaining the purpose of her trip. He stated that previously most Tata Scholars had gone to England for study, but that now "with the advent of the war and the increasing demand for scientifically competent people in India, its emphasis has shifted to the U.S.A."[9] The Tata Endowment's shift was representative of a larger shift in Indians studying abroad. In the earlier part of the twentieth century, far more Indians studied in Britain than in the United States, but by 1948 more Indians were studying in the United States than in Britain, a preference that would only grow over time.[10]

In the years after World War II, the Indian government had paid for some Indians to study engineering abroad and one of the purposes of Kharagpur was to render this unnecessary. But even with Kharagpur's opening, Indians still came to the United States and MIT, even as undergraduate engineering students. Forces both in the United States and India encouraged this trend.

In June 1952 U.S. Ambassador to India Chester Bowles sent an extraordinary letter to MIT President James Killian. Bowles, a former advertising executive, was far more positively inclined toward India than his predecessors and launched a campaign worthy of his previous career, to sell Indians on America. His efforts to dramatically increase U.S. economic aid to India were destined to fail in the face of a Republican Congress that saw India and Nehru as insufficiently supportive of the United States and its policies, but his proposal to Killian was one that would require no congressional sanction.[11]

Bowles wrote to Killian regarding Ajay Nehru, the son of R. K. Nehru, an Indian official in the Ministry of External Affairs. Bowles described the elder Nehru as a cousin of the prime minister and one of the "top two or three Indians who know the United States very well." The younger Nehru wanted to attend MIT, but needed financial assistance. Bowles argued that one of America's problems in India was the "lack of real firsthand knowledge of the United States" among Indian leaders, citing B. R. Ambedkar, untouchable leader and law minister, as the only Indian leader who had received his ed-

ucation in the United States. Bowles stated that the long-term interests of the United States would be served by "having many of the
sons of leading Indians get their undergraduate and professional education in the United States." Bowles asked Killian if he could find
any money to support the younger Nehru at MIT.[12]

Although Nehru had not yet even applied to MIT, Killian wrote
back stating that MIT would break with its general policy and offer
Nehru a full tuition scholarship for the first year. The scholarship
was, of course, contingent upon Nehru's admission to MIT. However, the fact that the letter was copied to the MIT dean of admissions and that Killian wrote that Nehru seemed "well qualified for
entrance to MIT" suggests that the normal admissions process was
all but bypassed.[13]

Whether or not the State Department and MIT made any more
formal efforts, the "sons of leading Indians" continued to come to
MIT throughout the 1950s and 1960s. High-ranking government
officials, including another Nehru, two governors of the Reserve
Bank of India, and a jurist who would later become the vice president
of India, saw their sons go to MIT. At the same time, as Chapter 8
will show, a number of leading Indian business families sent heirs
to MIT.

While Indians went to MIT, by the 1950s an earlier group had
returned. Nehru had a vision of the Indian state using science and
technology to create what he had called at MIT a "first class" nation. Nehru spoke of dams as "temples of modern India."[14] And if
dams were temples, then engineers, as Anant Pandya had said at the
Bengal Engineering College in 1939, were its priests. By this time
India had engineers educated in India, Britain, as well as American
institutions other than MIT. But the careers of its MIT graduates
exemplify the challenges that Nehru's vision faced.[15] They were key
figures in building dams, managing fertilizer plants, designing steel
mills, managing the development of atomic energy, and creating new
methods of processing food. India developed technical capacity that
it never had before and made a variety of technical achievements.
But as impressive as these individual technical skills were, they were
not sufficient to bring Nehru's dream into reality. Other barriers
stood in the way. A significant number of MIT graduates sought

work in private enterprise rather than within the state. Foreign assistance came with conditions that led to inefficiencies and squeezed Indian engineers out of jobs. The Indian bureaucracy proved to be a fierce opponent of engineers. One might question whether the goals that Nehru and his engineers pursued were the right ones for India, largely seeking development in the Western mold. Some dreams proved to be out of reach, while the developmental state became in part a national security state not unlike that of the United States. By the early 1970s, a disillusionment had set in among Indian scientists and engineers, government officials, and the middle-class public more generally.

Private Enterprise and the Developmental State

In January 1946 Ramesh Mehta wrote to Frances Siegel. Mehta, Siegel, and Pandya had been in an informal socialist cell group in the United States in the early 1930s. Mehta described to Siegel the career path of their friend: "Anant is a big GUN." He described how Pandya had left government service and had "entered on a big contract with a new firm of engineering contractors." Pandya was now "a successful man in the ordinary sense of the term," which had changed his "ideals and ideology."[16]

Pandya could have remained principal of the Bengal Engineering College for the rest of his career. But he was too driven, too ambitious, and too impatient for such a static job. After working in government service in munitions production during the war, he left to start his own company to seek the large construction projects that would inevitably come after independence. Pandya's goal was to develop an Indian firm modeled after large American contractors, keeping the work in India and developing Indian capabilities. His firm, working in a joint venture with another Indian firm, succeeded in winning large contracts. Their first major job was for the construction of a railroad tunnel in western India, then they won a large contract for building a five-mile-long water tunnel, designed to double Bombay's water supply. In 1948, at the request of Indian Deputy Prime Minister Vallabhbhai Patel, Pandya became the first Indian general manager of India's aircraft company, Hindustan Aircraft. During a nine-month stint, he reorganized the company and began the development of its first plane, a propeller-powered trainer.[17]

Thereafter Pandya returned to his contracting firm. Although Pandya's firm was private, his best customer, with the resources to undertake the large projects he sought, was the Indian state. Pandya's biggest coup was in winning a contract for the Konar Dam, part of the Damodar Valley project, a large-scale program often compared to the American Tennessee Valley Authority. Pandya's group beat out three European firms. Although Pandya's consortium had not previously built a dam before, the World Bank supported it after a German firm was included in the bid.[18]

Pandya attempted to create a reformed Indian contractor in a more Western mold, based on the belief that by replacing older techniques (such as carrying dirt in a basket placed on the workers' head) with modern ones, he would be able to build major projects more quickly and cheaply. One Indian colleague later said Pandya was "imbued with the go-ahead spirit of the West." Another said his personality was "almost atomic (in its new significance)."[19]

Anant Pandya was a young man in a hurry, with much building to do. In 1951, after a trip to Germany, one of his old friends asked how much more traveling he was going to do. Pandya replied, "Speed, great speed should be the great 'Mantra' of our land." Tragically Pandya, who had so much to do, would not be given much time in which to do it. On June 1, 1951, as Pandya was traveling at night from the Konar Dam site back to Calcutta (en route to Bombay), his driver fell asleep and ran into a truck, killing Pandya.[20]

At the time of his death, Pandya was not a widely known figure in India, or even Bombay. The *Times of India*, lamenting that his death had "excited comparatively little notice," claimed that India was a country where "too much attention is paid to politicians and too little to scientists." A later piece in the *Times* asserted that it was "impossible to realise what a great loss" Pandya's death was to the country.[21]

For many years India's MIT-trained engineers would have had a very limited public presence as "MIT-trained engineers." Even well-educated Indians would have had little reason to know of their existence. Pandya's death helped bring attention to MIT. In 1952 the Gujarati upper-class youth magazine *Kumar* published a special issue memorializing Pandya. It featured pictures of his time at MIT and his later career, as well as tributes from leading engineers in India and some of Pandya's professors in the United States. Pandya's education at MIT was a particularly prominent feature. Pandya received a level of acclamation rare among engineers anywhere.[22]

Some idea of what this issue of *Kumar* meant for young Gujarati men is suggested by the example of Kirit Parikh, who was in 1952 the seventeen-year-old son of a barrister in Ahmedabad. Parikh was considering the next step in his education after receiving his first college degree. Parikh, who had been raised in a Gandhian school that taught spinning on a charka as a graded course carrying as much weight as mathematics, physics, or English, said that seeing the story on Pandya in *Kumar* "created a dream that I wanted to go to MIT." He did, earning a doctorate in civil engineering from MIT in 1962. The editors of *Kumar* doubtless would have been pleased. Parikh would go on to have a distinguished career, combining engineering and economics, serving for many years on the Indian Planning Commission.[23]

Minu N. Dastur might be seen as Pandya's heir, seeking to develop an independent Indian contracting business, in this case focused on steel mill construction. His story suggests the frustration that may have befallen Pandya had he lived. A central tenet of Nehru's India was that the government would control major sections of the Indian economy. In 1948 the Government of India issued an Industrial Policy Resolution, which laid out the areas of the economy open to private enterprise and those restricted to the government. Atomic energy, railways, and arms were government monopolies. In steel and several other industries, the government had reserved the right to itself to make any new investments while leaving open the option of private-sector cooperation. Existing private operations in steel, such as Tata Iron and Steel, were allowed to continue.[24]

Steel production was one of newly independent India's biggest technological and economic challenges. Because of its essential role in so many other industries, advocates of industrialization commonly argued that steel was the keystone of the process. In 1951 India had the large Tata Iron and Steel (TISCO) plant and two smaller privately owned plants, which together produced 1.5 million tons of steel or roughly 8 pounds per person per year. By contrast, in 1950 the United States produced almost 97 million tons of steel or almost 1,300 pounds of steel per person. If India was going to industrialize and build machinery, factories, farm equipment, and railroads, it would need steel. Expanding India's steel production capacity was vital so that India would not continue to be dependent on foreign producers. Between 1953 and 1955, it made agreements with the Soviet, West German, and British governments to each build million-ton-a-year steel mills for India in Bhilai, Rourkela, and Durgapur, respectively.[25]

When Dastur arrived back in India in 1955, it seemed like a case of perfect timing. Dastur, born in 1916, was a Parsi whose father had worked as a clerk at the TISCO plant in Jamshedpur. The younger Dastur earned a bachelor's degree in mechanical engineering from Banaras Hindu University in 1938, then himself began work at the TISCO Works in Jamshedpur. There he took part in Tatas "loop course," where after working an eight-hour day, he attended evening classes. He advanced to a technical assistantship in the general superintendent's office. One of Dastur's brothers had earned a master's degree from MIT in metallurgy in 1941, and in 1945 Dastur himself entered MIT to study metallurgy, funded by a Tata Endowment loan scholarship. He studied with John Chipman, the head of MIT's metallurgy department and one of the country's leading metallurgists. Chipman focused on the production of steel, and he had many contacts in the American steel industry. Dastur finished his doctorate in 1949, and unable to get work with the big American steel companies, went to work for the New York consulting office of Herman Brassert. Brassert, a German émigré and veteran of Carnegie Steel and US Steel, was one of the leading steel consultants in the world whose work included the expansion of the TISCO plant in 1937, as well as several new mills in Germany during the Nazi period. By 1947 Brassert was involved in building new steel plants in Chile and Brazil.[26]

Working with Brassert gave Dastur the chance to see best practices in the steel industry throughout the world. At the same time he kept in touch with Indian government officials as he developed his plans to return to India. He wrote a detailed blueprint for the development of the Indian steel industry, which he sent to Arthur Lall, the Indian Consul-General in New York. His plans were catalyzed during a trip he took in 1954 to consult on a steel plant in Mysore, India. Then he met with T. T. Krishnamachari, the Indian Minister of Industry and Commerce. He spoke with J. R. D. Tata, the chairman of TISCO. Dastur later wrote that he received personal encouragement from Jawaharlal Nehru to come to India to establish a steel consultancy.[27]

Dastur's first job in India came from TISCO. His former employer needed a plant to produce ferromanganese, an important component of steel production. Although India had rich manganese deposits, its inability to produce ferromanganese required the alloy's importation. TISCO proposed to remedy this situation by building its own ferromanganese plant in Joda in the state of Orissa. Dastur had just a few people working with him to prepare a proposal in a competition pitting him against leading firms around the world, including Demag, the German firm leading the construction of the plant at Rourkela. The Tatas ultimately chose Dastur, who bid 30 percent less than his nearest competitor.[28]

The project faced many challenges. Dastur had to staff up from almost nothing. It was located in a remote area of Orissa that was essentially a jungle before construction began. The Suez Canal crisis occurred when vital equipment was being shipped. But the project was completed eight months ahead of schedule.[29]

Although Dastur started his firm by himself, he made a technological nationalist appeal to quickly build a first-rate staff. He offered the vision of Indian engineers working for an Indian boss in an Indian firm building a new nation, a technological nation, in the process showing the world that Indians were a technologically adept people. In a brochure publicizing its work on the Joda plant, Dastur identified eighteen other engineers with the company. One other engineer had a degree from MIT, while another had degrees from Stanford. But overall the common characteristic of Dastur's staff was that they had many years experience working in the Indian steel industry. Dastur's chief designer and lieutenant was D. S. Desai, who had a degree from Sheffield, then had worked for twenty-eight years for the Indian office of a British firm designing structures for the Indian Steel Works.[30]

In a letter to an Indian government official written prior to his return to India, Dastur had argued that there were two approaches to the construction of steel plants. The first approach, favored by foreign firms, was to import essentially the entire plant as a package unit. Dastur claimed that if the construction of the steel plant was overseen by a technically competent Indian instead, much greater use of Indian resources and capabilities would be possible. He estimated that this could lead to integrated steel plants being built for 60 percent to 70 percent of the cost of a plant imported as a package unit. He gave a specific example of how a Mexican firm, unable to get American or European help during World War II, built a steel plant for half of what it would have cost with outside help.[31]

In the late 1950s Dastur, in a series of letters to Indian government officials, continued these arguments, increasingly frustrated with India's foreign-managed projects and how little Indian capabilities were being used. Although he did not use the term, he implicitly argued that by handing over control of the construction of steel plants to foreigners, Indians were subjecting themselves to a new form of colonialism. Foreigners consistently underestimated Indian capabilities, hiring foreigners to do jobs that could be done by Indians. Furthermore they used their position of power to impose their will, saying that they could not guarantee the results if things were not done their way. They then imposed excessive consulting fees. Dastur provided a host of specific examples of how this foreign-controlled system was disadvantaging India. German carpenters were brought in when Indians could have done the job just as well. Excessive amounts of imported steel were used in construction when smaller amounts of Indian-made steel would have served adequately. Foreign consultants were engaged when Dastur's firm could have done the work. And worst of all this was done with little concern for the fact that India would have to pay for all these extravagances with foreign exchange.[32]

Dastur was not seeking national self-sufficiency in steel plant construction, for by the mid-twentieth century the industry was so specialized that no one country produced all the components necessary for a state-of-the-art plant. But what he was seeking was Indian control over the process, claiming that with such control the Indian content could be maximized. He did not argue that Indians would never make mistakes, but rather that under Indian control Indians could learn from their mistakes. A policy of contracting out to foreign firms for the construction of steel plants allowed these firms to gain the benefits of experience, leaving India in a state of perpetual adolescence.[33]

Indian Prime Minister Jawaharlal Nehru was an early ally of Dastur—in fact, more of an ally than the bureaucrats around him. In August 1958 Nehru wrote to his minister for steel, Swaran Singh, asking, "What has happened to Dastur?" Nehru gave Dastur an unqualified endorsement, saying that he "had a very high opinion of Dastur's ability," calling him "probably the most competent man for this kind of work in India." More subtly Nehru noted that Dastur had a "certain enthusiasm and some public sense, which is not always to be found among our experts"—Dastur was no bloodless technocrat.[34]

More fundamentally, Nehru showed that he accepted Dastur's main points about what an independent steel consultant could do for India. Nehru suggested "it may be that Dastur's advice might save us a considerable sum of money even in the plants that are being built." And he hoped that future steel plants could be built by "our own experts." Nehru implied that the Indian bureaucracy failed to make full use of Dastur writing: "It is odd that we employ large numbers of expensive foreigners and do not utilise the services of really capable Indians."[35] Later that month, Nehru wrote again to Singh, stating he had a "very uncomfortable feeling" about the new steel plants in Durgapur, Rourkela, and Bhilai, where India had to "rely entirely on foreign consultants and cannot check what they tell us," which made him anxious to use Dastur.[36] In December Nehru wrote to another cabinet minister, Vishnu Sahay, expressing the same feelings more powerfully, saying that Dastur was "probably better than the foreign consultants we have engaged at a very high price," and that it was "a great pity that we are not utilising him."[37]

Whether it was through Nehru's support or not, after this, the Indian government seemingly embraced Dastur. In 1959 Dastur submitted a proposal that his firm be given responsibility for building a plant to make alloy steel, a specialized, high-value steel. In doing so he could point to five alloy steel plants he had been involved with in the United States and Europe. He argued that awarding such a project to an Indian firm would be "psychologically" of "tremendous importance" to the people of India, who "will gain in confidence with the knowledge that a project of this nature was conceived, designed, and executed by Indians." While the government bureaucrats seemed hesitant about Dastur—at one point they questioned whether foreign firms would be willing to work under an Indian firm!—ultimately Dastur won the job.[38]

Getting the alloy steel plant contract was big, but the real prize was Bokaro, the fourth new steel plant to be built by the Government of India. It was to have an initial capacity of 1.5 million tons of steel a year, eventually expanding to 4 million tons a year. The Government of India had assumed the support of a foreign patron, expected to be the United States. And in fact the United States and India were in discussions over a program of assistance. The Kennedy administration, wanting to support Bokaro as part of its program to develop closer relations with India, ran into implacable opposition from some within the United States Congress who opposed U.S. aid to a socialist enterprise. Finally, in September 1963, the Government of India formally withdrew its request for American assistance for Bokaro, asserting its determination to go ahead on its own.[39]

Having the United States out of the picture seemed to work in Dastur's favor. American assistance would have invariably meant a large degree of American control, likely leading to the same issues Dastur had seen in the other plants. Dastur's proposal for Bokaro, released in July 1963, had asserted that the plant was "expected to be among the lowest cost producers in the world."[40] Bokaro had the potential to be a showcase, not of American technology, but of Indian technological prowess. Shortly after India had withdrawn its request for American support, Dastur came to Delhi and met with Prime Minister Nehru for over an hour on the Bokaro issue. Dastur later wrote to Nehru about the prime minister's "searching questions," "keen interest," and "extensive specific knowledge of the important issues." Dastur noted that India's decision to go ahead with Bokaro on its own would be "watched with wide interest here and abroad," and asserted his and his firm's confidence that they could do the job "completely and successfully."[41]

In late 1963 the Indian steel minister announced on the floor of the Lok Sabha, the lower Indian house of Parliament, that Dastur had been given the contract to design, engineer, and oversee the construction of the Bokaro steel plant, stating that this marked "a landmark in the history of advancement of Indian technology."[42] In response, Dastur's firm dramatically ramped up its hiring, putting together a team of 800 people with a wide range of engineering and business talents.[43] Instead of allowing foreign countries and technologists with their limited knowledge of India to provide a solution, a group of Indians would do so. And instead of having hundreds of foreigners coming to India, Indians would be doing the work, hiring foreigners only as necessary. The money spent on Bokaro would not only build up India's capacity to produce steel but also its capacity to produce steel mills.

However, things fell apart for Dastur from there. In Dastur's mind, the turning point came with the death of Jawaharlal Nehru in May 1964. Dastur felt that Nehru believed in him, a prime minister who had grown in the job having an affinity for an engineer who would grow as greater responsibility was entrusted to him.[44]

India's lack of capital played an important role in thwarting the plans for a swadeshi Bokaro. Dastur had assumed that half of the 5.5 billion rupees needed to build the plant would be provided by the Government of India, with the other half coming from a long-term loan. The government continued to seek foreign partners, and in May 1964 the Indian government and the Soviet Union announced an agreement on financing for Bokaro. But the Soviets provided more than financing: once they came into the picture, their desire to control the construction process meant that the role of Dastur would decrease.[45]

The Soviet design cost far more than the Dastur design and was twice as expensive for a given output as similar plants in Japan, France, or Great Britain. Dastur's vision was that as an independent consultancy, his firm would be able to choose from technologies around the world to give India the optimal steel plant. The vision was blocked by a Soviet program that tied India to one supplier.[46]

Ultimately the connection to the Soviets destroyed any idea of a swadeshi plant. Three-quarters of the design content came from the Soviets, with Dastur's contribution limited to outbuildings and auxiliaries. The Soviets supervised construction of the plant, although Dastur argued that his firm was capable of doing such work. The government commissioned Dastur to write a report on possible cost reductions at Bokaro, but when Dastur claimed that the plant could be built for a billion rupees less, the government rejected his report based on the principle that it would be "ungracious" to ask the Soviets to revise their plans, while the Indian ambassador to the Soviet Union suggested that the Soviets would not agree to be "mere suppliers of machinery."[47]

In Dastur's defeat, allies rallied to his side. An unnamed writer in the Indian journal *Economic Weekly* claimed that Dastur & Co. represented "the most outstanding achievement in designing and consultancy services," and that Dastur's team was "the best ever got up in a developing country." The writer went on to claim that Dastur has achieved an "immense stature" by international standards, proven by a Pakistani steel firm, funded by the Export-Import Bank, and supported by an American steel mill, which sought out Dastur's expertise.[48]

The press also saw the Indian government bureaucracy as complicit in Dastur's ouster, fearing the loss of power from having an engineering firm, even an Indian one, in control. Dastur had openly argued that technical projects should not be managed by nontechnical people. In 1968 the *Economic Weekly* carried a withering report on Bokaro, sarcastically observing that Dastur was not allowed to design the plant's foundations because "foundations for the important plant are considered too difficult to be designed indigenously." The report observed the project's labyrinthine structure, concluding "the career bureaucrat has succeeded in erecting suitably expensive hurdles and diversions in the path of yet another important project."[49]

Dastur continued to get work from the Indian government, but it was far less than what one might have expected. The Government of India established its own engineering and design group and Dastur ended up splitting work with it. Furthermore, when India established a fifth government-owned steel plant in Visakhapatnam, Dastur did some early design work, but India again turned to the Soviets for assistance, with the Soviets accounting for half the engineering cost.[50]

Ironically, although Dastur was a technological nationalist with a passion for steel, the firm got significant work outside of India. In its early years it did work for the National Steel Company of Pakistan. It worked for countries around the world, from Latin America to Africa to Asia. A German subsidiary, established in 1969, did early work that today would be considered outsourcing, winning contracts for the preparation of designs and drawings, which were then subcontracted back to the Indian parent company. Dastur's biggest job was overseeing the construction of a new integrated steel plant in Libya in the 1970s and 1980s.[51]

M. N. Dastur & Co., sometimes called Dasturco, was very much an extension of M. N. Dastur the person. When Dastur lost the Bokaro job, instead of summarily laying off employees, all (including Dastur) took a pay cut. Dastur's company embodied his humanitarian principles, offering healthcare to all his workers, with his physician wife serving as the de facto company doctor. One tribute to Dastur after his death noted the democratic spirit displayed on elevators and in company picnics.[52]

M. N. Dastur & Co. was as much a cause as it was a company, embodying Dastur's passion for steel technology and Indian self-reliance. Dastur's fierce commitment to those principles, at times violating norms of business diplomacy, won him both steadfast supporters and ardent opponents. Dasturco's family-like organization was a great source of strength, but also a limitation in a world where businesses were increasingly run on impersonal organizational principles.

The case of M. N. Dastur shows how a charismatic, technologically skilled India could build up a technological workforce. But Dastur's vision of a technologically nationalist steel industry, controlled by Indian technologists and designed to operate in India using the most suitable technology available anywhere in the world, was thwarted both by the politics of global aid and by an Indian bureaucracy unwilling to give up power.

Brahm Prakash, Atomic Energy, and Rocketry

A key part of India's program of technological development was the establishment of "display" technologies that were highly visible to the public both in India and abroad, with great symbolic value. Few nations had the ability to create these technologies, and to do so proclaimed a nation's technological sophistication to its citizens and the world. Between the 1950s and the 1970s, two such key technologies were atomic energy and rockets. Their dual use as both civilian and military technologies gave them a special valence, particularly as the military aspect remained at the surface.[53]

While atomic energy had a remarkably attractive power throughout the world, in some ways the most extraordinary atomic effort in the 1940s and 1950s belonged to India. Here Indian Prime Minister Jawaharlal Nehru and Indian physicist Homi Bhabha shared a vision of atomic power creating a new world, and in that new world, a new nation. India, lagging the West in most areas of industrial production, sought to develop nuclear power concurrently with Western nations. In mastering this most advanced technology, India would prove to the world that it was scientifically and technologically advanced. Nehru and Bhabha saw nuclear power catapulting India to a new stage of existence.[54]

In February 1948, after a long talk with Bhabha, Nehru wrote to his Defense Minister Baldev Singh that "the future belongs to those who produce atomic energy." While Nehru's public pronouncements emphasized peaceful uses of nuclear energy, here he was less circumspect, adding, "Of course, defence is intimately concerned with this."[55] This vision helps explain why, only one year after independence and before India even had a constitution, India's Constituent Assembly, acting at Nehru's behest, inaugurated an Atomic Energy Commission.[56]

The key figure in the Indian atomic program was Homi Bhabha, the perfect front man for a program where image counted for so much. A brilliant scientist, educated at Cambridge University in Britain, who was named a Fellow of the Royal Society of England at the age of thirty-two, Bhabha was widely known in the elite circles of physics and could count among his correspondents such leading physicists as Albert Einstein, J. Robert Oppenheimer, Niels Bohr, and P. A. M. Dirac. Bhabha was a charismatic extrovert with a wide range of interests and talents, ranging from art to music. Physicist C. V. Raman called him "the modern equivalent of Leonardo Da Vinci," an image reinforced by Bhabha's paintings and drawings.[57]

His elite social background was similar to that of Prime Minister Nehru's and, not surprisingly, they were friends. To a remarkable degree India's atomic program seemed to emanate from their personal visions rather than from an objective assessment of the technology's possibilities. And Nehru and Bhabha worked to keep it that way, clothing the program in secrecy. But Bhabha had a habit of over-promising and underperforming.[58]

Brahm Prakash was to be one of Bhabha's key lieutenants in India's nuclear program. While they respected one another, they were in many ways complementary figures. Although Bhabha was marked for greatness from an early age, Prakash had risen slowly. Prakash was an introvert. He was not flashy. But he would become one of the leading research managers in India, playing an important role in building up the managerial capabilities of the Indian atomic and space programs.

Prakash was the son of a railroad administrator who grew up in a middle-class Hindu family in Lahore. He studied chemistry at the Government College in Lahore and then went on to earn a doctorate in chemistry working with S. S. Bhatnagar, one of India's most prominent chemists. Prakash then took a job working with the Railway Metallurgical Laboratories in Ajmer. Even though he already had a doctorate, in 1945 at the age of thirty-three, Prakash applied for and was awarded a scholarship in the first round of the Overseas Scholarships, a recognition of the distance between Indian and American graduate programs.[59]

After briefly considering Columbia University, Prakash enrolled at MIT, where he switched from chemistry to metallurgy. At MIT Prakash worked with a group of scientists renowned in metallurgy who were working to expand metallurgy to a more scientific and general study of materials.[60]

When Prakash returned, Homi Bhabha selected him to be a metallurgist with the nascent atomic energy program. While physicists, such as Oppenheimer in the United States, were often most associated with atomic energy programs, metallurgists also played a crucial role in the program. At MIT Prakash worked with two metallurgists who had been involved in the Manhattan Project: John Chipman and Morris Cohen. But while Prakash was slotted into the position of chief metallurgist, the still-embryonic program did not yet require his presence. Prakash was loaned to the Indian Institute of Science, where he became its first Indian head of the department of metallurgy and worked to modernize its curriculum.[61]

India's atomic program put it on a global stage. Homi Bhabha was named the chair of the 1955 International Conference on the Peaceful Uses of Atomic Energy, showing the utility of being a nonaligned nation in a bipolar world. Prakash, who served as one of the conference secretaries, also presented a paper with his graduate student. Prakash formally joined the Atomic Energy Establishment in 1957 as director of the Metallurgy group. He had responsibility for a significant part of India's nuclear program, having the largest group at its main facility in Trombay.[62]

The economics of India's atomic energy program faced critics both in India and internationally. In 1958 British economist I. M. D. Little challenged Homi Bhabha's claims of the economic benefits of atomic energy for India, and noted that the longer India waited in developing its nuclear program, "the more free benefits she will get from the immense investment which has been poured into nuclear physics and engineering in the USA and UK." In 1961 the *Hindustan Times* asserted that the program "has never had the benefit of objective study," while the *Times of India* noted that economically, technological progress made in conventional fuels was making atomic energy less competitive. The United States and Europe also vastly overestimated the economic possibilities of atomic energy, but the opportunity costs of India's misjudgment were far greater.[63]

Of course atomic power was about more than generating electricity. Nehru and Bhabha played a double game here, often declaiming their interest in atomic weapons, but subtly making suggestions about India's capability of making them. The military applications provided both a justification for the program itself and also for shrouding it in a secrecy that kept the skeptics at bay.

In 1966 Homi Bhabha died in a plane crash and the leadership of the atomic program was taken over by Vikram Sarabhai. As the program developed, both politicians and the technical community debated whether it should focus on atomic power or whether it should aim to build an atomic explosive. Sarabhai had been one of the main figures not supporting the development of a nuclear explosive, and with his death in 1971, those who supported such a device were decisively in power. Prakash's sensibilities may be inferred by the fact that at that point, he left the atomic energy program to join the space program.[64]

In 1972 Satish Dhawan, the newly appointed head of India's space program, named Prakash his second-in-command. As mentioned in Chapter 6, Dhawan and Prakash had traveled to the United States together in 1945, with Dhawan going to Caltech to get his doctorate in aeronautical engineering while Prakash went to MIT. They taught together at the Indian Institute of Science, with Dhawan staying on and eventually becoming its director.[65]

Dhawan and Prakash made a very effective team. Prakash, the head of the Vikram Sarabhai Space Center in Trivandrum, was the inside man, with direct charge for the development of space vehicles. Under their leadership India developed the SLV-3 launch vehicle, which successfully put India's first satellite into orbit on July 18, 1980, making India one of only six space-going nations. Prakash, who had retired the year before, had served as a mentor to A. P. J. Abdul Kalam, the project manager for the SLV-3. Abdul Kalam later went on to lead the development of a family of ballistic missiles and serve a term as president of India.[66]

Prakash's career shows the development of India's technological capabilities, to the point where it could successfully complete the complex SLV-3. But at the same time it shows the rise of the national security state, where a significant portion of India's technical resources went into projects that had dual uses and where their military possibilities were significant drivers.

Darshan Bhatia, Government-Sponsored Research, and Coca-Cola

The Council of Scientific and Industrial Research (CSIR), a late colonial-era institution infused with new purpose and funding after 1947, was a key center for government research. So high was the enthusiasm about the possibilities of scientific research in the 1940s and 1950s that Nehru and other Indian dignitaries were often present at the laying of foundation stones for the laboratories or the opening of new facilities. After the opening of the National Physical Laboratory in New Delhi, a British scientist wrote that "it is astonishing the respect which the authorities in India give to science," noting an array of Indian dignitaries present, which he considered unlikely to be matched by equivalent English notables on an analogous occasion.[67] Nehru kept for himself the portfolio for scientific research and served as president of the CSIR, while attending almost all the meetings of the Indian Science Congress between 1947 until his death in 1964.[68]

Between 1947 to 1964 the CSIR opened twenty-seven laboratories that conducted research in areas ranging from chemistry and physics to electronics, aeronautics, and marine chemicals. From the late 1950s to the early 1970s within the Government of India, the CSIR's budget for research was exceeded only by the Department of Atomic Energy.[69] By the mid-1960s this enthusiasm for the role scientific research could play in building a new India had turned to disillusionment, as Indians found that their lavish research funding had provided precious little payback. At the same time researchers expressed widespread dissatisfaction with the conditions under which they labored.

Darshan Singh Bhatia's career illustrates the loss of faith in government-sponsored research. Bhatia, a Sikh, was born in Lahore in 1923 and attended the Forman Christian College. In 1945 he earned a master's degree in chemical technology from the Punjab University. He worked for two years for a military dairy before being selected in 1947 for a government scholarship for study in the United States. He left India in July 1947, on a U.S. troopship, with roughly 150 other Indian students coming to the United States for advanced studies. He left behind a pregnant wife (his son would be three years old before he would see him), knowing that a partition was coming, which would uproot his family. He entered the program in food science at MIT, earning his doctorate in 1950 for work on the effects of radiation on proteins.[70]

Upon earning his degree, he immediately came back to India and took up a position with the CSIR in the Central Food Technological Research Institute in Mysore, where he would became the head of the food processing division. His work there centered on increasing protein in the diet. He and other researchers worked on a scheme to develop a vegetable-based milk (from groundnuts), showing that it could be made roughly equivalent to dairy milk in terms of nutrients. Bhatia and his team conducted consumer trials on the favorability with which the vegetable-based milk was received.[71]

Bhatia was a global Indian, sitting on international committees and traveling throughout the world for meetings and assignments. In the 1950s he worked with a Swiss firm to attempt to develop a wheat-based "synthetic rice," while in 1962 he attended an international conference on food technology in London. In 1962 and 1963 he went to Saigon under the auspices of the United Nations to examine food production there.[72]

The period after Jawaharlal Nehru's death in 1964 was one of disillusionment in scientific research in India. This disillusionment, in part a reaction to an excessive faith in the ability of research to transform India, had analogs in the disillusionment American corporations experienced at the same time when they began to see that the millions of dollars they were pouring into scientific research was not leading to new products. However, India was trying to create a scientific research culture where none had existed before without having the same resources as Western organizations.

In 1965 Bhatia, frustrated by the politics of his CSIR lab, took a different direction: he joined Coca-Cola. Bhatia initially worked at the Coca-Cola Export Division's operation in New Delhi. His first major project was working with protein beverages, a continuation of his previous research, only now in the private sector. While at Coke, a United Nations committee selected him to be one of three food scientists to produce a report on the world protein shortage.[73]

As he stayed with Coca-Cola, corporate priorities took him further and further away from India and its priorities, both literally and figuratively. Bhatia was an outstanding researcher and manager, and perhaps inevitably, he was brought to the Coca-Cola metropole in Atlanta in 1972, initially leading a corporate protein group. Coke was working to develop protein drinks for third-world markets to help alleviate what was seen as a global nutrition crisis. The company test-marketed a soy-based protein drink in Brazil that contained 25 percent of a child's daily protein requirement and cost five cents. Later Bhatia became head of corporate research and development at Coca-Cola, and under his leadership Coke developed a wide array of new beverages, such as Diet Coke, New Coke, and Cherry Coke. In Atlanta, he helped start some of the city's earliest Indian cultural organizations.[74]

Bhatia was a mid-level manager when he left the CSIR lab and then India, and his departure went unnoticed by the press. However, in 1968 the failings of the Indian research system came to the fore at what might have been a moment of triumph, when the first Indian since independence was awarded a Nobel Prize in science. That year Har Gobind Khorana received the Nobel Prize in Physiology or Medicine for his work on protein synthesis. Khorana had been among the first group of Indians selected for study abroad in 1945, when he went to the University of Liverpool ostensibly to study agriculture. His Nobel Prize could be seen as a validation of both the strategy of sending Indians to study abroad and of the selection process used, except for one inconvenient fact: Khorana had essentially never returned to India and was a professor at the University of Wisconsin when he won the prize. Government ministers were forced to answer questions in the Indian Parliament about published reports that Khorana had previously been unable to procure a job in India, the implication being that the Indian government could not recognize scientific talent.[75]

Khorana's Nobel led to biting commentary from India on the state of its scientific system. The *Times of India* carried a satire, "Dr. Khorana Comes to India," which imagined that Khorana had returned and accepted a position as the head of a fictional National Genetics Laboratory, which was "designed by a famed Finnish architect" and "lavishly equipped with all the latest that money can buy" with no foreign exchange spared. Khorana was presented with his first day's schedule, where administrative trivia and the demands of a hierarchy overwhelmed the pursuit of science. The schedule included approving a grant of fifteen rupees, fifty paisa for the purchase of doorknobs and also the sanctioning of a scientific paper written by a research assistant, which had five coauthors, including the department head. After going through the schedule, he tendered his resignation to the prime minister, asserting his desire to work in a country "where there is no organised conspiracy to suppress science."[76]

The state of Indian research and Indian researchers came to the fore in a far more somber way in May 1972, when Vinod Shah, a University of Wisconsin-trained agricultural researcher, hanged himself in an act of protest against the injustices in the Indian research system. Shah had been turned down for an appointment in favor of a person he believed was less qualified. Shah wrote in a suicide note, "I think the time has come again when a scientist will have to sacrifice his life in disgust so that other scientists may get proper treatment."[77]

In response, a group of Indian scientists and engineers led a seminar on "The Best Use of Scientific Talent in India," which examined the failings of science and technology in India. One of the organizers of the seminar was Jagan Chawla, a Punjabi who had gone to MIT in 1944 to study aeronautical engineering. Chawla had then studied at Brooklyn Polytechnic Institute before going to work at Hughes Aircraft. In 1960 he returned to India, where he worked as the director of Technical Development and Production for the Indian Air Force.[78]

Chawla and his colleagues delivered a searing indictment of the Indian scientific research system. While huge amounts of money had been spent on scientific research in India since independence, there was little to show for it because of a corrupt administration system, which allowed for patronage, nepotism, and favoritism without providing adequate oversight. Chawla further complained that India had focused too much on "glamorising science and scientific research" instead of focusing on the more mundane requirements of developing and producing new technologies, noting Germany and Japan as examples of countries that did little research but had become wealthy. In addition, the government's essential monopoly on research circumscribed the researcher's freedom when disputes arose with management.[79]

Among the figures that Chawla and his fellow editors cited as examples of the sorry situation facing Indian scientists and technologists were M. N. Dastur and Har Gobind Khorana. The editors recounted Dastur's Bokaro ordeal, asserting that "his great expertise was given a slap in the face." It then pointed to Khorana, asserting that he "was offered a job on rupees 300 a month in this country and had to leave in disgust and go to America." India seemed hostile to scientific and engineering talent.[80]

The careers of Pandya, Dastur, Prakash, and Bhatia show Indians whose technological capabilities were recognized throughout the world in a way that would have been impossible to imagine at the turn of the century. However, each man's career shows the tension involved in bringing Nehru's developmental dreams into reality. In the case of Dastur and Bhatia, one might posit that the bureaucratic Indian was more powerful than the technological Indian, making it very difficult to work through the state.

In the course of Prakash's career, the national security state overtook the developmental state, with the work both on atomic energy and space increasingly shading into areas with military applications rather than the broader social applications originally envisioned. The peaceful nuclear explosion at Pokhran and the successful launch of the SLV-3 were technological nationalist events, showing India's technological capacity and putting it into exclusive ranks open only to a few nations. By the early 1970s, the idea of using the state to direct technological development to build a "first class" nation no longer inspired the confidence in India that had it had back in 1949. But Nehru's vision was not the only one about how an MIT education might be used in India.

8

Business Families and MIT

O N APRIL 21, 1961, John Kenneth Galbraith, recently installed
as U.S. ambassador to India, went with his wife to the Indian
prime minister's residence, Teen Murti Bhavan, for lunch. Galbraith
came to Teen Murti at a time when relations between the American
and Indian governments seemed to be undergoing a rebirth. Newly
inaugurated U.S. President John F. Kennedy had a long interest
in improving U.S. relations with India, proposing as a senator a
dramatic increase in U.S. economic assistance to India. Indeed, the
very presence in New Delhi of Galbraith, a Harvard professor, well-
known public intellectual, and personal friend of Kennedy, signaled
the importance the president attached to India.

Attending the luncheon were Nehru, his daughter, Indira Gandhi,
and Nehru's cousin, B. K. Nehru, an Indian diplomat serving in
Washington who was soon to become the Indian ambassador to the
United States. There were some uncomfortable moments as Indira
Gandhi "pressed" Galbraith on the failed U.S.-supported Bay of Pigs
invasion to depose Castro in Cuba. However, the afternoon also
included much more amiable conversation conducted on far more favorable grounds to Galbraith. One such topic was a discussion of "the
possibilities of MIT as an educational institution," which was of interest to all the Indians present. India and the United States were in
the midst of negotiating a program of assistance whereby MIT and

a group of U.S. universities would work to develop an Indian MIT at Kanpur (to be discussed in Chapter 10). But MIT was also a personal matter to the Nehrus: B. K. Nehru had two sons at MIT, while Galbraith reported that "Indira contemplates sending her son."[1]

Just days before, Galbraith had a parallel lunch with G. D. Birla, one of India's leading businessmen, and while the record of the meeting is not as extensive as that of the lunch with the Nehrus, it is a virtual certainty that MIT was a topic of discussion here too.[2] Birla had met James Killian, the former MIT president, the previous year, and in November Birla would visit MIT to explore the possibilities of his own plan to develop an MIT-like institution in India. Like the Nehrus, Birla also had a familial interest in MIT: later that year his grandson would apply to MIT. He would be accepted and start at MIT in September 1962.

Jawaharlal Nehru and G. D. Birla are two leading figures in the history of twentieth-century India, as consequential in independent India as they had been in colonial India. MIT was central in each of their visions for India, different though they were. For Nehru, MIT would be integrated into a system where the state controlled India's technical development. Initially MIT-trained engineers would help India build its dams, steel plants, and other infrastructure, but eventually the IITs, India's own MITs, would produce technical talent that would free India from dependence on Western powers in creating technology.

G. D. Birla saw MIT as part of his vision of a nonstatist India built by private enterprise. Even as Nehru's government sought to build an Indian MIT, Birla sought to build a private enterprise MIT in India. Furthermore, in sending his grandson to MIT, Birla sought to provide him with an American technological and managerial education in preparation for taking over the family business empire.

By the 1970s high-ranking government officials and business families in India were often seen as opposing one another; yet, ironically, in the 1950s and 1960s these two prominent groups were sending their children to MIT as undergraduates. Attending MIT as an undergraduate required more long-range planning and a better source of funds than attending MIT's graduate programs. Chapter 6 showed how American business and diplomatic interests sought to encourage the sons of Indian government officials to attend MIT. Chapter 7 showed how U.S. Ambassador Chester Bowles expressed the desire to have the sons of "leading Indians" educated in the United States. Bowles's desire was fulfilled in Indian business families. While the process may have had some later encouragement on the American side, it began in the 1920s with initiative from India. Between 1926 and 1973, nineteen children from Indian business families went to MIT to earn bachelor's degrees. Others earned graduate degrees. They came from some of India's prominent business families, such as Lalbhai, Kirloskar, Birla, Godrej, and Paul.

Business families whose children went to MIT saw that their continued success was dependent on professional education in modern management and technology. And in finding that education in the United States, they demonstrated a global outlook, which also suggested that they knew that for their continued business prosperity they would have to operate in an American-dominated world. The children carried on their family's global perspectives as they came into positions of leadership. Children educated at MIT were more an effect than a cause.

Like Birla, these business leaders had a vision for India that was dramatically different from that of Nehru and his daughter, and they pursued policies that put them at odds with their own government. Specifically, at a time when the Indian government was implementing a variety of trade restrictions aimed at keeping India from being exploited by foreign companies, these businessmen avidly sought to be part of global business.

Business Families in India before 1947

Shantanu L. Kirloskar was the first Indian associated with a business family to attend MIT. He came from a business family that was still in the making. As discussed in Chapter 2, his father, Laxmanrao Kirloskar, was a Maharastrian Brahmin who in the early twentieth century had used connections with the United States to start a plow and farm implement business. In 1922 Kirlsokar's son, Shantanu, and his cousin, Madhav, left India to attend MIT. The family intended that Madhav would study electrical engineering through the MIT cooperative program with General Electric and use that knowledge to help the family enter the electric motor business. However, on his arrival at MIT, he found that the cooperative program was closed to foreign students (although we have seen how T. M. Shah was later able to enter the program). To this misfortune was added tragedy, as Madhav contracted tuberculosis and died in the United States.[3]

Shantanu went on to study mechanical engineering. Although Kirloskar was the only Indian at MIT at the time, he was not alone. His family's business contacts in the United States helped him. In his memoirs Kirloskar recalls getting a letter at MIT from someone at Niles-Bement-Pond, an American machine tool producer. The letter writer had seen Kirloskar's name on the MIT student roster and inquired as to whether he was related to the Indian firm of that name, which was a customer of Niles-Bement-Pond. Through this connection Kirloskar was able to get summer employment from Niles-Bement-Pond's subsidiary, Pratt and Whitney.[4]

In most other respects, Kirloskar's experience at MIT was like any other student. He lived in the dorms and participated in the normal college pranks. He took part in military training where he learned to fire rifles, machine guns, and anti-aircraft weapons. He was commissioned as a Reserve Officer in the United States Army before he returned to India. When he got back to India, he joined the family business.[5]

The next Indian business leader to send a son to MIT was M. L. Dahanukar of Bombay. Dahanukar controlled a managing agency and also a sugar mill. In the 1930s Dahanukar served as president of the Maharashtra Chamber of Commerce and was a member of the board of directors of the Federation of Indian Chambers of Commerce and Industry, sitting alongside such leading business figures as G. D. Birla, Walchand Hirachand, and Kasturbhai Lalbhai. In 1938 Dahanukar sent his son, Shantaram, to MIT to earn a master's degree in civil engineering. The younger Dahanukar returned and joined the family business, concentrating in construction.[6]

The sons of other Indian business leaders would follow the younger Dahanukar's path. Kasturbhai Lalbhai was from one of India's oldest business families, which had occupied a position of prominence in Ahmedabad since the seventeenth century. In the late nineteenth century and the first decades of the twentieth century, Lalbhai and his family established a network of textile mills in Ahmedabad.[7]

In November 1944 Kasturbhai's two sons, Siddharth and Shrenik, came to the United States to enter MIT. Kasturbhai later said that an Oxford or Cambridge education, as had been pursued by the children of the Sarabhais (Ahmedabad's other leading textile family), was not what he wanted for his sons. Lalbhai said he wanted his children to be educated in chemical engineering or business, fields "that were far more developed in the States than in Britain." Shrenik, in a later discussion of his path to MIT, said that his father was determined to provide him and his brother with "the very best education." Given Kasturbhai's wide range of travels and contacts, he would have seen many options and MIT must have been a very deliberate choice. Two factors seem to have been important in his decision. First, the company's business had grown to the point where it required professionally trained managers. In 1931 Kasturbhai had hired B. K. Mazumdar, a graduate of the London School of Economics, and he became a trusted assistant. However Mazumdar's involvement in the nationalist movement meant he was lost to Lalbhai for years at a time: a family member would be a more reliable manager. Furthermore the business was considering more technologically sophisticated areas, such as dyestuffs. The clout that Lalbhai had is suggested by the fact that his two sons arrived in the United States on an Army transport plane at a time when most people came by boat. Shrenik earned a bachelor's degree in business administration from MIT before going on to Harvard for an MBA. Siddharth studied briefly at MIT before transferring to Brooklyn Polytechnic, where he earned a degree in chemical engineering. At the time his sons were in America, the senior Lalbhai was negotiating a collaboration with American Cyanamid for a dyestuff factory.[8]

S. L. Kirloskar

When S. L. Kirloskar rejoined his family's business in the industrial village of Kirloskarwadi in 1926, its products consisted of simple metal agricultural implements, such as plows and threshers. He designed new products and took an increasing role in managing the business as his father moved away from day-to-day involvement with the business. The senior Kirloskar sought to develop in his children the capabilities to take the business in new directions and new levels of sophistication. In 1934 he sent two more sons to the United States for education: Prabhakar to Cornell to study agriculture and Ravi to MIT to study electrical engineering. Under S. L.'s guidance, the firm would expand into a wider range of more technologically sophisticated products, often working through licensing agreements with foreign firms.[9]

One of the family's long-term goals was to make electric motors and diesel engines. Work had begun on engines in 1914 and continued with experimentation and trials into the 1930s. At end of World War II, S. L. Kirloskar set up two new companies: one for manufacturing oil engines led by him, the other for manufacturing electric motors, led by his younger MIT-educated brother Ravi. Each of these new businesses moved out of Kirloskarwadi, the industrial village where the agricultural implements were made. The oil engines would be made in Poona and the electric motors in Bangalore. Kirloskar would be an enterprise that made sophisticated technology in urban India.[10]

In 1949 Kirloskar began making oil engines. Jawaharlal Nehru, who visited the plant the next year, asked Kirloskar how his plant compared with plants in Europe and America. Kirloskar later recalled that he told Nehru that the plant was "better than some of those and not inferior to the top-ranking comparable units in UK and other parts of Europe."[11] Kirloskar and Nehru shared a vision of a technological India, operating on par with America and the nations of Europe.

However, the tension between Nehru's idea of a planned economy and Kirloskar's idea of free enterprise would soon manifest itself. Kirloskar built up the capacity to make engines, with a plan to produce 10,000 engines a year. But in 1952 the demand for Kirloskar's engines collapsed. Kirloskar blamed it on the Indian Planning Commission, which had underestimated the capabilities of indigenous manufacturers (like Kirloskar's company) and authorized the import of large numbers of engines, glutting the market.[12]

Kirloskar saw his business as a global one, both in its sources of technology and in its customers. In the early 1950s his business began exporting engines, with its first exports going to Cyprus. In 1955 exports began to Germany, an event that Minister of Commerce Morarji Desai called "a turning point in the industrial history of India."[13] By 1958 Kirloskar's company was exporting 150 diesel engines a month, roughly 20 percent of its production. Most of Kirloskar's customers were in countries bordering on the Indian Ocean, such as Australia, Malaya, Burma, Iran, and Iraq. It established an international dealer network. In 1958 Kirloskar had representation on a trade delegation seeking business throughout Africa. In trying to build an export business, Kirloskar was competing with the European, American, and Japanese companies whose much larger production runs gave them lower costs than Kirloskar.[14]

In 1957 Kirloskar went on a mission to the United States, led by G. D. Birla. The purpose was to seek American capital and collaborations. In the course of the mission Kirloskar traveled to Indianapolis to discuss a joint venture in India with executives from Cummins Engine. (The president of Cummins had been one of Kirloskar's classmates at MIT.) Cummins was seeking to establish international operations as a base for long-term growth. In visits to India to negotiate with Kirloskar, Cummins managers were impressed with S. L. Kirloskar's personal characteristics and became convinced that he was the best partner for Cummins in India. In 1962, after a prolonged period of negotiations between Cummins, Kirloskar, and the Government of India—where the major point at issue was the share of Cummins's and Kirloskar's stake in the venture—the two firms finally signed an agreement calling for a venture 50 percent owned by Cummins, 25.5 percent owned by Kirloskar, with the remaining 24.5 percent owned by the Indian public. The enthusiasm for this venture in India, combining a well-known Indian company with an American one, was shown at the Kirloskar-Cummins initial stock offering in India in May 1962, when 70,000 potential shareholders subscribed 2,000,000 shares, with only 36,750 shares available.[15]

Kirloskar-Cummins initially seemed like a match, if not made in heaven, then at least at MIT. However, when the two companies began to try to make engines, they saw how different their two worlds were. India did not have the network of local suppliers that Cummins could rely on in the United States. Moreover, what suppliers could be set up in India often produced goods that were not up to Cummins standards, with rejection rates three to four times higher than those in the United States.[16]

Worse yet the demand for the engines in India failed to materialize. Manufacture of the engine had been based on the assumption that it would be used for earthmovers and other equipment built in India. Several American equipment manufacturers who had been counted on as customers postponed their plans to build plants in India, while another manufacturer greatly decreased its production. The original plan called for 1,000 engines to be made in 1964, but in reality only eleven were made.[17]

Finally the American and Indian management styles were incompatible. The least unattractive option proved to be removing the American managers and technicians and allowing the Kirloskar team to run the operation. In spite of all the difficulties, both companies persevered with the joint venture. One of the uses Kirloskar-Cummins advertised for its diesel engines was power generation, capitalizing on the Indian government's inability to provide reliable electric power throughout the country. It was only in 1969 that the operation was able to declare its first dividend. This joint venture lasted until 1997 and survived changes in the Indian political realm that had forced companies such as IBM and Coca-Cola to leave the country.[18]

S. L. Kirloskar became an early and prominent critic of the Indian government's role in economic planning. In 1965 he was awarded the Padma Bhushan, one of India's leading civilian honors. If he had not been given the award then, his increasingly harsh criticisms of government policy may have disqualified him from receiving it later. That same year Kirloskar was elected the president of the Federation of Indian Chambers of Commerce and Industry (FICCI), India's leading business organization, and he used that position to become one of the loudest opponents of the Fourth Five-Year Plan, the government plan that laid out spending and production targets for the Indian economy. Kirloskar argued that it envisioned far too much public sector spending. In a 1966 article in *Fortune* magazine, Kirloskar was quoted as saying that his company "made money in spite of the [government] planners, not because of them."[19]

In March 1966 Kirloskar gave his presidential address to the FICCI. Traditionally the Indian prime minister was present and then delivered a response. At the time of the speech Indira Gandhi had been prime minister for less than two months, gaining that position only after the death of Lal Shastri in January. The notoriously blunt Kirloskar laid out his case directly to her. His speech was dedicated to explaining India's economic failures, which had left the Indian citizen "one of the poorest in the world," even as other countries had experienced great growth in the preceding decade. He first turned his attention to Indian business, saying it was inefficient, stuck in outmoded management practices, and not aggressive enough in pursuing exports. After acknowledging these failures, Kirloskar spent the vast majority of his speech focusing on the Indian government's responsibility for India's economic failings. He claimed that agriculture had been politicized to the point where "a model villager is a man who rides in a jeep and climbs the political ladder," rather than one who "improves his farm yield." The government had set up a series of regulations that constrained the Indian businessman. Taxes were too high and hindered development. Kirloskar claimed that to be successful, Indian business needed to be able to operate with the same freedoms that businesses around the world enjoyed. Kirloskar did not win over Prime Minister Gandhi, who took the podium after his address and, in the *Times of India*'s words, "ridiculed" the idea that a smaller government plan could lead to faster growth.[20]

If Kirloskar was out of step with the Indian government, he at least found a friend in the American press. India's position as a socialist-leaning and nonaligned country during the Cold War was not a comfortable one for America. But a 1966 story in the business magazine *Fortune*, "An MIT Man in Poona," carried the implication that Kirloskar was an American ally both in his technological sophistication and his commitment to capitalism. *Fortune* called Kirloskar "an outspoken critic of India's experiment with socialism, which he feels is doomed to failure." An earlier piece in *Time*, "Ancient Gods and Modern Methods," highlighted how Kirloskar could operate in two worlds, participating in ceremonies to honor the goddess

Lakshmi while also being a devotee of "modern and aggressive management."[21]

Kirloskar later wrote that with his attacks on the Indian government and the glowing stories about him in the American press, some in India saw him as "Indian by birth but American by training and thinking." But Kirloskar was more of a globally oriented capitalist than a closet American. By 1966 his businesses, which remained concentrated in high technology engineering, ranged from machine tools to motors, diesel engines, and agricultural implements. The businesses had sales of $64 million and employed 13,000 workers. In contrast to many Indian business families, which diversified into a wide range of areas, Kirloskar largely concentrated in engineering.[22]

Kirloskar laid the foundation for Poona to become an important center of automotive production in India. Despite efforts since the late nineteenth century to promote industry in Poona, the city had very little industry when Kirloskar began his Poona operation. In fact, Kirloskar's initial efforts to acquire land were met by opposition from government officials who did not want industry to destroy Poona's educational and cultural environment. In 1961 a Cummins executive had noted that Kirloskar was committed to improving Poona. By 1972 the journal *Commerce* was writing of a Poona "industrial complex," largely based on engineering industries. A 1975 article by *Commerce* asserted that Kirloskar had played a "prominent role" in the area's industrialization. Kirloskar had an explicit policy of working to build up local suppliers and by 1975 had 700 suppliers located in Poona. The Kirloskars established a foundation that provided loan guarantees for prospective entrepreneurs.[23]

The largest ancillary business midwifed by Kirloskar was Bharat Forge, which as of this writing (2015) is one of the world's largest forgings company. Neelkanth Kalyani was a Brahmin trader and agriculturist who was convinced by S. L. Kirloskar to start a forging business in Poona. Kirloskar needed forgings for his diesel engine business that were not available in India. An agreement with an American firm provided the initial technology to the new venture. Kirloskar served as the chairman of the board as well as one of the new company's biggest customers. Other customers came from the Poona area, such as Mahindra & Mahindra, Bajaj, and Firodia. By

1972 *Commerce* wrote that Bharat Forge's shop was "the largest of its kind in Asia," while Kalyani claimed the firm was making axle types produced only by two other firms in the world. By 1976 Bharat Forge had sales of 22 crore rupees (roughly $25 million at the 1976 exchange rate).[24]

Bharat Forge, like the Kirloskar's enterprises, was a family business, and Neelkanth Kalyani's son Baba played a large role in the company, ultimately taking control. He trained as a mechanical engineer, first at the Birla Institute of Technology and Science and then, with S. L. Kirloskar's encouragement, at MIT. Bharat Forge took advantage of the automotive businesses developing in Poona, but at the same time it suffered from the limitations of a slow-growing Indian economy. Although it had long sought to export goods, it only came into its own in the years following the 1991 liberalization of the Indian economy.[25]

G. D. Birla and the Birla Institute of Technology and Science

The draw of the socially modern, technologically ambitious S. L. Kirloskar to MIT and the United States was like metal to a magnet. However, the attractive power of MIT and the United States in years after World War II might be better seen by looking at those who might not be thought susceptible to their pull. No better case can be seen than the Birla family. Here an orthodox Hindu coming from a family with a long history in business looked to MIT. The Birlas brought a powerful ability to combine Indian tradition with the modern world.

The Birlas were Marwaris, a business community that had originated in the Rajasthan desert and was known for its Hindu piety, its insularity, and its frugality. In the nineteenth century, after the coming of the British had disrupted their traditional trading patterns, some Marwaris migrated first to Bombay and then to Calcutta, where they established trading operations. By the late nineteenth century these newcomers to the region had become the leading Indian traders in Calcutta, a status that earned them contempt from Bengalis.[26]

G. D. Birla was born in 1894 into a family that had small trading operations in Bombay and Calcutta. He entered the family business at the age of twelve and as the only family member who could read English was given an exceptional position of responsibility negotiating with the English brokers. At the age of seventeen Birla moved to Calcutta and set up a business with his brother. They prospered by being the first Indian business to import cloth from Japan and by joining an opium trading syndicate. In Calcutta, Birla faced a heavily entrenched English business community who subjected him to a variety of humiliations because he was Indian. While Birla smarted under these insults, he also developed an appreciation for certain British business skills.[27]

Birla made great profits in jute during World War I, the material being used for sandbags in the war's ubiquitous trenches. After the war, Birla sought to move from jute trading to jute processing, which up to this time had been a British monopoly. British businessmen tried in numerous ways to block Birla: their banks at first denied him loans, they preemptively bought the land he was planning for his factory, and they charged him higher freight rates on the railroads. Despite these difficulties, Birla persisted and developed a successful jute business. Later the family bought or started industrial ventures in sugar, paper, and textiles.[28]

Although G. D. Birla had only a grade-school education, education was to be one of his primary philanthropies. In 1929 he created the Birla Education Trust, which over the years supported a wide variety of institutions ranging from grade schools to universities throughout India. Birla concentrated his educational efforts in his ancestral village of Pilani, 150 miles west of Delhi in the desert of Rajasthan. In this controlled environment, Birla would create a distinct educational synthesis. As an orthodox Hindu, he had been one of the primary funders of Banaras Hindu University (BHU), an effort to create a modern Hindu university. In 1929, seeking to develop an educational system at Pilani similar to that at BHU, he hired S. D. Pande, a mathematics professor from BHU, to develop an Intermediate College there. In the 1930s Pande began combining compulsory craftwork with the academics at the college. Students spent two or three hours a week learning such subjects as tailoring, carpentry, carpet making, or handspinning. The emphasis on the moral aspect of manual labor might be considered Gandhian.[29]

During World War II, with G. D. Birla's permission, the Government of India established a center for training navy engine operators at Pilani, which after the war became the basis for the Birla Engineering College. In the 1950s the Birla Engineering College was one of a number of engineering colleges that received assistance through the U.S. government's Technical Cooperation Mission and an effort led by the University of Wisconsin. Several American professors taught engineering at Pilani.[30]

Birla had a long-term interest in India's relations with both the United States and MIT. He first visited the United States in 1931 and early 1932 after accompanying Gandhi to the Round Table Conference in London. He reported back to Gandhi that his visit was "entirely commercial," but "he did not like America much," finding the entire country "intoxicated with the 'Dollar Feeling.'"[31] (Birla did go to Battle Creek, Michigan, likely to the famous sanitarium there, to have his diet analyzed scientifically, the results of which he reported back to Gandhi.) As discussed in Chapter 4, in 1941 in response to Bal Kalelkar's letter, which called MIT "the best institute of its kind in the whole world," Birla had provided funds for him to study there. In 1945 Birla came to the United States as part of a contingent of Indian businessmen. Although Birla listed MIT on his planned itinerary and at least one member of the group did visit the institute, whether Birla visited or not is unclear.[32]

Before independence, Birla had carved out a role for himself as an emissary between Gandhi and the British, trying to explain each side to the other and to steer the Indian independence movement in a rightward political direction at a time when the Congress Party was pointing it leftward. After independence Birla took up an analogous role between India and the United States, serving as a valuable informal ambassador between the two countries. As an Indian capitalist he sought to bridge the tensions between a socialist-leaning and nonaligned India and a capitalist and anticommunist America. In 1948 he visited President Truman. The next year he came and examined textile factories making staple fibers. In 1956 he made a three-week visit to the United States, where he met with a variety of major business interests as well as with President Eisenhower. Birla wrote back to Nehru, telling him of Eisenhower's sympathy for India's nonaligned position. John and Nelson Rockefeller along with Paul Hoffman, the first president of the Ford Foundation, arranged a luncheon for Birla with seventy American businessmen. After the trip Birla stated in a letter for Indian Prime Minister Nehru that while the press emphasized the differences between the two countries, "it is not fully realized there is much more in common between the U.S. and India."[33] In 1957 Birla traveled to America in search of partners for an Indian aluminum operation, finally reaching an arrangement with Kaiser for a joint venture in India called Hindalco.

At some point in the late 1950s or early 1960s, Birla decided that the Birla Engineering College was inadequate for his goals. He may have seen the increasing role that technology was playing in business. The Indian government's establishment of the Indian Institutes of Technology meant that unless the Birla Engineering College refashioned itself, it would be relegated to second-class status. For help, Birla turned to MIT. In 1960 Birla had hosted former MIT president and Eisenhower science adviser James Killian at Birla House, when Killian was in India on MIT business. In November 1961 Birla came to the United States and met Killian at MIT. In a note following up on their meeting, Birla suggested several ways MIT might help him. First, Birla proposed starting a university and having MIT "give" him personnel to staff it. Another alternative he suggested was that MIT start a branch campus in India that he would support.[34]

Birla was a great entrepreneur, operating in the areas of business, politics, and education. While an entrepreneur starts new ventures and does things in new ways, Birla's ideas showed no knowledge of how an American university operated in the early 1960s and would have been complete nonstarters to Killian. Killian did not respond positively to Birla's suggestions. In fact, Killian did not respond to a series of notes from Birla over a period of months. But the indefatigable Birla kept the letters coming. He lowered his expectations by asking for a list of Americans who could help him start a technical institute in India. Killian first suggested Mansukhlal Parekh, Devchand's son who had received a doctorate in chemical engineering from MIT and who Killian said "was very highly thought of here."[35] Killian also gave him a list of American engineering educators, each of whom Birla then sent a letter stating that he was forming an institute of technology "on the lines of MIT," and that he was seeking "help and guidance." Birla couched his letter in the terms of a business proposition, asking his contacts "what terms" they required. The appeal was not based on idealism. By communicating as individual to individual, Birla was forgoing any possibility that the person's home institution could apply any suasion.[36]

Thomas Drew, a professor of chemical engineering at Columbia and a 1923 graduate of MIT, responded. In the summer of 1962 Birla hosted him during an inspection tour of Indian technical institutes. Even before he left for India, Drew told Birla that he thought something like MIT's chemical engineering practice school could be useful in India. MIT's practice school, dating back to the 1910s, was increasingly an anomaly at an MIT that by the early 1960s was privileging scientific theory over practice as the basis for engineering. In the practice school, students would leave the MIT campus for a semester and work at one of several "stations," industrial sites where they would work on real problems under the supervision of MIT faculty.[37]

Birla was making plans for his institute at the very time that the Indian government was finalizing its plans for an American program of cooperation at IIT Kanpur, led by MIT, as will be discussed in Chapter 10. Although the program was formally announced just days before Birla met with Killian at MIT, Birla was not at all daunted. In fact, he gave the government program a back of the hand compliment saying it was good "as far as it goes" while he asked Killian to "consider my suggestion also."[38]

The Indo-American program formed a reference point—in some ways a negative reference point for Birla. Even though Birla's businesses had grown in independent India, the increasingly socialist orientation of Nehru's government confined Birla's great ambitions. It kept him out of the steel industry and led to a government takeover of his airline. It must have galled Birla to see India's government trying to build an industrial India of steel mills, hydroelectric power plants, and fertilizer plants, sidelining India's leading industrialists in favor of bureaucrats with no business experience. The resulting cost overruns and schedule delays were anathema to the way Birla ran his businesses. Here was one area where he could show the superiority of his way.[39]

Birla's aims are further suggested in a May 1963 letter to Jawaharlal Nehru describing his observations during a three-week trip to the United States. The previous eight months had not been good ones for either Nehru or India, and perhaps Birla thought Nehru would be open to rethinking some of his policies. The Chinese invasion of India in October 1962 had been a crushing blow. In the face of the near total collapse of the Indian Army, Nehru had been forced to call on the United States for military assistance, revealing that his belief that India could occupy a nonaligned position in the world was an unrealistic illusion. Whether there was a causal connection or not, many who saw Nehru after that time saw him as a broken figure.[40]

Part of the letter showed Birla to be the loyal soldier, supporting Nehru's policies even when they went against his own interests. Birla reported on his lobbying for U.S. assistance for an Indian steel plant, trying to argue to Americans who opposed supporting a socialized industry that the failure to build the plant would hurt private industry. (This was an argument that must have caused Birla some pain to make because he had been denied permission to build a private steel plant in India.)[41]

Birla, ever the practical businessman, was looking at America as a potential market for Indian goods. But he saw India failing to take advantage of opportunities, as when he noted shirts from Hong Kong being sold in the United States, due to Hong Kong's "first class mechanised equipments," which India did not have. Birla was blunt with Nehru here, calling into question one of Nehru's most cherished ideals: that the state could build up a technological India, reducing poverty and raising India's position in the world. Birla mercilessly pointed out the contradictions between what Nehru talked about and what in fact the government was doing, saying that while Nehru stated "we should use technology to help production," that did not reflect the actions of the government. Birla castigated both India's "primitive and out-of-date methods of production" and also government policies and practices that made exports difficult. In a statement that must have cut Nehru to the quick, Birla said that "technology is not a popular item with us. If some businessmen desire to resort to it, Government is not favorable."[42]

Birla had a proposed remedy, but as he must have surely known, it was an illusory one. Birla stated that India needed a "vigourously pursued" policy and program. He wrote to Nehru saying "no action, I fear shall be taken unless you personally direct it." That was something Nehru, tired and in failing health, could not do. This final call on Nehru, who had made building a technological India such a personal crusade, showed that he could do no more. The mantle would have to be taken up by others.[43]

Birla was able to get officials from both MIT and the Ford Foundation interested in his program, but there was one major hurdle. He had to reorganize his college so that its administrative structure and curriculum would reflect that of an American university. Birla colleges in Pilani had been set up under the typical Indian fashion where they were teaching institutions existing under an often-distant degree-granting institution. Degrees were awarded based on examinations administered by the degree-granting institution, which set the curriculum. The teaching institution had no authority over the curriculum, which only changed slowly. Thomas Drew, the Columbia University chemical engineering professor who had agreed to work with Birla, claimed that the chemical engineering curriculum was twenty years out of date. When Drew and a colleague, Howard Bartlett of the MIT humanities department, came over to India in 1963, their first priority was to draft an American-style organization plan for the institute.[44]

One of the first documents that Drew and Bartlett prepared when they got to Pilani was called "The MIT Plan," which was intended to describe to the existing staff what exactly MIT was. Whenever there were questions from Birla or the Indian administration about some proposed organizational structure, a simple reference to MIT's practice was enough to end all debate. Even as the name MIT was bandied about, neither Birla nor Drew imagined a literal effort to replicate MIT, with its research orientation and its new and increasingly science-oriented curriculum (as will be discussed in Chapter 10). But they saw that the association with MIT could be the vehicle for a new model of engineering education.[45]

The reformed school took up the name the Birla Institute of Technology and Science (BITS), and in a 1964 letter Drew provided a clear rationale for its existence, asserting that "no small part" of the educational advances in the United States had come from private institutions such as Harvard, Columbia, and MIT. Drew claimed that these schools, "being free of control of well-meaning government committees, have been able to quickly adjust their programs to the needs of their country and her industry and to experiment with new techniques of instruction." Drew stated that the Indian private sector should provide "a strong independent school, as free of government control as may be."[46]

The MIT faculty supporting IIT Kanpur were apoplectic about the BITS proposal, considering it an unconscionable diversion of resources since its remote location and existing faculty constituted impossible impediments to it ever becoming a first-class engineering school. But Drew, in his writings to Birla, suggested that rather than being a pretender, BITS was the rightful heir of MIT. In a letter to Birla, Drew stated that "our State Department persuaded MIT to form a consortium . . . for the purpose of aiding the IIT at Kanpur," implying that MIT acted due to government coercion rather than of its own accord. He also stated that some (among whom he included himself and MIT President Julius Stratton) believed that BITS, with its greater freedom from government interference, could actually assist Kanpur's development by establishing precedents that Kanpur could later follow. Drew wrote to Birla that "all here are very much attracted by the fact that BITS is a private sector operation."[47]

Birla himself was not as dogmatic about the private sector nature of BITS. Birla had spent his career working seamlessly between the government and the private sector (as, of course, had several prominent MIT administrators). At one point he wrote to Drew, "Whatever autonomy we have, please do not forget that the Government is a Government and there are a hundred and one things for which we shall have to approach them and therefore whatever we may do, we shall have to do with their full blessings and good wishes."[48]

MIT's support of programs both at Kanpur and Pilani raised questions about the nature of MIT itself. MIT had been a beneficiary of the enormous growth of the federal government and had supplied leaders for government service. But at the same time MIT's leaders had expressed ambivalence about the power of the government even as they had helped bring it into the university. MIT's leaders might be best described as Eisenhower Republicans, concerned about the growth of the federal government, but not looking to roll it back. Drew was suggesting that while MIT officials were supporting both IIT and BITS, their hearts were with BITS, and while his suggestions were obviously self-serving, they were not entirely fatuous.[49]

Birla ran BITS with a practical businessman's eye for costs. He had originally considered building a new technological institute, but studies convinced him that it would be cheaper to convert the Pilani colleges. In a letter to one of his associates in 1962, he said, "We are not the government with unlimited funds at our disposal." When some at MIT criticized what was going on at Pilani, Birla fired back that "Cawnpore [Kanpur] has not impressed India," with some people saying that the Kanpur students, trained at a "huge expense," were no better than students at the Soviet-sponsored IIT Bombay or Pilani itself.[50]

The Ford Foundation-BITS-MIT program ran for a total of ten years. During this time the Ford Foundation provided $3 million in grants, which were used for visiting American faculty (some, but not all, from MIT) and to pay for equipment and library books. Fifty-six people came over from the United States as visiting professors, contributing twenty-one person-years of effort.[51]

The effort to establish an "MIT of India" at Pilani could be seen as a combination of modernity and tradition. Pilani was tradition, an isolated ancestral homeland, dominated by one family. MIT was modernity, committed to using science and technology for ceaseless innovation. G. D. Birla believed that these worlds could be brought together. Perhaps the best symbol for the catholicity of his views was the Saraswati temple he built at Pilani. Among the images there were Henry Ford, Lenin, and Albert Einstein. Pairing MIT faculty and administrators with G. D. Birla and the existing BITS staff was perhaps just as unlikely as pairing Albert Einstein, Henry Ford, and the Hindu goddess of knowledge, Saraswati.

The first director of BITS, the incumbent from the older engineering college, had only a bachelor's degree from an engineering college in Madras, received in the 1920s. And there were scores of other faculty and staff members who would be more comfortable keeping the old ways. It was a sign of Birla's ambivalence toward a new way of doing business, where personal loyalty yielded to other values, that it took him four years to replace his longtime director, and only after both the Ford Foundation and MIT gave him an ultimatum. The director who finally proved to be acceptable to all parties was himself an Indian graduate of MIT, C. R. Mitra, who had earned his master's in chemical engineering there before earning his doctorate from Columbia University. More importantly, Mitra had experience doing exactly what Birla wanted to do at Pilani, for he had already modernized an existing technical school, Harcourt Butler Technical Institute, into an autonomous technological institute. Harcourt Butler, located in Kanpur, sat almost literally in the shadow of IIT Kanpur, and Mitra had transformed it even though he had few resources available.[52]

At BITS, Mitra developed a distinctive educational synthesis, taking ideas from MIT, but also moving beyond them. The hallmark of Mitra's BITS became the practice school, which intended or not, had resonances with the industrial training that Pande had introduced in the 1930s at the Birla Colleges. Mitra expanded it across the institute so that it not only covered all engineering disciplines, but also sciences, humanities, and social sciences. Each student did two stints in practice school: a two-month term in the summer after the student's second year and a five-month term in the student's final year. Students worked in small interdisciplinary teams with faculty members. By 1980 BITS had over 700 students participating in sixty-eight practice school stations throughout India, at sites such as government research laboratories, steel mills, textile plants, banks, museums, newspapers, and rural development centers.[53]

Although BITS never gained the recognition of the IITs, it was more effective at producing engineers who would and could work in the Indian environment. A 1974 report to the Ford Foundation asserted that because it had private support, BITS had a greater measure of independence than other universities, which it had used for "experimentation and innovation." The report asserted that BITS had a "degree of flexibility unmatched at any university or IIT in India." In a 1994 article, Mitra claimed that while BITS had the "highest fees structure" in the country, covering 50 percent of the cost of operation, at the same time BITS had a dramatically lower "unit cost" per degree and per student than the IITs.[54]

Aditya Birla

At the very time that G. D. Birla was working on establishing the MIT connection with his engineering college, he had another project involving MIT: his grandson, Aditya, born in 1943. Although Aditya Birla proved to be one of the great Indian businessmen of the late twentieth century, there is reason to believe his career took a different trajectory than his grandfather had anticipated. Aditya's career was a response to the policies of Indira Gandhi and the limitations she put on business. Aditya looked outside India, but not to the United States. Aditya moved in a different direction—toward Southeast Asia.

Although the attribution of agency is difficult within a business family, G. D. Birla's enthusiasm for MIT clearly had something to do with Aditya going there. In January 1962 the elder Birla cabled Killian to check the status of his grandson's application. Killian replied in an indirect way that must have provided some comfort to Birla, telling him that "my interest in your grandson is well known here."[55]

Sending Aditya to the United States was a very bold move for the religiously conservative and orthodox Birlas. No family member had studied abroad before. Aditya's parents were worried about what would happen to him at MIT. The family was based in Calcutta. There he had all the resources of his family at his disposal: a personal servant, a driver, anything he needed. He would be giving all that up by going to Cambridge. Ultimately Basant Kumar Birla, Aditya's father, persuaded the parents of Ashwin Kothari and Om Bhalotia, two of Aditya's classmates at St. Xavier's Collegiate School in Calcutta, to also send their sons to MIT. The Birlas were sending Aditya to get an American education, but he had obligations in India to maintain. He was expected to come back and participate in the family businesses. He had been engaged at fourteen to a girl who was ten at the time of the engagement. The Birlas would very carefully manage his experience in the United States. His father admonished his son that there should be no smoking, dancing, or drinking. Grandfather G. D. Birla sent Aditya a letter with a long list of taboos, mostly based on personal safety, such as "Never dive into a swimming pool," and "Don't drive a car yourself."[56]

Aditya Birla, Ashwin Kothari, and Om Bhalotia, the "three musketeers," as Basant Kumar (and others at MIT) called them, stuck together. They all majored in chemical engineering. They lived together in an apartment that a Birla executive had picked out beforehand. They cooked and maintained their vegetarian diet. None of them were stellar students, but under the watchful eye of the Birlas, they completed their degrees in four semesters, having received credit for previous college work done in Calcutta.[57]

Aditya had studied engineering, not management at MIT. Management was in the family's blood. When Aditya returned to India after graduating, his father put him through a six-month course in accounts where he learned the Marwari partha system of accounting. Aditya's first venture was setting up a small cotton spinning operation in West Bengal.[58]

Aditya globalized the Birla industries, but not in the way one might have expected from an MIT graduate. While he was finishing up at MIT, his family began negotiations with DuPont, the American chemical giant, over the possibilities of a joint venture to manufacture acrylic in India. Aditya participated in the negotiations from the United States, and this seemed like the kind of venture for which the Birlas had sent Aditya to MIT. DuPont was the epitome of the high technology American chemical company and itself had close relations to MIT. However, DuPont wanted to build a larger plant than the Birlas did and more importantly, DuPont wanted a level of control that the Birlas found unacceptable. The Birlas ultimately turned down the venture. The Birlas would not simply be the Indian front for an American-controlled operation. The Birlas were managers. Aditya's partners would not be Americans.[59]

At the same time, shortly after Aditya Birla came back to India, the atmosphere for business radically changed. G. D. Birla had helped to lay the groundwork for Indira Gandhi's 1966 visit to the United States, which at the time had appeared to be a great success. She had expressed her support for private investments in India, while President Lyndon Johnson had supported an increase in food aid. Gandhi made the concession of expressing sympathy for the American position on Vietnam. In June 1966 India devalued the rupee, apparently as part of a series of agreements reached during Gandhi's Washington meeting.[60]

The devaluation of the rupee marked the apex of both G. D. Birla's and American leaders' influence with Indira Gandhi. Things quickly went downhill from there. Devaluation was seen almost universally by Indians as a national humiliation. Furthermore, America seemed not to have kept its share of the bargain by increasing aid to India. To make matters even worse, the failure of two monsoons made food aid an urgent matter, but Lyndon Johnson provided it only in the most piecemeal matter, seeming to take delight in inflicting maximum humiliation on India and its leaders. Gandhi turned sharply to the left, and in doing so she turned against those, including G. D. Birla, who had advised her to look to the United States. Gandhi's government instituted a number of anti-big business policies, including the nationalization of large banks, restrictions on the expansion of large companies, and limits on foreign investment.[61]

While Aditya Birla kept his domestic businesses, most of the expansion would come from abroad. He established production facilities in Southeast Asia. Indira Gandhi's license raj, where starting or expanding a venture was predicated on getting government permissions, made expansion in India difficult and often a matter of bribing or otherwise placating government officials. In 1969 Aditya began working on establishing a synthetic fibers operation in Thailand. Birla was able to attract Indian investors living outside of the country and from Thais, who were happy to give Birla managerial control of the company. Birla later expanded into Malaysia, the Philippines, and Indonesia. In 1978, when a plan to expand Birla's Gwalior Rayon plant was tied up awaiting government approvals, a plant was built in Indonesia instead. In these countries he found more receptive governments. India's license raj had led to a complacent Indian industry with little incentive to try to expand, knowing that any potential competitors faced a crippling thicket of government restrictions. Aditya Birla's multinational industries grew up in a completely different environment, facing open competition from businesses in Japan and other parts of Southeast Asia. Birla stayed within a narrow range of technologies, mostly intermediates, including such unglamorous products as textiles, caustic soda, acids, carbon black, and cements. The strength of Birla was not new technology or new products, but a relentless focus on costs and schedules.[62]

In 1991 India liberalized its markets, removing the restrictions that had hobbled Aditya's businesses in India, Tragically just two years later, at a time when Aditya was at work developing the new business opportunities now possible, he was diagnosed with prostate cancer. It was a final sign of his family's global orientation that his managers did a search for the best hospitals in the world for the treatment of his disease. Birla went to Johns Hopkins, but unfortunately the cancer had spread and no effective treatment was possible. He died in 1995. His son, Kumar Mangalam, a chartered accountant, with an MBA from the London Business School, carried on.[63]

The career of Shailendra Jain, who came to be a key lieutenant of Aditya, provides another perspective on the Birla operations, showing the role that personal contacts played in bringing an MIT graduate into the family business and also the way the Birlas managed the transition between generations. Jain was from a business family, and his grandfather had established textile mills in Ujjain and Indore in central India. However, the succeeding generation was not as successful in managing those mills. Jain earned a degree in electrical engineering from Victoria Jubilee Technical Institute in 1963 and then won admission to MIT. Funded by his family, he studied electrical engineering, graduating with a master's in 1965. After graduation, Jain returned to India for a brief vacation before planning on returning to the United States, where he had several job offers, including one from Xerox. Due to a scheduling mix-up, Jain missed his flight to the United States, whereafter he stopped at the Bombay office of D. P. Mandelia, a family friend. Mandelia was not just a family friend, but also the right-hand man of G. D. Birla. Jain met Mandelia at Industry House, Birla's Bombay base of operations. Birla walked in on the men and queried Jain about his plans, challenging him to stay and work in India, going so far as to offer him a job. Jain ended up working at Birla's Grasim Viscose Staples Fibres plant in Nagda, only sixty kilometers from his home in Ujjain.[64]

At Nagda, Birla was establishing a large integrated operation for the production of rayon. Jain stayed with G. D. Birla's businesses, but he had interactions with Aditya Birla's businesses as well. As Aditya expanded his operations outside India, he was able to use his grandfather's existing businesses for support. They designed the plants that Aditya would build in Thailand and the Philippines. On the death of G. D. Birla, the Nagda plant fell to Aditya Birla, and Jain became one of his top lieutenants.[65]

Other Business Families

An Indian business family sending a child to MIT was a marker of the preexisting orientation of the family, with individual careers taking a wide variety of shapes. Ramesh Chauhan, born in 1940, was from the second generation of an extended Bombay-based family that had started a biscuit and soft drink business targeting the Indian middle class. Chauhan and three others from his generation went to MIT. Chauhan, who graduated in mechanical engineering, built his career trying to resolve the forces of globalization and nationalism so that they reinforced each other and benefited his company. He sought to convince Indians to become members of the global community of soft drink consumers, but to forgo international brands, such as Coke, in favor of Indian-made products. Some believed Chauhan had a role behind the Indian government's campaign against Coke, and when Coke was forced from India, one of his products, Thums Up, became India's most popular soft drink. But Chauhan was nothing if not adaptable: after Coke reentered India in 1993, Chauhan sold it the bulk of his beverage business.[66]

Adi Godrej was from the third generation of a Bombay-based Parsi business family that made such consumer products as safes, furniture, and soaps. Godrej's soap business put it in direct competition with the multinational Hindustan Lever. Adi, whose interest in MIT was sparked by the fact that his relative M. N. Dastur had studied there, graduated with a degree in management in 1962. Upon his return to India, he and Keki Hathi, a Godrej manager who had gone to MIT's business school as part of a mid-career program, worked to introduce professional management practices into the company, such as cost accounting and human resources management. Godrej began recruiting summer trainees from the newly established Indian Institute of Management in Ahmedabad, which was supported by Harvard Business School. Hathi developed a formal management structure at Godrej. Adi's younger brother Nadir earned a bachelor's degree in chemical engineering from MIT in 1973. After receiving a master's at Stanford and a Harvard MBA, he returned and worked on the animal feed and chemical side of the business.[67]

While these were some of the most widely recognized business families in India, families that had much smaller businesses also sent their children to MIT. Firoze Sidwha was a Parsi lawyer in Bombay who had first developed a ceramic tile business and then a grinding wheel business called Grindwell. Sidwha, who had no sons, decided after examining college catalogs at the American library in Bombay that his daughter Almitra, born in 1936, should go to MIT to study ceramics. Her education was paid for by the company, which required her to sign a bond promising five years of work on her return. In 1958 Sidwha became the first Indian woman to earn a degree in engineering from MIT. On her return, she worked in a small research group in the company. Her husband, a chemical engineer trained at the University of Michigan, joined the company upon their marriage.[68]

Both Nehru and a group of Indian capitalists looked to MIT to build their India, but they were substantially different Indias. While Nehru sought to use the state to create his vision of a prosperous India, a group of Indian capitalists sought to use MIT to develop capabilities to expand their businesses globally and to protect their businesses from foreign competition. Of all the Indian capitalists, G. D. Birla had the most ambitious vision: to create a private MIT. Winning MIT as a partner to build an Indian MIT in the desert was in the words of the BITS official history, *An Improbable Achievement*. The business families that sent their heirs to MIT had a global vision, reflected in their companies, albeit in diverse ways. For Aditya Birla, it was by expanding into Southeast Asia. For Ramesh Chauhan, it was by competing with Coca-Cola. For Godrej, it was using American management methods to compete with multinationals.

The Birlas and the Tatas were the biggest business families in India. In the 1960s, the House of Tata, perhaps the most globally oriented of Indian businesses, would use a group of MIT graduates to start a most unlikely venture in India. This venture would show the tension between the state and private enterprise.

9

The Roots of IT India

IN THE 1950s nothing better demonstrated the fundamental challenges of contemporary Western technology in India than the computer. It was a fragile, capital-intensive, labor-saving device that could only be operated in an environment where temperature, humidity, voltage, and dust could be precisely controlled. India was a country that had a scarcity of capital and an abundance of labor. Moreover, a reliable electrical supply was difficult to come by, and the computer's other environmental requirements were almost an impossibility. How could the computer make sense in India?

This was the question Kirit Parikh faced in August 1958. In that year Parikh, who had been inspired by Anant Pandya's story, received an offer of admission to MIT to study civil engineering. However, because Parikh had applied late, MIT could not offer him financial aid. Parikh's father insisted that the young man go to MIT, but money was going to be an issue. Once he got to MIT, Parikh received word that a graduate assistantship had opened up in the computer center. Parikh was torn. He had attended a Gandhian school in Ahmedabad that taught spinning and he was committed to using his MIT education to build up India. He was also convinced that computing would not be a relevant technology in India for many years. Parikh took a weekend to consider what to do before finally saying to himself that $283 a month was $283 a month, and that as a

Gujarati bania (merchant), he could not refuse the money. He took the position.[1]

This anecdote suggests the tension that MIT and the computer involved for India. In the 1950s and 1960s, computers at MIT were everywhere and difficult to avoid. While few, if any, Indians went to MIT at that time with the intention of studying computing, a substantial portion of those who went were introduced to the technology there. And of those, a small cadre played a crucial role in computing in India. They did so first of all by their precocious exposure to the technology. Through that exposure they began to see the computer's possibilities. They envisioned a far different role for the computer in India than that envisioned by the Indian state. At a time when the state put a high priority on self-reliance and import substitution, these Indians accepted that computers in India would operate in an American-dominated environment. India would use computers made in the United States as a way to pursue business internationally. Although this vision was in conflict with the Indian government's, the Indian polity provided (begrudgingly and barely) these Indian entrepreneurs the space to explore a variety of possibilities for exploiting computer technology. Through this exploration Indian entrepreneurs developed a variety of businesses, some of which were evanescent while some proved to be lasting. By the early 1980s, while these computing businesses were still small, they had given Indians a foothold in doing computing work overseas. Those with imagination saw that this business had the potential for great growth. And by the late 1980s, the Indian government was even providing support for these ventures.

The Computer at MIT

The modern electronic computer emerged out of World War II, with different incarnations appearing in the United States and the United Kingdom. In the United States, engineers at the University of Pennsylvania developed the ENIAC, an electronic programmable calculator designed to calculate ballistic tables for the Army. The rapid development of computing in the United States in the two decades following World War II owed much to the substantial funding provided by the U.S. military at a time when the market for this new and unproven technology would not sustain large investment by private enterprise.[2]

Penn's work was soon overtaken by MIT, which had far greater technical capabilities and was willing to engage in computer research and development efforts that at the time were unusual for a university. Just as important were the remarkable entrepreneurial abilities of MIT engineers in securing military funding. Starting in 1945 and over eight years, MIT engineers transformed a small project to produce an aircraft simulator, estimated to cost $200,000, into Project Whirlwind, an $8 million behemoth, the largest computer of its time. Whirlwind in turn was transformed into Project SAGE, an $8 billion program led by MIT to provide a computer system that would be used in defending the United States against air attack from the Soviet Union. Although the Soviets' development of intercontinental ballistic missiles made SAGE obsolete before it even entered service, engineers working on SAGE and Whirlwind pioneered such key computing technologies of the time as magnetic core memory, computer networking, and computer graphics.[3]

With funding from the Defense Department's Advanced Research Projects Agency (ARPA), in 1962 MIT began Project MAC, aimed at developing new ways to use computers. Project MAC would develop timesharing, artificial intelligence, and computer communication, leading to the threshold of the Internet. The project's annual budget peaked at $4 million a year and its staff reached 400. Computers were not only limited to these large research projects: by 1962, over a thousand MIT students were using computers in their classes. Computing was where both the money and the action were at MIT, making it almost inevitable that the Indian students there, who were in many cases looking for financial support, would be drawn to the technology.[4]

The Computer in India

Global connections meant that even in the earliest days of the computer, some of the same stimuli acting in the United States were also felt in India. In 1948, as Homi Bhabha worked to establish an institute for fundamental research in India, he corresponded with computer pioneer John von Neumann about the possibilities for computing. In 1955 R. N. Narasimhan, working at Bhabha's institute, the Tata Institute for Fundamental Research, and relying on logic diagrams from a University of Illinois computer project, began work on what would be India's first indigenously designed computer, formally named the TIFRAC by Nehru in 1962.[5]

Military considerations provided a further stimulus to electronics. Following India's disastrous performance in the 1962 war with China, Nehru commissioned Bhabha to study India's electronics industry. The report emphasized developing a largely indigenous, self-reliant electronics and computer industry. However, the rapid development of electronics and computing in America caused by its massive spending on defense meant that a truly indigenous Indian program would lag far behind the United States.[6]

In any case, the technically oriented in India maintained a certain ambivalence about computing. In a country with so many needs such as food production, electric power, steel, and education, computing was well down the list of priorities. Even the directors of several Indian Institutes of Technology were circumspect about having them in their programs, with the director of IIT Bombay declining a Soviet offer of a computer in 1961, believing that the money could be better spent elsewhere.[7]

The Tata Computer Centre

The House of Tata was one of the first Indian organizations to explore how the computer could be used commercially in India. Here three Indians, trained in computing at MIT, received the support of India's largest business group. Tata's support for computing was far more than financial, however. Tata had the business experience, the patience, and the management talent (including another MIT graduate) that were crucial in translating a technology into a viable business.

The development of a separate computer enterprise within Tata began with three young MIT students and some auspicious timing. Lalit Kanodia, a Marwari, was from a well-to-do business family in Bombay; his father had been the director of the Bombay Bullion Exchange. The younger Kanodia enrolled at IIT Bombay, where he graduated with a degree in mechanical engineering in 1963, in the second batch to graduate. He was accepted to MIT for graduate work in industrial engineering, a program housed in MIT's business school.[8]

While at MIT Kanodia saw his first computer (he had had no exposure to computing at IIT Bombay). The computer would become the center of his career, but at first it represented something very practical to him: a means of support. Although Kanodia's family funded his initial studies at MIT, he sought to support himself and soon became deeply involved in computing. He held several assistantships, eventually working with Professor Donald Carroll on Project MAC. Through his MIT connections, Kanodia also worked doing consulting with Ford Motor Company and Arthur D. Little.

In 1965 Kanodia had an unintended extended stay in India. He had proposed marriage, but the auspicious date for the marriage was six months in the future. He needed something to do during this time and, through a neighbor, he secured a meeting with Rustom Choksi, a Tata executive. Choksi gave him an assignment acting as an informal consultant for Tata, and in the course of his work, Kanodia suggested three ways that Tata could apply computers to its businesses. Kanodia proposed two ways that computers could be applied in the electric power business, to make load dispatching and billing more efficient. He also suggested that Tata could advantageously set up a computer operation for its businesses.

The Tata organization offered Kanodia the opportunity to start up and lead the last of his suggested enterprises. Kanodia then enlisted two other Indian MIT students to join him, Nitin Patel and Ashok Malhotra. Patel, the son of an engineer, was a master's student in electrical engineering who had also been involved with Project MAC and had written a thesis on computer timesharing. He had then worked with a Texas firm in using computational techniques in oil exploration. Malhotra, the son of an accountant in Delhi, had done his undergraduate degree at MIT and then completed a master's both in management and electrical engineering. He had worked at MIT in a defense-oriented laboratory and then also worked on Project MAC. Kanodia returned to MIT to complete his doctorate, and then came back to India to start the new venture.[9]

It was unheard of in India for three young men in their mid-twenties to be essentially given charge of a new business requiring a major capital investment. (It would have been unusual in the United States also, with the seminal Silicon Valley event, the founding of Fairchild Semiconductor by eight young scientists and engineers, being the closest parallel.) A more senior Tata executive, P. M. Agarwala, provided supervision and guidance. These three young men, newcomers to Tata, were to develop a center in Bombay that would provide computing services to all of the Tata companies.[10]

Kanodia, Patel, and Malhotra through their association with Project MAC had an unparalleled knowledge of the state of the art in American computing as well as a clear insight into where leading researchers were trying to steer the future of computing technology. But they had almost no experience with business in India. The company they set up reflected this in a number of ways. Their model was that of an American management consulting firm, along the lines of Arthur D. Little or McKinsey, with the idea that they could justify the same types of expenditures those firms did. They had a library. They had a seminar series every Monday. They planned to publish a magazine, *Tata Computer Review*, to "disseminate information regarding computers and their benefits to management." They planned to transfer all the latest computer programs available in the United States such as PERT, a computerized management tool originally developed for the U.S. nuclear submarine program, as well as dynamic programming, queuing theory, and statistical programming. It would be as if the Tatas had their own private branch of McKinsey & Company.[11]

The three young MIT graduates gave a series of talks on "EDP [Electronic Data Processing] and the Management Revolution" to Tata managers at the Tata Management Centre in Poona. They were talks that would have been right at home at MIT's business school, but they showed no knowledge at all of any local business culture. While Kanodia gave a talk on the basics of computing, the most striking lecture in the series was his concluding one, "Frontiers of Electronic Data Processing." Here he presented a vision far out into the future of what computers might become—the vision that MIT researchers had in Project MAC. Although the computer was in the mid-1960s a technology used by only a very small technical elite in the United States and a minute few in India, Kanodia confidently stated that there was "no doubt" that the "common man will be interacting with the computer in his daily life within the next decade or two."[12] While from the perspective of the twenty-first century this vision has proved prescient, it was a vision that would have had little direct relevance for Indian businesses in 1967.

Using Harvard University's library as a taking off point, Kanodia sketched out the possibilities of an automated computer library, where books were put onto magnetic tape. The library would then be accessible through a wide number of "remote centres" throughout the country. (The country in question was implicitly the United States.) Kanodia predicted such a library would exist in thirty years at a cost of $100 million. Kanodia went on to suggest the possibilities of automatic language translation, possible in twenty to thirty years, and automated money, which was a little more speculative and remote, with "some scholars" anticipating it within fifty years. At the same time one can only imagine the dissonance this talk would have caused among Tata managers as they compared what they heard with their daily business activities.[13]

In April 1966 a memo formally announced the creation of the Tata Computer Centre, requesting that all Tata companies "take full advantage of the proposed computer installation."[14] In September 1967 the Tata Computer Centre began operations with an IBM 1401 computer. Tata central management required all the Tata companies to use the computer for the reconciliation of stock shares, and this was initially the main work for the computer centre.[15]

While the Tata Computer Centre was a new enterprise that was allowed to bring in new ideas and operate with some freedom, the head of the Tata group, J. R. D. Tata, had a long-term interest in the field of computing. Through his position on the board of the Tata Institute for Fundamental Research, he received briefings on the state of the field by Homi Bhabha and his successor, M. G. K. Menon, two of India's leading scientists.[16]

Furthermore, while the House of Tata was known from its earliest days under J. N. Tata for its long-range vision and its willingness to adopt technology from other countries, there never was any question that the House of Tata was a business and that ultimately its units were expected to make money. It would not tolerate a fanciful waste. This was suggested by an incident early in the history of the Tata Computer Centre, when it held an open house attended by senior Tata executives. Computer Centre employees had arranged typical computer demonstrations of the period, with games and the printer set up so that in its printing rhythm, it would play music. As one Tata director, J. J. Bhabha, played one of the games on the computer, J. R. D. Tata pulled out a slide rule and asked if Bhabha knew how much his game was costing.[17]

The Computer Centre lasted less than a year in its original form. The Tata companies, operating as part of a decentralized system, were giving it less business than expected. At the same time, Tata managers came to believe that a market existed for their business outside. In July 1968 the Tata Computer Centre was transformed into a new venture with a new strategy. Tata Consultancy Services (TCS) would focus on providing management consulting and computing services to clients outside of Tata enterprises. Early brochures produced by TCS continued to emphasize the advanced technological capabilities of the organization and the technical prowess of its consultants. One brochure gave detailed biographies of nineteen of its consultants, while also providing minute details on the capabilities of its IBM 1401 computer. Another brochure was filled with mathematical equations and flow charts.[18]

Anti-Automation

As Kanodia, Patel, and Malhotra shared their vision of the computer, another group in India was sharing a far darker vision of what the computer would mean for India. They saw the computer as a "job-eater" that threatened the well-being of Indians. This opposition to computing and automation was centered in Indian labor unions, particularly those associated with the Life Insurance Company of India.

At a union conference in 1963, leaders reported on an article in *The Hindu* which they understood to mean that the Life Insurance Company of India (LIC) planned to abolish many of its offices. Further research by employees led them to believe that these closings would come about through computerization. Comparisons with the American situation amplified the threat. The Indian union reported that with automation, American firms handled over 120 times the business of Indian firms with only four times the employees, raising the possibility that over 95 percent of the company's employees were at risk because of the computer. The next year the LIC union demanded that the company rescind its automation plan. In 1965 the insurance company union led the organization of a National Conference against Automation. The next year employees at the LIC held a fifty-minute strike in opposition to automation and union leaders presented the government with over a million signatures on petitions opposing automation.[19]

The nature of the struggle between opponents and advocates of automation might be seen in two publications issued in 1968. The leftist publication *New Age* published an article "Ram Raj or Computer Raj," asserting that the computer was being ushered in with "teargas shells and police batons." It noted that Honeywell had made a gift of computers to the Government of India and wondered if these computers had been "wired" for the CIA. It further noted that President Mobutu of the Congo was introducing IBM computers and suggested that he was the type of customer computer firms were looking for. It acknowledged beneficial uses of computers, but saw them being introduced in India primarily to replace clerks. It called for February 23, 1968, to be observed as a "National Day against Automation."[20]

The Employer's Federation of Bombay published a 1968 brochure titled "Automation: Blessing or Curse," whose foreword was written by Naval Tata, a family member who was one of the senior executives in the Tata group. Tata titled his essay "The Inevitability of Automation," a fact that he said was proven by how automation was part of broader technological changes and also how it was embraced by both capitalists and socialists. Tata criticized trade unions who opposed automation for their "emotional outburst" and their "biased and bigoted appraisal of automation." Most profoundly, Tata claimed that opponents of automation threatened to deprive India of "its legitimate share" of progress and allow India to be "stigmatized as backward and undeveloped." Tata acknowledged that computing would likely produce temporary unemployment, but that the disruption would soon dissipate.[21]

This fear of automation affected the Tata Computer Centre in several ways. Nitin Patel recalled seeing black flags, which had been raised by workers in protest against computers, as he walked past the LIC building in Bombay on his way to work at the Tata Computer Centre. In its early days, the Tata Computer Centre had to apply to the Bombay government for special permission to operate its computers twenty-four hours a day. Given the computer's cost and the difficulty of buying new ones, it was essential to squeeze as much work out of each machine as possible. The government granted Tata permission to operate twenty-four hours a day, but Tata had to agree that the computer would result in "no displacement of any labour whatsoever."[22]

Tata was also an indirect beneficiary of this anti-automation movement. The British computer company ICL brought two computers into Calcutta, one intended for sale to the LIC, the other intended for sale to Calcutta Electric Supply. At Calcutta Electric Supply, workers conducted a gherao against automation, an Indian protest tactic of blockading company management inside their building. Workers, supported by the leftist West Bengal government, were, over a period of years, able to keep the company from operating the computer. In a parable about growing technological ascendancy of western India, Tata Consultancy Services bought that computer and moved it from Calcutta to Bombay.[23]

TCS 2.0

TCS had reached a pivotal point in 1969. The consulting company was making money and developing new lines of business. However, it had not been integrated into the overall Tata culture. It lacked a formal management structure. It had no long-term strategy. The three founders, who had been given an extraordinary degree of freedom, had interests that were more academic and research oriented. Kanodia, from a business family, had a long-term goal to start his own company. That year Kanodia left the company and, in

September 1969, Tata managers brought in F. C. Kohli, then forty-five years old, to domesticate TCS. Kohli had a background that spanned several disparate worlds, which suited him well to take a complex business that had been started on an experimental basis and make it a thriving ongoing venture.[24]

Kohli, born in 1924, had grown up in the city of Peshawar (now part of Pakistan), where his father ran a drapery and apparel store that catered to the needs of the British. The younger Kohli earned a physics degree from Punjab University and then was selected for a government overseas scholarship. He first went to Queens University in Canada, where he earned a bachelor's degree in electrical engineering. He then went to MIT, where he earned a master's degree in 1950. At MIT, while he had nominally studied power engineering, he had imbibed the new fields of control and systems engineering. After six months of training at utilities in the United States, Kohli started work at Tata Electric.

By the late 1960s he had advanced to one of the top two operational positions in Tata Electric. A sign of how important Tata senior management saw the computing enterprise was that they took one of their senior executives in one of their larger businesses to head it. As a sign of how new and experimental the venture was, Kohli received an agreement that he could return to his old job if the new one did not work out. Kohli had familiarity with computers, helping to introduce them into Tata's electric power utility.[25]

Given that computers were so closely associated with the United States, Kohli's connections with America would also prove important. In addition to having gone to MIT, Kohli played an integral role in establishing a Bombay section of the Institute of Electrical and Electronic Engineers (IEEE), the American-oriented professional society. Furthermore, Kohli was friends with the first director of the American-supported IIT at Kanpur and made regular visits there, serving on its electrical engineering faculty selection committee.[26]

Kohli was appointed general manager and managing director, titles that had never existed before, making a statement that this was a formal business that would have a formal organizational structure. The Monday seminars were gone, as were the brochures that seemed to emphasize the brilliance of individual computer scientists and their impressive foreign education. Patel and Malhotra stayed on for several years, but eventually left to complete doctorates at MIT. As TCS developed, it would require competent people, but MIT degrees would become less salient.[27]

TCS had developed its business within the constraint of limited access to computers. Kanodia and his group would have liked to acquire a large IBM computer from its new System/360 line, but IBM would not sell them in India, limiting them to older obsolescent 1401 models. Restrictions by the Government of India made acquisition of new computers difficult.[28]

In 1969 TCS submitted a quotation for generating bills by computer for the Bombay telephone system. It won the job, and the system became operational for the system's 140,000 subscribers in 1971. The system allowed calls to be debited to an account within ten days while the manual system had taken up to two and a half months. TCS estimated that the system would lead to 720,000 rupees a year in revenue. TCS also won a contract to computerize the compilation and printing of the Bombay telephone directory.[29]

While TCS pursued a variety of businesses, its top managers had a particular interest in the overseas market. A 1970 report on TCS's performance shows its global vision as well as the attention given to the business by the head of the House of Tata, J. R. D. Tata, whose remarks were sprinkled throughout the report. TCS was already pursuing the software business in the UK through advertising in British newspapers. It was "urgently" trying to assess the business possibilities in computing in Australia in an effort to keep ahead of the Japanese. TCS was looking to establish an operation in Singapore as a way to bypass the difficulties of acquiring computers in India. At the same time TCS reported that it was trying to get business from IBM. In spite of this enthusiasm, the report also recognized that there were "serious problems of communication (both technical and commercial)" that needed to be resolved before the foreign business could truly develop.[30] (And indeed it appears that most of the plans discussed in the report never came to fruition.)

The tone of the report made clear that the attraction of the software export business was more than the immediate money it might bring in. Global connections would provide ancillary benefits, with the report claiming that software development for export would "provide job satisfaction to many of our brighter staff" and would "also attract talented Indians working abroad."[31]

In 1972 Kohli was elected to the board of directors of the IEEE, the first Indian ever to hold such a post. Part of his responsibilities were to attend periodic meetings in the United States. Kohli later said that whenever he had to attend an IEEE meeting, he would also make a visit to Detroit to seek business from Burroughs. Burroughs was an American computer company, one of what wags at the time called "the seven dwarfs." But Burroughs was a dwarf only in comparison to gigantic IBM; it still had a substantial computer business.[32]

In July 1973 Burroughs gave TCS its first contract. It was a small job, valued at only $30,000, requiring a turnaround in less than three months, a schedule no American company was willing to commit to. Ironically because TCS did not own a Burroughs computer, its first job was to build an emulator that would allow it to run Burroughs software on its ICL machine. TCS completed the work for Burroughs successfully. The emulator itself became a successful product, helping companies converting to Burroughs equipment.[33]

This was the beginning of a small but steady flow of business to TCS based on Burroughs products. At times TCS would get jobs through Burroughs for businesses that were converting from a competitor's system to a Burroughs system. Other times they got jobs for extensions to existing software packages or for new ones. TCS's 1976 statement of operations provides a sense of TCS's business. Its overall revenues for the year were 30 million rupees, roughly $3.4 million. Many of the jobs were "body-shopping," sending its employees to work at the clients' sites, but TCS bragged in 1976 that it was doing some foreign work in India.[34]

The 1976 report provides a sense of the kinds of foreign work TCS engaged in. It had an eight-member team working on developing a hospital information system, while four TCS employees went to Detroit to work on a version of a Burroughs bank information system for a new line of hardware. It had a nine-member team working to convert the Maryland Department of Transportation from an IBM system to a Burroughs system, while four TCS employees worked to convert the Casco Bank of Portland Maine from a Honeywell system to a Burroughs system. TCS had similar jobs through Burroughs with clients in New Zealand, the United Kingdom, and the Netherlands. Furthermore, TCS had jobs, not linked to Burroughs, in Iran.[35]

From its origins in the 1850s when it traded with China and England, Tata had been a global enterprise, getting ideas from a wide range of places and doing business across boundaries. When TCS began, the Tata Group already had a New York office, and its manager, Naval Mody, became an integral part of the TCS operation. He frequently sent back to India clippings from newspapers and magazines apprising Tata managers on the state of things in the United States.[36]

Doing business in America was not without its problems. In 1974 Burroughs and Tata began negotiating with Ford for a job converting computer programs to run on a new Burroughs machine. The job would require eight people working over a six- to eight-month period. While TCS submitted a bid for $98,000, which was $20,000 less than the closest American competitor, ultimately Ford chose the American firm, claiming that TCS was not registered to do business in the United States and that its employees lacked the proper visas. Burroughs itself seems to have been complicit in Ford's decision, with an executive writing to TCS managers that Ford was "too sensitive a customer to take any chances." Tata managers quickly worked to comply with American regulations.[37]

TCS's business grew slowly for a variety of technological reasons, but another impediment to the growth of the business was the environment created by the Indian government. The government imposed tight restrictions on purchases that required foreign exchange. Whenever TCS sought to expand and buy new hardware, a protracted series of negotiations with the government was required centering around the question of foreign exchange. Typically TCS would have to guarantee that a given computer would generate foreign exchange worth twice the value of the computer over a five-year period.[38]

As discussed in Chapter 8, in the late 1960s Indian Prime Minister Indira Gandhi, in the face of a struggling Indian economy combined

with political challenges, turned sharply to the left, adopting increasingly populist policies that put restrictions on business. This
was the license raj in full force. Among these policies was the
Monopolies and Restrictive Trade Practices (MRTP) Act, which
required government approval for any industry expanding its facilities by 25 percent or more. This stipulation frequently came into play
for TCS when it sought to buy new computer systems, requiring another laborious approval process. Thus, in 1978, TCS was forced to
argue that its purchase of a Burroughs 6800 computer system "would
not be prejudicial to the public interest and would not result in a
concentration of economic power to the common detriment."[39]

On top of these restrictions, there were an assortment of complex
import duties to be paid on any new computer entering India. In 1975
when TCS bought a Burroughs B1728 small computer, TCS managers thought the import duties would amount to 75 percent, but
when they sought to pick up the computer, the government officials
who went through the byzantine process of calculating duties told
them that the duties in reality amounted to 101.25 percent. In this
situation TCS's position as part of the larger Tata group became
helpful as Kohli wrote a letter to other Tata companies that were
TCS clients and asked them to prepay their expected charges a year
in advance to cover the additional duties.[40]

Both the requirement for guaranteed foreign exchange and the
MRTP Act meant that it could take years to get government approval
to acquire a new computer. There was more than a little irony in the
fact that for the year 1976–1977, the Government of India recognized
TCS as being India's top engineering exporter. (It gives a sign of how
small India's engineering exports were that TCS's 1976 exports
amounted to only 5.9 million rupees or a little less than $700,000.)[41]

Creating a computing business in India involved creating an entire infrastructure, and a large part of Kohli's work went beyond
transactional business for TCS. Kohli served a term as the head of
the Computer Society of India, where he helped write the organization's constitution and began its transition from an academic to a
business orientation. In his presidential address given in 1975, Kohli
invoked the concept of systems engineering he had learned at MIT,
saying that a systems approach would enable society to deal with

increasingly complex problems. Kohli concluded his speech by laying forth both an opportunity and a challenge to India: "Many years ago there was an industrial revolution. We missed it for reasons beyond our control. Today there is a new revolution—a revolution in information technology, which requires neither mechanical bias nor mechanical temperament. Primarily, it requires the capability to think clearly. This we have in abundance. We have the opportunity to participate in this revolution on an equal basis—we have the opportunity, even, to assume leadership in this revolution."[42]

In 1976 Kohli led the organization of a South East Asian regional computer conference in Singapore. The conference brought together computer professionals from Japan, Hong Kong, Singapore, Australia, New Zealand, and India. At this conference Kohli and a colleague gave a paper on "Software Development for Self-Reliance in Computers," which laid out the TCS strategy. However, the paper's title was misleading, for while the words "self-reliance" were almost a mantra in India, ultimately the paper was about computing as a global technology, where countries such as India would not be self-reliant but able to carve out a profitable business in an interdependent world. The vision Kohli presented was one very much at odds with the more autarkic policy of the Indian government.[43]

Kohli outlined in detail the way the computer was the product of the United States and the distinctive characteristics that applied there. He noted the role that American research universities had played in the computer's development, the massive funds provided by the defense department, the high degree of capital intensity of the business, and the rapid rate of obsolescence. Kohli stated that the "technological momentum generated by funding on such a scale is truly staggering."[44]

What Kohli saw was the impossibility of competing head-to-head with the American computing industry, a realization that technocrats in the Indian government were only starting to come around to. But trends in computing were starting to favor developing countries. As software became a larger part of the cost of computer systems, the high cost of skilled labor in the developed countries provided an opening for developing countries. If they took advantage of the freely available technical information throughout the world and developed

software skills, they could use their advantage in labor costs to cap-
ture business from developed countries.

Crucially, this model assumed not competition between India
and the United States but what Kohli called "a mutually beneficial
relationship and close cooperation." Kohli's paper described the relationship that Tata had developed with Burroughs. It also described the
relationship that other India firms were developing with American
computer companies.[45]

The *Tata Review* for 1977 carried an extraordinary story. The period was one of turmoil for India. Indira Gandhi had suspended
democratic governance in India from 1975 through March 1977. The
economy was in the throes of what became known as the "Hindu
rate of growth," a 3 percent annual growth rate that would do little
to raise India's standard of living. As India celebrated its thirtieth
anniversary of independence from the British, Jawaharlal Nehru's
dream of a democratic and technological India that was a "great nation" was in some ways no closer to realization and perhaps even
harder to imagine. At this time the *Tata Review* described how a
group of European computer personnel had traveled to Bombay to
receive training from TCS to manage the transition to a new Burroughs computing system. A representative of a Swiss insurance company asserted that TCS was a crucial factor in their decision to buy
the new Burroughs equipment. He added that his firm would send
more people to TCS for training to "bring more knowledge back to
our country."[46] In the almost 500 years since Vasco Da Gama had
arrived on India's shores, Westerners had come to India in search of
many things, ranging from treasures to conquest to spiritual enlight-
enment, but perhaps this was the first time Westerners had traveled
to India to gain knowledge about a technology that had originated
in the West.

If Kanodia, Patel, and Malhotra's precocious vision of computing
developed at MIT was critical in getting TCS started, and if Kohli's
connections with the United States were essential in building business, the larger Tata organization contributed much to TCS's success.
It provided an organizational stability as well as a willingness to fund
and nurture a new business. It provided an organizational structure
that could grind through all the Government of India's requirements

for acquiring new computers. Tata managers provided a strategic vision to help TCS navigate through uncertain passages. TCS was more than just four MIT-educated engineers.

Lalit Kanodia and Datamatics

Kanodia left TCS in late 1969. He would continue on in computing, although now without the resources of Tata behind him. He formed two new companies: one, Datamatics Staffing, was a recruiting firm, while the other, Datamatics, leased computer time in order to offer courses in computing in Bombay. By 1972 Datamatics had a "Datamatics Institute of Management" that was offering a postgraduate diploma course in computing as well as a course in business. Advertisements for the business program noted Kanodia's doctorate from MIT and claimed that the program was "patterned after advanced marketing and advertising programmes in the United States."[47]

One fundamental problem Kanodia faced in starting a computing business was finding a computer to use. He initially rented time on other computers in Bombay. Only in 1975 did Kanodia's firm have its own computer. While India's complex regulations made buying a computer difficult, it was possible for an Indian returning from abroad to bring a used computer with him without paying extravagant duties. Kanodia found such a person who brought in an old IBM 1401—the same computer TCS had been chafing under nearly a decade before. Kanodia used this computer as the basis for starting a new venture, Datamatics Consultants, which concentrated on doing financial data processing for Indian businesses.[48]

In 1977 IBM, faced with a requirement by the Government of India that it cede majority ownership of its Indian operation, left India. The departure of IBM, which had by far the largest share of the Indian market, encouraged smaller American computer companies to enter India. At the same time the requirement in India that foreign-purchased computers generate foreign exchange prodded these American companies to send programming work to India. By 1981 Datamatics was advertising a connection with U.S. minicomputer maker Wang Laboratories, using its equipment and helping to promote its products. Like TCS had done with Burroughs, it also began doing programming work for its American partner. In a 1981 advertisement claiming to be "India's leading data processing company," Datamatics Consultants stated it had a staff of 300.[49]

In the late 1980s Datamatics Consultants began programming for AT&T, the American telephone giant, and in 1989 Datamatics and AT&T's subsidiary Bell Labs opened a satellite link connecting their facilities in India and the United States. The satellite link allowed Datamatics to work directly on Bell Labs computers from India, eliminating the need for sending Indians to work in the United States and also eliminating the need for Datamatics to buy the computers on which its employees worked, reducing problems of government red tape and foreign exchange. In its announcement of the link, Datamatics claimed to be the first Indian software company to have a satellite link for software exports. Although this represented a major breakthrough, Datamatics was still a small company. While Kanodia had been running Datamatics for twenty years, the company's annual billings were roughly $6 million. But like TCS, it had established itself and proved to American partners that it could do the required work.[50]

Patni Computer and the Road to Infosys

MIT graduates shaped two other important Indian IT firms, one directly and one indirectly. Narendra Patni was a Marwari and the son of a textile mill owner. The younger Patni, born in 1943, studied electrical engineering at Roorkee, where he was encouraged to go to MIT by a faculty member who had earned a master's degree there. How little either his mentor or Patni himself knew about the contemporary situation at MIT is seen by the fact that Patni applied seeking to study electric power, only to be told by MIT that that field was no longer offered. He was nevertheless offered admission, where he did a master's thesis related to gyroscopes, a key technology for guidance systems for missiles and space vehicles. He stayed on at MIT, first to earn a master's degree in management, later to serve as an assistant to Jay Forrester. Jay Forrester was MIT's entrepreneur par excellence, who had led the development of Project Whirlwind and SAGE. After that Forrester moved into the school of management where he extended systems engineering into nontechnical areas in an effort to model society as a whole. Forrester had a private consulting business and frequently referred prospective clients to Patni.[51]

Patni also worked for Arthur D. Little, which in the early 1970s was developing the LexisNexis database. This legal and news database required entering a huge amount of data in digital form. In 1972 Patni began a series of experiments in Cambridge exploring the possibilities of doing data entry work in India for American companies. In his Cambridge apartment working with his wife, he labeled one room "India" and one room "United States." The purpose of this experiment was to see if it was possible, using only written instructions, to provide sufficient information such that jobs originating in the "United States" could be performed in "India." Patni used the positive results of this experiment to begin a business converting information into electronic form. He had typists in India working to input text into paper-tape punching machines. The paper-tapes would then be shipped to America, where they would be uploaded into a computer. The main skill that India provided was data entry, and Patni built a business with 100 typists working in Poona.[52]

One salient feature of Patni was his transnationalism. He lived in the United States and traveled frequently between the two countries. In the late 1970s Patni made an agreement with Data General, analogous to those made between Tata and Burroughs and Datamatics and Wang, to sell their minicomputers in India. Because of the need to earn foreign exchange to support sales of computers in India, the Patni arrangement included writing software.[53]

Patni had put together a team of computer programmers, the leader of which was Narayana Murthy. Murthy, the son of a teacher, had studied engineering in Mysore. Like Patni, Murthy had intended a career in electric power, although his interests ran to hydroelectric power. But when he went to IIT Kanpur for his master's, which he earned in 1969, he got hooked on computing.[54]

When Murthy was completing his master's degree, Jashwant Krishnayya from the Indian Institute of Management (IIM) came to Kanpur looking to hire computer programmers. Krishnayya, the son of an educator, had gone to MIT and then was recruited by Vikram Sarabhai to come to the newly established IIM to help set up a new Hewlett-Packard computer the school was getting with a grant from the Ford Foundation. Krishnayya hired Murthy and several other master's students from IIT Kanpur for the work. After a few years Murthy left the Indian Institute of Management and then worked at a consulting firm Krishnayya had started, Systems Research Institute (SRI), doing consulting work for nonprofits. The sterility of the work, which entailed preparing reports, discouraged Murthy, and he left to work for Patni.[55]

In July 1981 Murthy and six of his colleagues left Patni to start their own company, Infosys, which would eventually become one of India's largest IT companies.[56] The founders of Infosys were known for having been educated in India rather than abroad. But MIT-educated Indians played a crucial role in Murthy's early career and thus the founding of Infosys. He was introduced to computing at IIT Kanpur, and his two mentors there, V. Rajaraman and H. N. Maha-bala, had either a degree from MIT or spent time there. MIT-educated Krishnayya gave Murthy his first computer job in India and then worked with him at SRI. Finally, through Patni, Murthy began doing business with American firms. Patni had the connec-tions enabling him to win business in the United States, connections which Murthy initially lacked but developed through his work with Patni. Without Rajaraman, Mahabala, Krishnayya, and Patni, it is doubtful there would have been an Infosys.

By the late 1980s the environment for writing software in India for export had changed dramatically. Its possibilities were clear to American companies, Indian entrepreneurs, and even the Indian government. The Government of India, as part of the more positive policies toward business initiated by Indira Gandhi in her restora-tion to the prime minister's post in 1980 and continued by her son Rajiv, started providing assistance to Indian firms by giving special consideration to computer work done for export. By the late 1980s there was a wider movement to promote software development in India. At a 1987 Boston conference on software development in India, twenty Indian companies and sixty-five American companies partici-pated. A year later the Indian government sponsored a conference on the same topic in San Jose. By 1989 the *Times of India* was noting that "Indo-U.S. cooperation in the field of computing is so extensive that it is difficult to name a major U.S. company which does not have some sort of association with India, that is except IBM and Apple."[57] In 1991, as India stood on the threshold of liberalization, the Indian IT magazine *Dataquest* published a list of the top-twenty Indian software exporters. Five of the top-ten had MIT in their genealogy.[58]

In the earliest stage of the development of the Indian IT industry, Indians who had been to MIT played an outsized role. They had a precocious exposure to computing. They had contacts in the United States, which they could exploit. They had a sense of the areas where India had a competitive advantage and those where it did not. Indian government bureaucrats make easy targets in this story, but most basically they did not know how to operate in a global, capitalist, American-dominated environment in the same way that Kohli and others did. Of course an MIT education was not the only factor: for TCS, also vital were Tata's deep pockets and far-reaching vision.

Liberalization of 1991 removed some of the tangle of regulations that Indian entrepreneurs had to deal with. Much faster electronic connections between India and the United States made it much more feasible to do work in India. As software became an increasingly large component of the price of computer systems, the advantages of software development in India became increasingly clear. By this point the MIT connection had become less important.

MIT had been important in giving Indian students exposure to American technologies, an exposure that had helped them develop firms that could pull business from the United States to India. But in the 1960s the exposure to the American system of engineering education would also pull in the opposite direction.

10

From India to Silicon Valley

DURING INDIAN PRIME MINISTER Jawaharlal Nehru's visit to the United States in November 1961, the United States announced that an agreement had been reached whereby nine U.S. universities, anchored by MIT, would provide assistance to the new Indian Institute of Technology at Kanpur. President Kennedy issued a statement hoping that Nehru would consider this program a "souvenir" of his visit to the United States. The United States spent $13.5 million on IIT Kanpur, and with American professors and professional staff spending a total of 188 person-years there. However, over time, with the students from Kanpur and the other IITs going to the United States, where they were often to spend their entire careers, some questioned who had been the giver of the gift and who had been the recipient. In 1984 the British science journal *Nature* quoted an Indian academic as saying that the IITs were "India's most generous gift to the United States."[1]

These results were not something that either government wanted, even though each government's actions contributed to them. The Indian government decided to establish and munificently fund institutes of technology offering an engineering education based on American models. At the same time, American national policies funding its graduate programs in engineering made them accessible to many Indian engineering students. Finally, in 1965 America's changed immigration policies made it possible for Indian engineers to stay in the United States in ways that had been impossible before.

In the end, the actions of individual Indians were decisive. Although Nehru's dream was to use the technological elite created by an Indian MIT to build a new nation, ultimately many individuals refused to be a part of that dream. Instead, they committed their lives and talent not to the Indian nation, but to a technological system based in America. They would join that system as individuals, where they would become professors, researchers, and entrepreneurs, creating a technological Indian in the United States.

A New MIT

At the very time that IIT Kharagpur was starting, the experience of World War II followed by the early years of the Cold War led MIT to redefine both itself and engineering education more broadly. Before World War II most of the relatively small amount of money for engineering research came from private foundations. During World War II, academic researchers, working with government funding, had played a vital role in the development of a wide range of technologies with military applications, including the atomic bomb, radar, servomechanisms, and the computer.

As World War II ended and the Cold War with the Soviet Union began, the U.S. military saw the continued need to support academic research that might have relevance to military technologies. MIT, whose research contracts were not even mentioned in the summary statements of the Institute's finances that appeared in the President's Reports before the war, had research contracts by 1948 that accounted for over half of its budget. Research had become MIT's central function.[2]

At the same time, the experience of World War II, along with the military's continued funding of research, led engineering administrators to rethink the engineering curriculum. Engineering professors like Gordon Brown at MIT and Frederick Terman at Stanford saw that scientists with a stronger education in fundamental science and math had taken the lead in the great World War II research projects, relegating engineers to the background. Although MIT's curriculum had long had a significant scientific component, Brown and others developed an engineering curriculum branded "engineering science," which emphasized higher mathematics and science and de-emphasized training in existing technologies. In 1952 Brown gave a paper at a meeting of engineering educators: "The Modern Engineer Should Be Educated as a Scientist." In MIT's ensuing reforms it discarded its traditional teaching of electric power and machinery, the fundamental but mature technology underlying much of American society. MIT scrapped its electrical machinery lab, where earlier generations of students had received their induction into the field. MIT was the leader, but other American universities followed, both in the move to federal research funding and in the curricular reforms.[3]

By the late 1950s, Indian government officials were beginning to grapple with what the American transformation of engineering education meant for India. In 1959 a committee set up to review IIT Kharagpur, led by British engineer Willis Jackson, issued a withering critique, finding Kharagpur wanting according to the new American engineering standards. Jackson's report noted that the previous decades had been characterized by a "fantastic range and pace of scientific discovery and technological development," including computers, transistors, nuclear power, and many new materials. The report asserted that these developments necessitated a new type of engineering training that put a greater emphasis both on fundamental science and on scientific thinking. Although the committee acknowledged that Kharagpur's rigorous selection process and well-equipped labs were yielding better quality engineers than most Indian engineering colleges, on the whole the committee found that the curriculum at Kharagpur was too much like that at other Indian engineering colleges, asserting that men trained along the lines needed for the new technological era "have not yet been produced within India."[4]

Jackson's report reflected a consensus that top Indian officials with responsibility for technological education had already been coming to, and in the late 1950s, they made an all-out effort to get MIT involved in providing assistance to the new IIT at Kanpur, which was already slated for assistance from the United States. In 1958, citing Indian officials' interest in an "MIT-type" institution, the International Cooperation Administration of the State Department (ICA, the forerunner of the U.S. Agency for International Development) requested that MIT prepare a proposal for the development of Kanpur. MIT declined, citing lack of personnel.[5] In the absence of MIT's participation, a group consisting largely of retired American engineering educators went to India and wrote a report outlining their plans for the new institution. P. K. Kelkar, the man who would become director of IIT Kanpur, later wrote that their report would have produced an IIT Kanpur that was "just one more Engineering College added to the long list." The ICA then began negotiations with Ohio State University, a good enough engineering school but no MIT, to support Kanpur.[6]

But Indian officials would not give up on MIT. In the summer of 1960, Government of India (GOI) officials repeatedly met with various MIT officials pressing their case. An MIT professor wrote that M. S. Thacker, a leading Indian official with responsibility for technical education, expressed "the willingness of the Indian Government to meet almost any conditions to persuade MIT to take on this task."[7] Later an American official stated that top officials with responsibility for technical education in the Government of India believed that "only MIT can successfully break through old customs and traditions and start technical education on [a] new course essential in India."[8] The Government of India's belief in the indispensible role of MIT was so strong that as American State Department officials entered into the final stage of negotiations with Ohio State, GOI officials requested that the negotiations be broken off, so that they could have one last try at enlisting MIT.[9]

Before any American program of assistance was in place at Kanpur, Indian officials appointed a director, P. K. Kelkar. Kelkar, born in 1909, had a doctorate in electrical engineering from the University of Liverpool and taught at the Victoria Jubilee Technical Institute in Bombay. Just prior to his appointment at Kanpur, Kelkar had been the chief planning officer for IIT Bombay, where he had worked closely with Soviet engineers. While Kelkar had experience with both the British and Soviet systems of engineering education, he later described an epiphany he had when he read the proceedings of a American conference detailing the reforms launched by MIT to reorient American engineering programs around the idea of engineering science. Kelkar later wrote that the report had a "traumatic psychological impact" by showing him a new approach to engineering education, which he believed to be clearly superior to India's existing system. Kelkar (and, one imagines, other Indian officials) became a convert to the new American system of engineering education.[10]

The Indian officials' persistence ultimately led MIT to send three professors to India to visit Kanpur and explore the possibilities there. In Kelkar, they found a kindred spirit, and largely based on his enthusiasm and the commitment of Indian officials to the new American curriculum, they recommended that MIT support the venture at Kanpur. The MIT professors believed the task was too big for any one university, and so proposed a consortium of American universities, informally anchored by MIT, that would provide support to Kanpur.[11]

In 1961 the Indian and U.S. governments along with nine American universities reached an agreement to provide support to Kanpur through what became known as the Kanpur Indo-American Program (KIAP). Ironically, although Indian officials had wanted an "MIT-type" institution, the structure of the program, with nine American universities (eight of which were not MIT) contributing visiting professors, meant that politically it could no longer be known as a program to establish an Indian MIT. The program was reframed as an effort to bring an American model of engineering education to India.

Kelkar and the American professors formulated a plan to develop an American-style engineering curriculum, with a heavy emphasis on engineering sciences. While this curriculum had been established in the United States to enable engineers to work on technologies such as nuclear power, semiconductor electronics, new materials, and aerospace engineering, its introduction in India was justified in two ways. First, India would need its own programs and its own workers in these new technologies. Second, the conceit of engineering science was that it provided fundamental tools that would enable Indian engineers to tackle their society's challenges whatever the specific technology might be.

From the American perspective, KIAP was a New Frontier program, where young idealistic professors took up President Kennedy's challenge to "pay any price, bear any burden" to assure "the success of liberty." In this case it was to establish a first-rate engineering institute that would both help develop the Indian economy and point that new nation toward the United States and away from the Soviet Union. It had a far higher profile than the Illinois program at Kharagpur had. U.S. Ambassador John Kenneth Galbraith kept up on its progress and visited Kanpur in 1962. University presidents gave it their support. Professors were willing to halt their research programs for a year or two in order to go to India, putting their career trajectories at risk.[12]

Indians brought a similar idealism to the program. Young Indian professors gave up promising jobs in the United States, not just to come back to India, but to come back to an unattractive part of India, in almost all cases far away from their home regions where they could have easily found jobs. They came to Kanpur with the goal of building a technological India.

In parallel with the program at Kanpur were programs at IIT Madras (with West Germany), IIT Bombay (with the Soviet Union), and IIT Delhi (with Great Britain). While the different national patrons for the IITs suggested India's nonalignment, the Americans at Kanpur had a Cold War perspective, fearful that the Soviet-supported IIT Bombay would come to play a dominant role in Indian technological education. What had shaped up to the Americans as another battle in the Cold War proved to be no contest. Language differences kept the Indians and the Soviets apart. The Soviet program at Bombay was conducted at more of an arm's-length distance than the American program, with the Soviet professors largely confined to graduate courses.[13]

More fundamentally, actions of the Indians, not the Soviets or the Americans, were decisive. At each of the IITs were Indian faculty who had gone to American graduate schools, and this American-oriented Indian technical elite partially decoupled the IITs from the global politics of the Cold War and worked to align each of the IITs with the American model of technical education. This could be seen most dramatically at IIT Bombay in the years 1971 and 1972, when formal Indo-U.S. relations were at an all-time low.

During those years the personal enmity between Richard Nixon and Indira Gandhi was exacerbated by strategic considerations that pointed each nation away from the other. One of President Nixon's signature issues, opening relations with China, depended on using Pakistan as an intermediary. In 1971 India signed a friendship treaty with the Soviet Union. The crisis in East Pakistan, which would eventually lead to the creation of Bangladesh, led to war between India and Pakistan in December 1971. Things hit rock bottom when Nixon ordered the aircraft carrier *Enterprise* into the Bay of Bengal, in what Indians saw as a threatening gesture aimed at them. There seemed to be no doubt that the United States had tilted toward Pakistan and away from India, while India had tilted toward the Soviet Union and away from the United States. At IIT Kanpur, Americans and Indians had envisioned a second-stage program, which would continue past the original 1972 end date. However, poisonous formal relations between the countries' governments made a second stage impossible.[14]

In 1970 Kanpur director P. K. Kelkar was named director of IIT Bombay, returning to his hometown. As the officials who appointed him must have known he would, he sought to reform the curriculum and introduce aspects of American engineering education practice. In March 1972, near the peak of anti-American sentiment in India, the faculty committee that Kelkar appointed made its final report, proposing a series of reforms whose overall effect was to bring the curriculum more in line with American engineering education standards. And to make completely clear what these reforms signaled, Suhas Sukhatme, an MIT-trained IIT Bombay faculty member and member of the curriculum committee, inserted into the beginning of the report quotations by MIT founder William B. Rogers and former MIT president Karl Compton on the goals and methods of engineering education. The faculty at IIT Bombay had made its own tilt.[15]

Kanpur and each of the other IITs, as they had developed by the late 1960s, were small institutes, emphasizing quality over quantity. They had undergraduate populations of between 1,200 and 1,600 spread out over a curriculum of five years, and roughly 500 more graduate students. The IITs together produced roughly a tenth of the engineering graduates in the nation at the time, but they were to be India's elite engineers. The IITs represented an enormous investment in the top of India's educational pyramid. A 1964 study showed that the IITs were spending 16,400 rupees per student per year, as compared to 7,000 rupees a year for Regional Engineering Colleges and 3,000 rupees for State Engineering Colleges. In turn, each of these numbers was enormous when compared with expenditures on school education. India spent 24 rupees per student in primary education, 65 rupees on those in middle schools, and 230 rupees on those in secondary schools.[16]

The tuition at the IITs was extremely low (200 rupees per year at IIT Madras in 1973). Given that most students were from the upper middle class, the IITs represented a substantial subsidy by a poor nation to one of its wealthiest groups, in the faith that it could lift the entire country up. At the time, with India's per capita national income of 425 rupees per year, the cost of sending a student to an IIT was a national investment of the per capita income of almost forty people.[17]

With the full implementation of the IITs, engineering education in India consisted of a series of interlocking hierarchies, with a substantial momentum pushing students toward an apex that lay not in India, but in the United States. The first and most basic hierarchy was a hierarchy of professions. Among the Indian middle class, the two most coveted career choices were engineering and medicine. In industrializing India, engineering jobs were expected to be abundant and well-paying. While some entered engineering from a genuine interest in technical things, for others there was a simple calculus behind their decision: if their grades were sufficient to get them admission to an engineering school, then they should go.[18]

In parallel with the creation of the IITs, the Indian government created a series of Regional Engineering Colleges (RECs). The expectation was that these RECs would provide the more workaday engineers who would staff steel mills and power plants. The previously established engineering colleges, such as Bombay's VJTI, Calcutta's Bengal Engineering College, or Roorkee, occupied the next place in the hierarchy, with the IITs at the top.[19]

The hierarchy was further elaborated with the instrument of admission to the IITs, the Joint Entrance Exam (JEE). This exam, introduced in 1961, was developed as a solution to the problem of devising a standard, objective method for determining entrance to the IITs. The exam originally had sections in physics, chemistry, math, and English. In 1963 18,000 students took the exam.[20]

The exams produced a ranking, known as the All-India Ranking. The student with the highest score on the exam got the choice of any seat in any major at any IIT. The student with the second highest score had the choice of any seat, except that already occupied by the first student. This procedure went on down the rankings until every seat was occupied. The JEE manifested and also played some role in producing a hierarchy among the engineering majors. This hierarchy had some of its origins in perceived career prospects, but became self-perpetuating over time. By the mid-1960s, the hierarchy was electrical engineering, mechanical engineering, chemical engineering, civil engineering, metallurgy, naval architecture. (When computer science programs were introduced, they would occupy the top spot in the hierarchy.) The electrical engineering seats would be filled first, then the mechanical engineering, and so on. The student's inherent interest in the field was often a secondary consideration.[21]

The hierarchy of JEE rankings was closely connected to the family's income and also to the student's precollegiate education. Students who performed better on the JEE were more likely to be from wealthier families and to have been educated in private, English-medium schools. Students who scored lower on the JEE were more likely to have gone to government-supported schools and to have a Hindi-medium education. To know a student's major was in a sense to know their All-India Ranking on the JEE and also have a hint about the family's social status.

The next level of hierarchy manifested itself upon graduation. The distinction between the Regional Engineering Colleges and the IITs was that the IITs were to train students for more technically demanding, research-oriented jobs. In truth, such jobs were rare to nonexistent in India, but the conceit suggested that IIT students were perhaps above the jobs that REC students might do. Furthermore, these jobs working in steel mills or hydroelectric plants neither paid particularly well or offered a promising career trajectory. Graduate school in America, offering a stipend equal to a salary one might receive in India, was an attractive choice.

Each of the IITs put a great deal of effort into developing graduate programs. But ironically, while American engineering professors worked to build up the graduate program at IIT Kanpur, the larger American system of engineering education undercut those efforts. The prestige and the funding of American graduate programs were too much for any of the Indian IITs to compete with.

While few of IIT Kharagpur's original graduates had gone overseas for training, that pattern changed in the next decade. Of the students who graduated from IIT Kharagpur in 1962, the top student in the class as well as the top student in metallurgy went to MIT. The top two students in electrical engineering went to the United States for graduate school. The top student in mechanical engineering had options to go to America, but decided to stay in India. In the first half of the 1960s, before his institute was even ten years old, IIT Bombay director S. K. Bose complained repeatedly in his annual reports about what he called the "craze" his students had for foreign (read American) graduate degrees. In his 1966 report, he noted that one-third of the graduates had gone abroad for studies, in preference to Bombay's own programs. IIT Kanpur's 1975 annual report showed that out of its first eleven classes, roughly 20 percent of undergraduates went abroad (which in the overwhelming majority of instances meant to the United States) for graduate study.[22]

As this book has shown, the IITs did not cause Indians to come to the United States for graduate engineering education. And in spite of India's hierarchical educational structure, students came to MIT and the United States from a wide range of colleges in India, from the IITs to the long-established engineering colleges to the new Regional Engineering Colleges. Surveys conducted by the Institute of International Education showed that the numbers of Indians pursuing graduate studies in engineering in the United States rose from 1,344 in the 1963–1964 school year to 3,364 by the 1968–1969 year.[23]

The IITs would operate as part of a global hierarchy in engineering education, where American institutions, and MIT within them, stood at the top. By and large, IIT undergraduates would shun IITs for graduate education in favor of American schools. It became common at each IIT for the winner of the President's Gold Medal, the award given to the student graduating with the highest academic average, to then go to an American school and work for a doctorate. The IIT graduate programs would largely serve students who had studied at schools lower down in the Indian hierarchy, such as Regional Engineering Colleges (although even students from these colleges would apply to American graduate schools).

Both from the Indian side and the American side, the IITs standardized the route to American graduate schools. Prior to the full development of the IITs in the early 1960s, while there were forces acting to draw Indians to the United States, at the individual level the process typically involved some unusual event. Very rare was the student or family who had a long-term plan to come to MIT. Particularly important during this time were the programs and people of the United States Information Service (USIS), which as part of its efforts to fight the Cold War with the Soviet Union provided information that worked to recruit Indians into the American technological empire.

Tapan Kumar Gupta, born in 1939, was from a family that had been displaced and impoverished by the 1947 partition of India. His father was a lawyer, who was unable to reestablish his practice when his family was forced to flee East Pakistan on short notice. The younger Gupta described his educational career as essentially moving from one school that would give him a scholarship to the next. After getting a degree in chemistry at a small local college in West Bengal, he moved to the University of Calcutta for a master's degree in applied chemistry, which he received in 1960. He and some friends regularly went by the United States Information Service library in Calcutta. One day they went in to inquire about the possibilities of going to American graduate schools. The staff at the library helped them through the process of applying. Gupta was accepted and given an assistantship at MIT. Needing only to borrow money for his passage to America, Gupta began his studies at MIT in the fall of 1961.[24]

Pramud Rawat was the son of a lawyer and had graduated from the third batch of IIT Kharagpur in 1958 with a degree in naval architecture. He subsequently went to Hamburg, Germany, for a training program. While he was there he attended a program put on by the U.S. consulate's office discussing automation. After the program Rawat approached the speaker and asked for additional reading on the subject. The speaker replied that if Rawat was interested in automation, he should go to MIT. Rawat then sent MIT a letter asking for application information, addressed "MIT, Massachusetts, United States," not even knowing the school's city. In 1959 he was accepted and began his studies at MIT.[25]

John F. Kennedy famously put the American space program in the context of a competition with the Soviet Union having global implications, emphasizing "the impact of this adventure on the minds of men everywhere, who are attempting to make a determination of which road they should take." Although Kennedy's claim may seem exaggerated, for some Indians, space did help them determine the road they took. Ani Chitaley's father was an editor of a legal magazine in Nagpur. The family received both the magazine published by the Soviets and the magazine published by the USIS, *Span*. Chitaley recalled the Soviet magazine as being printed on cheap paper while the American magazine had beautiful pictures, showing what he later called a "wonder land." When the Soviets first launched Sputnik, Chitaley got interested in space and made a plan to go to the United States for higher studies. Satya Atluri was the son of a lawyer, studying in Andhra University. In 1961 or 1962, as he listened to a Voice of America radio broadcast, he heard one of President Kennedy's speeches on America's plan to go to the moon. Atluri at that point decided to focus on aeronautics and to try to come to the United States for further education.[26]

Thomas Kailath provides an example of the idiosyncratic process of getting to MIT in the pre-IIT days. Kailath, born in 1935, was a Christian whose family's roots were in the southern state of Kerala. His father, trained in biology, worked for a company selling vegetable and flower seeds in Poona. After going to a Jesuit school in Poona, the junior Kailath entered the Poona College of Engineering, majoring in telecommunications. He read an article in an American publication on Claude Shannon's information theory and became fascinated with the subject. A family friend, Stephen Krishnayya, had earned a doctorate in education from Columbia University and worked in India as an inspector of schools, but made regular visits to the United States. (In fact in 1952, Krishnayya had written a guidebook for Indian students, *Going to USA*.) Krishnayya encouraged Kailath to apply to American graduate schools, at a time when, in Kailath's words, such action "was not on my horizon." Kailath imagined that he might have a career working for All-India Radio. (And many of his professors doubted the chances that Kailath's application would be successful.) However, one of Kailath's professors, S. V. C. Aiya, was a graduate of Cambridge University and had published papers in an American technical journal, a rarity for an Indian professor. Aiya instilled confidence in Kailath and wrote him a letter of recommendation, which Krishnayya personally delivered to MIT. In 1957 Kailath was accepted to MIT with a fellowship.[27]

One of the effects of the IITs was to produce students who were recognized as being compatible with the American system of engineering education. Electrical engineering was a discipline that had undergone vast changes at MIT, with the advent of the computer, the rise of electronic devices, and the development of information theory. An application from India might face skepticism about whether an Indian engineering education provided adequate preparation. In Kailath's case, support from Aiya and Krishnayya seems to have overcome that hurdle, and he became the first Indian to earn a doctorate in electrical engineering from MIT. Prior to 1965 Indians who earned doctorates in engineering from MIT were more clustered in older and less glamorous fields of engineering such as metallurgy, civil engineering, and ceramics, rather than in the fields that had been more revolutionized by the introduction of engineering science. After 1965 Indian engineers at MIT were much more widely distributed in such mainstream engineering fields as electrical, mechanical, and chemical engineering.

An IBM 1620 computer installed at IIT Kanpur in 1963 serves as a parable for the ambivalent nature of the connections between the United States and India fostered by the IITs. Both Americans and Indians were eager to have a computer at Kanpur, and the IBM 1620 was installed as soon as the buildings could accommodate it. Both Indian and American IIT faculty conducted seminars introducing a wide array of institutions to the machine's possibilities. A large number of Indian undergrads got hooked on the computer. But getting hooked on the computer also meant getting hooked into an American technical system. American universities had far more resources to support work on computing, and most of those under-graduates involved in computing ended up in the United States. By contrast, the master's degree students at Kanpur, most of whom had come from lower-ranking colleges in India, were more likely to stay in India. We have already seen how Narayana Murthy worked at the Indian Institute of Management and had several other posi-tions before he started Infosys. Other IIT Kanpur masters students went on to work for the Indian Institute of Management or for Tata Consultancy Services. The computer at Kanpur helped to develop India's capabilities, but there was a tax for doing so, payable in talent to the United States.[28]

A broader picture of how Indian students ended up at MIT is pro-vided by the 1967 report of the adviser to foreign students at MIT. An entire section of the report was devoted to India, the only country receiving such detailed attention. The report noted that with the ex-ception of Canada, India provided more students (82) and more ad-missions queries (2,500) than any other country in the world. At this time, the total population of Indian students in the United States was 7,500, the second largest foreign contingent from any nation other than Canada. Of these, 2,300 were pursuing graduate degrees in en-gineering. At this time, India's annual production of engineers with bachelor's degrees was roughly 10,000, of which roughly 1,000 came from the IITs. For Indians even thinking of coming to the United States to study engineering, the first step was often to write to MIT.[29]

To manage the many admissions queries from India, MIT insti-tuted a two-step application process. It first sent a preliminary ap-plication to determine whether the would-be applicant had sufficient credentials to merit receiving a formal application. The 1967 report asserted that MIT received from India a total of 460 applications for graduate admission to MIT the previous year, a number which, based on the proportion of engineering students in the total Indian MIT student population, suggests that MIT received roughly 380 appli-cations from engineering students.

MIT admitted only forty-six of these applicants, but surprisingly only 40 percent of the admitted students enrolled. In spite of MIT's attractiveness, money was often decisive. Each department at MIT had its own policy, but on the whole MIT seems to have been less generous in its offers of financial support than other American graduate schools. MIT typically expected most incoming foreign graduate students to pay for at least the first semester of their education, offering support after the students had proved themselves. (Each department had some degree of freedom and often seems to have made decisions on a case-by-case basis.) Unless a family had significant resources, if the student was not offered support, MIT would be out of reach, no matter how much a student wanted to go to MIT.[30]

In 1964 a member of the MIT Club of India wrote to MIT's dean of engineering expressing his disappointment that a graduate of IIT Bombay, who was the son of an MIT graduate, was denied admission to MIT. MIT's foreign student adviser, Paul Chalmers, responded by laying out the admissions situation Indian students faced. He began by noting, as MIT commonly did, that its 900 international students out of a total student body of 6,000, gave MIT "the highest percentage by far" of international students of any major American university.[31]

While Chalmers denied that MIT had a "strict quota" of Indian students, his letter showed the informal quotas that inevitably played a large role in the admissions process. The civil engineering department had an admissions allotment of thirty slots, of which "it was thought fair" to allocate three to Indian applicants. After the first stage of the MIT admissions process for Indian students, there were still forty-five Indian applicants in civil engineering who were considered potentially qualified (qualified Indians could have filled every seat). The student who ranked first in civil engineering at IIT Bombay was admitted, while the student denied had been ranked second. Only one of the forty-five applicants seems to have gone to MIT, but the other forty-four may have well ended up in other American graduate schools. If MIT had actually accepted three Indians into its civil engineering program, two accepted students likely declined for financial reasons.[32]

As the IITs developed, students intuited the calculus that American graduate schools faced and engineered their own processes to maximize their chances to go to American graduate schools. In various degrees of formality, students at each IIT would find out the graduate school choices of the students in their major. They would then seek to spread their collective applications over as wide a range of American universities as possible and avoid collisions with their fellow students. The top students would be expected to limit their applications to particular schools, made known to all, so that lower-ranked students could avoid them. This system avoided wasted application fees, which were not trivial either in their amount or in the time it took to get the foreign exchange from the bank. Some students used less prestigious American universities, more willing to provide support and even waive application fees, as a stepping-stone to more prestigious schools like MIT.[33]

Paths in America

Indian graduates of MIT began staying in the United States in significant numbers in the 1960s. This section follows those MIT students who earned doctorates in engineering to provide a fine-grain analysis of that process. This group left a rich paper trail, allowing their history to be followed in detail. The 305 Indians who received doctorates in engineering from MIT in the twentieth century represent an extraordinary investment in human capital, both in India and in the United States. They reached the pinnacle of engineering achievement, becoming the quintessential technological Indians. Their careers tracked changes in both India and America, but particularly in the years after the early 1960s, their careers were more tightly coupled to American society than to Indian society.[34]

Chapters 4, 5, and 7 have detailed a group of Indians who received their doctorates from MIT in the 1930s, 1940s, and early 1950s, people like Anant Pandya and Minu Dastur. In these years, most Indians had significant support from sources in India and then returned and spent their careers in India. Factors in both India and the United States were responsible. On the Indian side was an enthusiasm for building up the new Indian nation and the desire to return to one's country and people. On the American side was a set of discriminatory immigration laws for those from the "Asia Pacific Triangle," which up through 1946 blocked Indians from residing permanently in the United States and then limited the number of

Indians allowed to permanently immigrate to the United States to 100 per year.

The first group of those with doctorates in engineering from MIT who stayed in the United States did so for personal reasons. Going to the United States and to MIT represented an openness to new experiences and to what a world beyond India could offer. Not surprisingly, a number of these young men (the first woman earned a doctorate in engineering in 1976) met women (either American or European) while they were at MIT and later married them. In many cases this had important consequences for the men's careers. U.S. immigration law allowed spouses of a U.S. citizen (or European immigrant) permanent residency status. At the same time, both members of the couple typically recognized the challenges that going back to India would present. Would the women be accepted by the man's family? By other Indians? Would she be able to play the role expected of her? How would she adapt to the climate and the different economic conditions? These women had made commitments to Indian men, not necessarily to India. At the same time, the very fact that an Indian would come to the United States and marry a European or American woman indicated a willingness to break with tradition.

Although it is not now possible to identify every spouse of every Indian student at MIT, the trend of Indians marrying Europeans or Americans and subsequently staying permanently in the United States was so pronounced that sometime in the 1950s, when J. C. Ghosh, the first director of IIT Kharagpur, was on a recruiting trip to the United States, he commented positively on C. R. Mitra's Indian wife and what it meant for Mitra's career: he was likely to return to India.[35]

These cross-cultural relationships could cause tension. Was an Indian student in America an individual free to make his own decisions, or did he have some larger loyalty to his family and to his country to consider? In 1949 the Harvard-trained Indian mathematician D. D. Kosambi corresponded with longtime friend and MIT professor Lawrence Arguimbau about an Indian student who was considering marrying an American woman. Kosambi planned to confront the student during a trip to the United States and try to talk

him out of it. When Arguimbau objected to Kosambi's condem-
nation of the proposed marriage, Kosambi quickly claimed that his
only objection to the relationship was economic. The man was in no
position to support a wife, and it would be irresponsible to get married. Kosambi later wrote that the student was "letting his family
down badly," and cited him as an example of Indians who "believe
they owe nothing to anyone," and whose "personal worries over-
shadow the miseries of millions from which they themselves make a
nice living." Here Kosambi suggested that his real objection was the
way that the marriage involved putting individual considerations
above family or country.[36]

When Jamshed Patel came to the United States, he met his future
wife, Eileen, an immigrant from Ireland, whom he married in 1951.
At MIT Patel had gotten involved in a sophisticated experimental
study of the properties of materials using the high-pressure experimental apparatus of Harvard Nobel Laureate Percy Bridgman,
equipment not available in India. After he left MIT, Patel stayed in
the United States, working for a series of companies performing
studies of the characteristics of semiconductors. This work put Patel
on the leading edge of technology and brought him widespread recognition, but would have been impossible in India. Patel, who spent
most of his career at Bell Labs, embraced the new world open to
him, both technically and socially, and never looked back.[37]

By coming to MIT, Indians were entering the heart of the U.S.
technological system. And in the late 1950s and 1960s, that heart was
government, and often military, funding. And even though these students were not citizens, the degree to which Indians were integrated
into the military/government system is striking. A 1949 dissertation in
metallurgy was the first one mentioning funding by the U.S. military, in this case the Army Ordnance Department. Sixty-six Indians
earned engineering doctorates between 1959 and 1971. Of the sixty-
four dissertations I have been able to examine, fifty-four mention
a funding source. Fifty of those mention funding by a U.S. government agency, and of those, twenty-seven mention funding by an
agency of the military. Funders included NASA, the Atomic Energy Commission, as well as branches of the U.S. Army, Navy, and
Air Force.[38]

MIT students were involved in government projects in a variety of other ways. Avinash Singhal was a civil engineering student learning computer modeling methods at MIT, and during two consecutive summers in the early 1960s, he worked at Douglas Aircraft on issues related to the Saturn rocket, and then contributed to the design of the wing of the F-111 fighter aircraft. Anil Malhotra worked for a consulting company owned by professors, where he did modeling work for the Apollo spacecraft. Ashok Malhotra did research as part of a U.S. Air Force contract. Not being a United States citizen only became an issue when he was halted at the gate at Wright-Patterson Air Force Base as he came with a group to present the results. (He was eventually let in.)[39]

Although the United States maintained its discriminatory legislation, by the early 1960s, some graduates of MIT without American or European spouses were beginning to stay in the United States. Immigration law had given preference to those whose services were urgently needed in the United States. Now, with the increasing need for technical manpower in the post-Sputnik world, American universities and large corporations were willing to take some Indians on.

One of the first such cases was Thomas Kailath. After coming to MIT, Kailath began working in the new field of information theory, which had been essentially created by MIT professor Norbert Wiener and by Claude Shannon, who went to Bell Labs after earning his doctorate at MIT. In the following years, MIT continued to dominate the field. Kailath's work, funded by all three branches of the military, dealt with communications: specifically, receiving signals when an enemy was attempting to disrupt those signals, a topic of obvious military interest. Kailath was identified in his graduate career as someone with extraordinary potential, being invited in 1960 as the only graduate student to give a paper at one of the leading conferences in the field, the London Symposium on Information Theory. A Stanford professor saw him and targeted him as a possible hire. At the time Stanford, which had built up its electrical engineering program in many areas, was well behind MIT in information theory. The legendary Frederick Terman, then provost at Stanford, made a strong effort to recruit Kailath, personally spending an hour with him to explain Stanford's great future and why it was preferable to MIT. When Kailath turned down an offer of an assistant professorship, Terman offered to make him an associate professor. Kailath came to Stanford in 1963 and quickly fulfilled the potential people had seen in him, being promoted to full professor at Stanford in 1968.[40]

In the 1950s it had been rare for an Indian graduate of MIT to stay in the United States. In the first half of the 1960s, almost equal numbers of those who earned doctorates stayed as returned. By the second half of the decade it became uncommon to return. A number of factors were responsible for this change, with American firms and universities recruiting Indian engineers as never before.

In 1965 Lyndon Johnson signed into law an immigration reform bill, which eliminated the features of the previous law that had discriminated most blatantly against Indians. The new law set a quota of 20,000 for each country in the Eastern Hemisphere, and had additional provisions for accommodating family members. While MIT graduates may have found positions in the United States in any case, the bill changed the official tenor of the American attitude toward Indians: now they were welcome in the country. Immediately the number of professionals coming from India spiked. In 1965 fifty-four Indians immigrated to the United States in the technical and professional workers category, a number which rose to 1,750 the next year.[41]

The removal of the barriers to Indians staying in the United States brought out the fundamental contradiction facing Indians studying at MIT in the 1960s. They were being socialized into a technical and economic system that was increasingly incompatible with India's. American industry knew exactly what to do with these people; where they would go in Indian society was unclear. At MIT they worked in an environment where they had access to the best-equipped laboratories and were connected to teams of similarly minded researchers both at MIT and around the nation or around the world. In contrast, most had few professional contacts back in India, and oftentimes there was no similar work being done. A common belief was that success in India required a high degree of social capital, primarily in the currency of personal contacts to help navigate one's career. An MIT degree by itself was not seen as sufficient. Returning to India required a willingness to start anew, applying principles learned at MIT to the Indian environment. Momentum tended to keep them in the United States.[42]

The social dimension of this momentum can be seen in the career of Tapan Kumar Gupta. After finishing his doctorate from MIT, Tapan Kumar Gupta went back to Calcutta to visit his family. There he got a letter from the college roommate of his MIT dissertation adviser, now working as a manager at Westinghouse Research and Development, asking him to explore the possibilities of a job. Gupta did, and accepted a job.[43]

At the same time, if getting a job in America was the logical next step, Indians who stayed in the United States did so by an incremental process, not necessarily imagining that they would stay forever. But each commitment drew them further into American society and the American technological system, making leaving harder.

The 1950s and 1960s were the golden years of the American industrial research lab. Large vertically integrated companies, primarily based on the East Coast, flush with money and seemingly facing little competition from either America or abroad, were willing to invest large amounts of money on long-term research projects with little thought to immediate returns. The research lab provided a university-like environment both in its campus-like setting and in the freedom it offered to individual researchers in their work. While research work was creative, it was highly routinized, with each researcher given a specialized area of responsibility and a wide range of supporting resources. It provided engineering doctorates, whether Indian or American, a close approximation of the world they had occupied in graduate school. Very often upon receipt of their doctorates, Indians entered American industrial research labs, working at places such as Bell Labs, U.S. Steel, Westinghouse, Union Carbide, or IBM. An ancillary advantage to such a job was that the companies typically sponsored the worker's request for a green card, taking care of the legal expenses. As with Americans joining these labs, many spent their careers as researchers, publishing papers and gaining patents. A few joined the ranks of senior management.[44]

Praveen Chaudhari stands as an exemplar of the Indian in American industrial research, becoming one of the first Indians to occupy a senior technical managerial position in a large American corporation. Chaudhari, born in 1937, was the first bachelor's graduate of an Indian Institute of Technology to earn a doctorate from MIT. Many more would follow. His life might be best characterized as being one that did not recognize boundaries, a strategy that required a high degree of self-confidence to be executed successfully. Chaudhari's family was from the Punjab, where they had been wealthy landowners. (His grandfather had owned a village.) Chaudhari's father was an attorney who then moved into business, first selling army surplus goods and then owning a small steel mill. The younger Chaudhari went to La Martinere, an elite English-medium boarding school in Lucknow. From there he studied metallurgy at IIT Kharagpur. A summer position at the Tata Iron and Steel Works at Jamshedpur convinced him that he did not want permanent work there. After graduation, he considered taking a temporary position with a Scottish firm that would allow him to travel, but his father told him about MIT. The senior Chaudhari had a friend whose son had gone there. The junior Chaudhari wrote letters inquiring about admission to metallurgy professors at three schools: MIT, Cambridge, and Berkeley. The professors at both Berkeley and Cambridge welcomed Chaudhari but could provide no support. Only the MIT professor, John Chipman, offered funding. Chaudhari's family could afford to pay, but he wanted financial independence and he chose MIT.[45]

Chipman was a metallurgist, but Chaudhari entered MIT just at the time the metallurgy department was being transformed into a department of materials science. (It wouldn't be until 1970 that the department would make the formal name change.) Materials science was interdisciplinary, combining metallurgy, physics, chemistry, and other engineering fields. For his doctoral dissertation, which was supported by NASA, Chaudhari worked with a metallurgist, a physicist, and an electrical engineer. He married a Danish woman and spent a year working with the Danish Atomic Energy Commission. Upon completing his doctorate, he considered jobs with GE and RCA, before choosing IBM's facility in Yorktown Heights, New York.[46]

Later in his career a senior IBM manager told Chaudhari that he could not figure out Chaudhari's work at IBM, implicitly wondering what common themes held it together. Chaudhari denied that he was either a microscopist or a spectroscopist or even a scientist or an engineer. Chaudhari characterized himself as a problem solver who was willing to follow a particular problem in whatever direction it led. This strategy often involved moving into territory claimed by particular disciplines. In the 1970s, while Chaudhari had been working on amorphous semiconductors, he was asked to take over the management of IBM's bubble memory research. While bubble memory research never led directly to a product, Chaudhari and several other scientists discovered amorphous materials that led to the recordable compact disc. In the 1980s Chaudhari was appointed the director of science within IBM Research. The fact that Chaudhari, with a degree nominally in the field of metallurgy, was given responsibility for managing teams of scientists—physicists and chemists—represented a departure from the normal hierarchy, where scientific knowledge was seen as more fundamental to that held by engineers, and he could not have gotten that appointment without the implicit support of the IBM scientific community. Under Chaudhari's leadership IBM won two Nobel Prizes in Physics.[47]

Of the 305 Indians who earned doctorates in engineering from MIT in the twentieth century, only nine were women, with the first woman earning her doctorate in 1976. Women have been underrepresented in engineering in both the United States and India throughout the twentieth century. In the United States, the percent of women earning bachelor's degrees in engineering rose from less than 1 percent in 1970 to 10 percent in 1980, and then hit a plateau of 15 percent in 1987. In India less than 1 percent of graduating engineers were women in 1975, a number which grew to almost 9 percent by 1988.[48]

As might be expected, the case of the first woman to navigate both the Indian and American educational systems to earn a doctorate from MIT in engineering was an exceptional one. Uma Dalal was born in Bombay in 1947. Her father, Rasiklal Maneklal Dalal, was a member of the Visa Oswal Jain Gujarati business community. He had moved from Ahmedabad to Bombay and established a successful stock brokerage business. Fulfilling a wish of Mahatma Gandhi, he had donated money to build a public library in Ahmedabad. He was an associate of G. D. Birla and Kasturbhai Lalbhai, serving on the board of directors of their companies. Rasiklal put a priority on his children's education, sending a son to Case Institute of Technology in the United States to study mechanical engineering. Uma went to the elite Queen Mary High School in Bombay, and although her father tried to get her interested in medicine or fields seen as more appropriate to a woman, such as music, art, or dance, she developed an interest in math and science. She then went to Elphinstone College, the leading liberal arts college in Bombay, before earning her bachelors degree in physics from Bombay's Institute of Science in 1968.[49]

It became clear to Uma she was disadvantaged by being a woman in India and she sought opportunities elsewhere. As it had for many other people before, the United States Information Service Library in Bombay proved an important resource. There she got catalogs of American universities. She applied to a number of top American universities for graduate studies in physics, earning acceptance to Caltech with a fellowship. For a young Indian woman to go to the United States by herself was a departure from the normal conventions of Indian middle-class society, but her father allowed her to go.[50]

She arrived at Caltech seeking to do particle physics at a time when Caltech, with Nobel Laureates Richard Feynman and Murray Gell-Mann on its faculty, was one of the world's leading centers for physics. But at Caltech, Uma's interests moved away from the highly esoteric world of theoretical physics, as she met a professor there who encouraged her to consider work in materials science.

At the Institute of Science in Bombay she had developed a relationship with a chemistry student there, Vinay Chowdhry. Chowdhry had earned a master's degree from the Indian Institute of Technology in Bombay and then gone to the University of Michigan to pursue a doctorate. Dalal moved to Michigan to join him and they got married, but after a short time, they moved to Cambridge, with Uma pursuing a doctorate in materials science at MIT, and Vinay a doctorate in chemistry at Harvard. At MIT Uma concentrated on ceramics, working with Robert Coble.[51]

After receiving her doctorate, Uma took a position at the DuPont Company's Research Labs in Wilmington, Delaware. She never seriously considered returning to India, believing the opportunities for women in the United States to be far better. She worked on a variety of materials science–related areas, ranging from batteries to catalysts to ceramics for electronics. In 2006 she was appointed DuPont's chief scientist and technical officer.[52]

Reconnecting to India

Roughly 80 percent of the Indians who earned doctoral degrees in engineering from MIT between 1961 and 2000 stayed in the United States for their careers. The most common track for those who returned to India was to work at American research lab for a brief time before returning to India, usually one of the IITs or the Indian Institute of Science. These were the positions in India that most directly used the skills developed at MIT, even if these institutes could not match American research budgets. It was not uncommon for Indians who had taken up jobs in the United States to either consider returning to India or to actually return to India. In most cases, especially when they involved working for Indian businesses, such efforts met with frustration. In the 1970s Suhas Patil had discussions with people working in the computer division of DCM, a company originally known as Delhi Cloth Mills, but by this time a conglomerate. DCM wanted Patil to develop a clone of a popular Digital Equipment Corporation minicomputer. To Patil, this was insulting work, presupposing that he could do no better than copying. He also had discussions with the Tata group. He ended up staying in the United States. In 1972 Avinash Singhal, who had worked at defense contractors such as GE and TRW, went to Engineers India Limited, a large government-owned civil engineering firm. Singhal was hired to be the head of its systems division. He lasted only two years, before frustration with the position and the Indian bureaucracy drove him back to the comparative stability of the United States. Sanjay Amin had earned a doctorate in chemical engineering from MIT in 1975 and then worked for Upjohn. In 1979 he returned to India to work for a new venture of an Indian textile company to make dyestuffs. He saw

that one of his main tasks would be working with the Indian bureaucracy, and so in less than fifteen months, he quit and returned to his
job at Upjohn. Other people had similar experiences.[53]

A highly atypical route, both to and from MIT, was followed
by B. C. Jain, born in 1949. Jain, a Marwari and one of fourteen children, came from a small town in Rajasthan called Bijainagar, where
his father had a business as a cotton trader. He had gone to Birla Institute of Technology and Science (BITS) in Pilani, eight to ten hours
from his hometown, and graduated in 1971. BITS students at that
time were divided into "mustards" and "castors," based on the type of
oil they used on their hair. Mustards were typically from rural backgrounds and had been educated in Hindi-medium schools. Castors
were from urban backgrounds and were more often educated in
English-medium schools. Jain was a mustard. He was at BITS in the
midst of the MIT-Ford Foundation program of assistance and had
several courses with MIT professors. He applied to several American
schools and was accepted at MIT. He had gotten married just after
graduating from BITS and had planned to leave his wife in India as
he started out. No one from Jain's village had ever gone to the United
States before, and some men in the village warned his father about
the perils of the young man abandoning his wife while in America.
Subsequently Jain's father forbade Jain from going. Jain's brothers
then colluded in a plan to spirit Jain out of India without the father's
knowledge.[54]

Jain then completed a doctorate in mechanical engineering with
research on modeling engine behavior, funded by the U.S. Army and
the U.S. oil company Texaco. The oil shocks of 1973 had given Jain
a vision of developing renewable energy in India. Upon the comple-
tion of his dissertation in 1975, he decided to return to India. He later
described a scene where in the month before he, his wife, and
daughter were scheduled to leave to return to India, friends would
visit his apartment nightly, trying to convince Jain to stay in the
United States. They argued that if he would not do it for himself,
then he should at least do it for his daughter. They claimed that given
the low wages paid in India, he would have to make sacrifices. If he
gave his daughter milk, he wouldn't have enough money to pay for
fruits. Jain returned to India anyhow.

In his search for a suitable position, Jain found something remarkable: a technology-oriented company, working in renewable energy, led by an Indian graduate of MIT. Nanubhai Amin had graduated from MIT in 1942, having been there with Mansukhlal Parekh and being part of the group of Indians at MIT described in Chapter 4. Upon his return, he started a company, Jyoti Limited, which began under the umbrella of his family's chemically oriented company, Alembic. Jyoti was a medium-sized company that made electrical equipment such as motors, pumps, and switchgear.[55]

Amin had a technologist's fascination with new areas and he made Jain the head of a renewable energy group and, in 1979, established a research center in solar energy in the Gujarati village of Tandalja, on the outskirts of the city of Baroda. Two years later the *Times of India* published a lengthy profile on the work conducted under Jain's leadership. Jain had a team of forty engineers and scientists working on solar water heaters, solar cookers, and windmills. After nine years with the company, Jain left to start his own business, Ankur Scientific. Jain's strategy was to develop renewable energy technologies that would not be dependent on foreign technologies. With support from the governments of India and Gujarat, his company developed a series of biogas converters, having success on a small scale.[56]

Even among those who spent their careers in the United States, there were cases of significant backward linkages to India. In 1969–1970 Kailath, by this time a full professor at Stanford, spent a sabbatical year at the Indian Institute of Science in Bangalore. While he was there, his mentor at Poona Engineering College, S. V. C. Aiya, now a professor at the Indian Institute of Science, connected Kailath with officials from the Indian Air Force, who wanted his judgments on a proposal from an American contractor on a radar system. Kailath recommended that India develop a university-based research program similar to the Joint Services Project the U.S. military had with American universities. Ultimately the Indian government approved a program at the five Indian Institutes of Technology and the Indian Institute of Science, replicating the military-university connection that had been so strong at MIT and Stanford.[57]

In the 1980s the Indian government developed a rhetoric of "brain bank" to counter that of "brain drain." The claim was that

Indian engineers abroad were strategic assets that India could tap for its benefit. In 1985, during a state visit to the United States, Prime Minister Rajiv Gandhi, more technically oriented than his mother, announced a plan for India to select ten high technology areas as focal points for research. Gandhi envisioned recruiting Indian scientists and engineers working in the United States to take leading positions in these efforts. In the fall of 1988 Rajiv Gandhi and his advisers invited leading Indian-American scientists and engineers in the United States to come to India to help chart out a possible path of technology-driven economic development in India. This group included MIT graduates Praveen Chaudhari of IBM and Satya Atluri of Georgia Tech. The effect of this visit is difficult to determine. Gandhi already had a technology adviser with an American education and work experience. Would the American group have different counsel? Would they have more prestige? In spite of Gandhi's desire to recruit senior Indian technologists and scientists back to India, none of those on this trip went back, and one has to wonder how effective they would have been working in the Indian environment after having spent so many years in the United States.[58]

The Indian Entrepreneur

While in the 1950s and 1960s industrial research labs were the principal commercial habitat for those with doctorates in engineering, Indian or not, over time technological entrepreneurship became increasingly prominent, with Route 128 outside of Boston and Silicon Valley in California being the most widely known centers. By the 1980s and 1990s, Silicon Valley had developed as an ecosystem that had surpassed Boston's Route 128 in the opportunities it offered for technology entrepreneurs. The most successful Indian entrepreneurs would start their companies there.[59]

Thomas Kailath used his research done at Stanford, one of Silicon Valley's hubs, to start several firms as part of a career that was remarkable even for a Stanford electrical engineering professor. The first, Integrated Technology, began in 1980 as a contract research firm, but later sold software for computer-aided control systems.

Kailath later founded with a graduate student Numerical Technologies, a company that commercialized work they had done on using
mathematical techniques to break what had been a barrier to the
semiconductor industry's continued drive to reduce the minimum
feature size printed on integrated circuits. This continual reduction
of feature size, a crucial component of Moore's Law, played an important role in the industry's ability to produce electronics with
continually greater capabilities.[60]

Kailath's path to entrepreneurship was one not unusual for a professor at Stanford or MIT, but not necessarily reproducible by those
without his standing. The first Indian MIT doctorate to be widely
known as an entrepreneur and who helped to establish a path to Indian entrepreneurship in Silicon Valley was Suhas Patil. Patil, the
son of an executive at Tata Iron and Steel in Jamshedpur, developed
an interest in radio through his father, studying a manual produced
by the American Radio Relay League. Patil went to IIT Kharagpur,
where his interest in electronics developed further using American
textbooks and the classic series on electronics produced from the
MIT Radiation Laboratory. In 1963 a professor from the University
of Illinois, Donald Bitzer, had come to Kharagpur to install an IBM
1620 computer and to teach the students and faculty how to use it.
Bitzer had brought with him several transistors and Patil used them
in an early project to build a digital counter. Through Bitzer and the
IBM 1620, Patil received his introduction to computing. Patil graduated from Kharagpur in 1965, and although he got admission to the
University of Illinois and had planned to go there to work with Bitzer,
a later offer from MIT changed his mind.[61]

Now in the computer metropole at MIT, Patil continued his work
on computing. He became a part of Project MAC, MIT's pioneering
project on Machine Aided Cognition, where he worked on computer
architectures. Patil was invited to join the MIT faculty, and after
some years there and an abortive attempt to return to India, he moved
in 1975 to the University of Utah. At Utah, Patil's work on developing software to design integrated circuits was funded by the
electronics company General Instruments. Eventually General
Instruments offered Patil a contract to start a company to design

complex integrated circuits, and on that basis he started Patil Systems in 1981. In 1984 he reengineered the company, changing its name to Cirrus Logic and moving the company to Silicon Valley. There he got connected into Silicon Valley networks and hired an experienced CEO. In 1989 the company went public, earning a profit that year of $4 million.[62]

In 1993 Patil and a group of Indian entrepreneurs founded TiE, The Indus Entrepreneurs, as a way to mentor entrepreneurs. Whether intentional or not, it also helped to raise the profile of Indian entrepreneurs. Cirrus Logic was perhaps the most widely recognized firm headed by an Indian, but by this time a handful of Silicon Valley firms had been started by Indians, often IIT grads. In 1994 TiE held a seminar to promote entrepreneurship. Patil gave the keynote address, and other Indian entrepreneurs shared their experiences in starting new ventures. Five hundred people attended, of whom seventy-five were not Indian. One of the main goals of the organization was to develop Indian entrepreneurial networks.[63]

If Kailath, Chaudhari, Chowdhry, and Patil represented stages of the technological Indian in America, Vivek Ranadive, born in 1957, represented another: the more than technological Indian, the Silicon Valley entrepreneur who broke through, becoming both a wider technological evangelist and a celebrity, appearing not just in the pages of business or technical journals, but also in the *New Yorker* and *Esquire*. Ranadive had a lifelong gift for knowing where the action was and putting himself in the middle of it. Going to MIT as an undergraduate could be a sign of extraordinary privilege or the sign of extraordinary drive, and in Ranadive's case, it was primarily the latter. The son of a pilot for Indian Airlines, Ranadive had gone to a school in Bombay that had emphasized science and technology. He had listened to the Apollo moon landing on a radio in 1969 and decided at that point that he wanted to come to the United States. Later Ranadive saw a documentary on MIT and was fascinated by the idea of student research and resolved to go there. Although he started at IIT Bombay, Ranadive applied and got admission to MIT. At the time, India had severe currency restrictions on those seeking to do undergraduate study outside India, and Ranadive had to appeal directly to the head of the Reserve Bank of India and was only allowed to leave India with $50. While at MIT he supported himself through loans, scholarships, and his own entrepreneurial efforts. He started a small job shop, taking advantage of the cheap pool of labor he had access to: MIT students. He searched the *Boston Globe* want ads for small businesses seeking to hire computer programmers. He then contacted them and offered to meet their requirements by hiring an MIT student on a part-time basis.[64]

In 1986, after several jobs and a Harvard MBA, Ranadive started a company: Teknekron Software Systems. It was not a true independent start-up but operated under the umbrella of a parent which itself had been founded in 1968 with the goal of incubating new businesses. The parent's philosophy was to have new companies develop a niche product and expand from there. Ranadive was charismatic, ready to take risks, both technical and personal. When he began his company he called up Hewlett-Packard and asked to speak to its founder, Dave Packard. When the secretary put him through, he introduced himself to the legendary figure and asked if he had any advice. Packard spoke to him for a few minutes and gave him some contacts.[65]

Technology entrepreneurs face the challenge of positioning their ventures between an appealing technological vision and a profitable product. Many entrepreneurs have visions with a technological logic to them, but which are ahead of their time. On the other hand, a profitable business without a technological vision risks being quickly overtaken by new technologies. Ranadive had a technological vision, but he was able to trim that vision to what was doable in the 1980s. Ranadive's big idea, which he called "the information bus," was to have an interface for various types of software that would be as simple and modular as the interface for hardware. His ideas to provide fast access to information offered the biggest payoffs in the financial industry, and Ranadive targeted it, developing a system that would give Wall Street traders integrated access to information from a wide variety of sources on a single workstation. His company won contracts with Fidelity, Salomon Brothers, Goldman Sachs, and others, coming to be widely known in the industry. In 1993 Reuters, a competitor, agreed to buy the company for $125 million.[66]

Ranadive stayed on with the company, which was reincarnated as TIBCO (The Information Bus Company), and continued to evangelize for his idea. The router manufacturer Cisco Systems was an investor and supporter of its products. Ranadive became a Silicon Valley personality, in 1999 writing a book, *The Power of Now*, enunciating his philosophy of the power of real-time information and the information bus. Ranadive's renown expanded beyond technology in 2009 when the *New Yorker*'s Malcolm Gladwell chronicled his work as a middle school basketball coach. Ranadive, coaching his own daughter's team, developed a distinctive strategy, which enabled the team to achieve remarkable success, even though it lacked traditional basketball skill. Ranadive followed up by buying the Sacramento Kings, outmaneuvering Microsoft's Steve Ballmer and becoming a hero in Sacramento for his promise to keep the NBA team in the city.[67]

In the late 1980s, Suhas Sukhatme, professor of mechanical engineering at IIT Bombay, began a series of studies of India's "brain drain." Sukhatme's work showed that from 1973 to 1977, roughly 30 percent of those receiving undergraduate engineering degrees from IIT Bombay ended up in the United States. He also showed that in the 1980s, the overall number of engineers who left India for the United States per year was roughly 1,700, or 7 percent of the engineers India produced annually, a number he called "quite low."[68]

Sukhatme's study was in part motivated by a popular perception in India that almost all IIT graduates migrated to the United States. From that perspective, his work showed that the IITs contributed far more talent to the Indian economy than was commonly perceived. But even if 70 percent remained in India, for a system designed to produce engineers for India the loss of one-third of its input was a remarkable kind of engineering inefficiency. Sukhatme professed to be unconcerned about the absolute numbers of engineers who left India. What concerned him was India's loss of its most brilliant young people to the United States. He called this "the real brain drain" and proposed a number of measures to arrest it.

Ironically, his family embodied "the real brain drain." His father Pandurang Sukhatme, who came from a poor family in a village in

Maharashtra, had earned a doctorate in statistics from London University in 1936. He worked for the Indian Council of Agricultural Research before moving in 1951 to the Food and Agriculture Organization of the United Nations in Rome as the head of its statistical department. Suhas had earned his doctorate in mechanical engineering from MIT before returning to India. His younger brothers Vikas and Uday would come to MIT also, receiving bachelor's degrees with financial support from their father's employer, the United Nations. They would each go on to earn doctorates in physics, but unlike their brother, their careers would be in the United States, with brother Uday teaching physics at American universities before becoming a university administrator, while brother Vikas turned to medicine after his physics doctorate, earning his medical degree from Harvard, where he became a professor.[69]

In its efforts to build a technological nation, by the late 1960s the Government of India had developed a system of technical education that produced engineers trained to the highest standards. However, the design of this system offered a frictionless connection with the American system of graduate education and then the broader American technological system. To many newly trained Indian engineers, that system was more attractive than the Indian one. In adopting an American model of technical education, the most perfect example of that model would always be in the United States.

Kailath, Chaudhari, Chowdhry, and Patil are exceptional figures, largely unknown to the American public. In this history they stand in for other Indian graduates of MIT responsible for technological work that is less well known but fundamental to American society, ranging from pumps to gas liquefiers to methods of computation. They also stand in for the tens of thousands of other Indians who did not go to MIT, but made the decision to come to the United States for engineering training and then made their careers in America. From the mid-1960s onward, Indian engineering students, some from IITs, some from older engineering colleges, some from Regional Engineering Colleges, flocked to American graduate schools: MIT, Stanford, and Berkeley, as well as SUNY-Stony Brook, Purdue, and Ohio State, and lesser known schools like the

University of Wisconsin-Milwaukee. By the 1990s no one with any familiarity with technology could doubt that there was such a thing as the technological Indian, but these technological Indians, who had come to the United States, had undercut Jawaharlal Nehru's dreams of a technological India.

Conclusion

E VEN IF SUKHATME's and others' studies showed that only 20 percent to 30 percent of the students went to the United States in the 1980s and the 1990s, there was ambiguity at several levels about whether the IITs were Indian or American institutions. A history of IIT Bombay asserts that in the 1990s "standard dress code for students was T-shirts emblazoned with the crests of American universities, speaking of allegiances being forged long before the flights were taken." In 1994 a student at IIT Kanpur wrote of how the coming of the GRE, the examination necessary to gain admittance to American graduate schools, changed the entire tenor of the campus, with students so busy in their preparations that they stopped attending lectures. Even two former IIT directors questioned the institutions' place in India, writing in 1993 that the "IIT graduates are the only high tech products in which India is internationally competitive." They were forced to admit that there was no "satisfactory answer" to whether India was ready even in 1993 for an "MIT type" institution or if using MIT as the model for the IITs had been a wise choice.[1]

Ironically, they wrote at a time when the environment for the IIT graduate was changing dramatically. The IITs as they developed in the 1960s through the 1980s were based on a fundamental contradiction. They offered an autarkic India an education developed for an American society. For a significant part of the student body, the

303

resolution to this contradiction was to go to the United States, first for graduate school and then for employment. In 1991, acting under duress because of a foreign exchange crisis, the Indian government liberalized its economy, removing a host of regulations that had hobbled Indian businesses and kept foreign enterprises at bay.[2] The 1991 liberalization allowed for a wide array of new connections between India and the United States, with American firms able to operate in India with an ease that had never been possible before and with Indian firms able to make alliances with a range of foreign firms that had never before been possible. Gradually these connections opened up new job possibilities in India for students at the IITs (and other engineering schools). If the IITs had been exceptional in India for their implicit connection to the United States, that was no longer true in the 2000s. American technology and global capitalism had come to India in a variety of forms.

A quarter century on from liberalization, India's technological globalization continued apace, becoming so associated with 1991 that its much older origins are lost. Pune, whose complete absence of industrialization had been the subject of the laments of the editors of the *Mahratta* and the *Kesari* for years and years in the late nineteenth century, has built on the foundation laid by S. L. Kirloskar to become one of India's leading centers for automobile manufacture while also being (as are most large Indian cities) an important IT center. M. M. Kunte's 1886 dream of having factories covering "the Bhamburda and Chatursingi grounds," four miles from the Poona Central Railway Station, seen at the time as a "romantic" idea, has in 2015 proved far too modest, with the area (and miles and miles further) being gobbled up by industrial sprawl. Outside Pune, the Paisa Fund Glass Works, where MIT-educated Ishwar Das Varshnei and a group of Japanese workers began producing glass according to modern methods, today sits in the shadow of an enormous General Motors plant.[3]

While MIT graduates such as Anant Pandya, M. N. Dastur, Brahm Prakash, Lalit Kanodia, and F. C. Kohli had played central roles in the technological development of India in the two decades after independence, in the post-liberalization world MIT graduates no longer have that same role. In the years after independence, MIT graduates stood out in a country with relatively few engineers and fewer still with advanced training. Today India is awash with engineers, coming from the IITs, the Regional Engineering Colleges (now rebranded NITs, National Institutes of Technology), and private engineering colleges. One survey concluded that India produced 220,000 engineers or computer scientists with four-year degrees in 2005–2006, compared to roughly 130,000 produced by the United States at that time.[4]

Furthermore, although the first generation of postindependence MIT graduates came back to India for their careers, later generations did not. In the twenty-first century, IIT graduates are much more likely to stay in India than they had been previously.[5] However, the current career tracks of Indian MIT graduates are less clear. An analysis of Indians who earned engineering degrees of any type from MIT between 2001 and 2003 (the last years that commencement programs allow for identification of graduates' hometowns) shows that of those who could be identified, roughly 80 percent had careers in the United States in 2015.[6]

One of the effects of liberalization has been an increased flow of Indian students more generally to the United States. Today, a greatly enlarged Indian middle class, with the capacity to foot the bill themselves, has created a demand for high-quality technical education that India has not been able to meet. While in 1991, the U.S.-based Institute of International Education counted 32,000 Indians studying in the United States, by 2001, that number had grown to 54,000 and then by 2013 to 103,000. A 2014 Brookings Institution study showed that a large percentage of Indians came to study what are called in the United States the STEM fields (science, technology, engineering, and mathematics), with 70 percent of the Indian students studying in these areas compared to 48 percent of the domestic American student population. And for Indians, STEM overwhelmingly means information technology or engineering, with 80 percent of Indian STEM students studying in those areas.[7]

Liberalization may have opened up new employment opportunities in India, but many Indian students clearly come to the United States with the expectation of using their education to land a job in America. The U.S. H1B Visa program, which allows high tech workers to work in the United States temporarily, is dominated by Indians, who often hope to use the H1B Visa as a stepping stone to permanent residency status in the United States.[8]

A group of Indians has continued to find a home, both technological and physical, in the United States, and their advance up the American technological hierarchy has continued apace. While in the early 1960s, Thomas Kailath was one of very few Indian engineers in what would become known as Silicon Valley, in 1999 AnnaLee Saxenian showed that Indian and Chinese entrepreneurs ran one-quarter of the firms there.[9] Although graduates of MIT represent a minute fraction of the Indian engineers in America, a glance at their careers gives a hint of how central Indians have become to the American technological system. Three individuals—Praveen Chaudhari, Rakesh Agrawal, and Thomas Kailath—have been given the nation's highest technical honor, the National Medal of Technology or Science. In 2010 President Barack Obama named Subra Suresh, a graduate of IIT Madras with an MIT doctorate, as head of the National Science Foundation, whose annual budget of $7 billion plays an important role in shaping the direction of American scientific and engineering research. Three years later Suresh went on to accept the presidency of Carnegie Mellon University, one of America's premier technological institutions. Thousands of other Indians occupy positions of major responsibility throughout the American technological infrastructure.

The children of those Indians who came to the United States have been a new type of technological Indian, the second-generation Indian American. The first American-born person of Indian descent to attend MIT was Amar Bose, whose father had immigrated to the United States in 1920. Bose graduated from MIT in 1952, joined the MIT faculty, and later started an audio equipment company. In the years since 1965, as the number of Indian Americans living in the United States has increased, reaching 3.2 million in 2010, so have the number of Indian Americans at MIT, reflecting a continuing preference for engineering careers. The best-known representative of this group is Salman Khan, the creator of virtual learning environment Khan Academy. Khan, the son of a Bangladeshi father and an Indian mother, grew up in Louisiana. After earning three degrees from MIT, getting a Harvard MBA, and working at a hedge fund, he transformed his series of online math lessons for his cousins into an entire online learning system, accessed by millions of people around the world. Khan is one of the second-generation Indians whose career has linked him back to the countries of his parents. But for all the talk of Khan Academy's democratizing effects, Khan's software is used far more in the United States than India.[10]

MIT and American systems of technical education have had a particularly strong attractive power in Asia. The story of how India looked to MIT has parallels in Japan and China and sheds light on how each country sought to industrialize and meet the challenge of the West. As Meiji Japan sought to modernize its society, it sent officials throughout the world looking to examine institutions in industrial societies, and MIT was one of those places. In 1874 Japan provided MIT's first foreign graduate, Aechirau Hongma, who then worked in Japan as a government engineer.[11] China similarly looked to the United States and to MIT. In the five years prior to World War I, Chinese students made up the largest foreign contingent at MIT, even outnumbering Canadians. In 1999 Chinese Premier Zhu Rongji spoke at MIT, noting that China's Tsinghua University was known, when he studied there (as it still is today), as "the MIT of China." He said that during his days at Tsinghua, students used copies of MIT textbooks. Rongji also recalled that while he was at Tsinghua, he wondered if he might ever be able to study at MIT.[12]

Even today, the three largest contingents of graduate students at MIT come from China, India, and Korea, respectively, with Taiwan and Singapore sending delegations that outnumber the major European countries. This pattern is replicated on a national scale, with China, India, and South Korea being the three largest senders of international students to the United States and with seven of the top-ten sending countries being from Asia.[13]

And the effects of these students are felt in the United States as well as Asia. From the 1960s onward, Asian students have been an important factor in providing the students that make many American engineering schools viable. Today, the many Asian students who pay their own tuition are responsible for a significant transfer of money to the United States. And through the Indians, Chinese, and others who have stayed, the last fifty years have seen a significant Asianization of the American technical workforce.[14]

MIT and the Mahatma

The empire of technology was another empire that Gandhi fought against, but this was a struggle he lost. The decisive struggle over India's future is often portrayed as between Gandhi and Nehru. But we might instead imagine critical points being in T. M. Shah's, Bal Kalelkar's, and the other young Gandhians' choice of MIT and modern technology over the Gandhian ideal of the charkha used in village service. Even if he spoke of a personal industry, they found the appeal of another kind of industry more attractive.

This book has told the story of remarkable developments over nearly a century and a half that have made India a participant in the world of high technology and brought Indians to the pinnacle of American technology and industry. But that is not the whole story. It is apposite to give Mahatma Gandhi the last words, reminding us of the costs of a global system of technology-based capitalism, often hidden.

The people in this book are a tiny elite, and while the years since independence have seen the remarkable growth of an Indian middle class, human development statistics arrestingly show that the India of high tech engineers, shopping malls, and consumer society does not represent the entire country. When the Indian child receives an average of 4.4 years of schooling, when the adult literacy rate is 74 percent, when up to half of India's children are malnourished, the limited ability of the technological Indian to transform India is all too clear. Although the technological vision of Gandhi has long seemed archaic, in a country where half of the population defecates out of doors, a man who had an obsession with the proper disposal of waste surely still has some relevance.[15]

Although it would be overly simplistic to suppose that there were binary opposite choices, the MIT-based model points to a pyramidal structure, suggesting that the focus should be on the top. Gandhi attacked that model, which he said had an "apex sustained by the bottom." He contrasted this with a model of an "oceanic circle."[16]

His Wardha scheme of seven years of compulsory education, using craft skills as the base on which to build other areas of instruction, such as mathematics, history, or geography, represented a different incarnation of MIT's motto "Mens et Manus"—Mind and Hand. If the establishment of the IITs is seen as one of Nehru's achievements, his inability to establish a system of universal education must be seen as a great failure.[17]

Gandhi's language and (at times) his actions regarding technology could seem hyperbolic and grossly exaggerated. In *Hind Swaraj*, he had written, "I cannot recall a single good point in connection with machinery," and claimed that machinery represented a "great sin."[18] The global system of technology-based capitalism that MIT graduates participated in was capable of producing remarkable technological developments, generating fabulous wealth, and increasing the standard of living of many people. But that system was also capable of exacting horrific costs, frequently incurred by those not enjoying its benefits. At those times, Gandhi's words could seem prescient.

Throughout this book, Indians who studied at MIT have served as a divining rod, pointing to important sites of technological activity. While they have pointed to the birth of the Indian IT industry, they also point to the worst industrial accident in world history. On the night of December 2/3, 1984, methyl isocyanate gas leaked from the tanks of Union Carbide's money-losing Bhopal pesticide plant, which had been subjected to safety-compromising cost-cutting. The disaster, which so overwhelmed the Indian public health and legal infrastructures that no definitive toll has ever been established, killed 3,800 immediately and perhaps 10,000 more in the following days, with estimates of those permanently injured ranging past 100,000. Most of those killed lived in the slums near the plant. In the early decades after Indian independence, American companies doing business in India not infrequently hired MIT graduates for key positions in their Indian operations, and this was particularly true for Union Carbide. The manager of the plant, Jagannathan Mukund, whose father had been director of the Reserve Bank of India, had earned a master's degree from MIT. Union Carbide India's nonexecutive chairman was Keshub Mahindra, an executive at Mahindra & Mahindra who had attended MIT during World War II.[19]

But the actions of another MIT graduate stood as the clearest statement of how the intersection of capitalism, nationalism, and technology could play out in the late twentieth century. Ashok Kalelkar, born in 1943, was the nephew of Bal Kalelkar, and had started college in the United States while his father, an Indian diplomat, was posted in Washington. The younger Kalelkar had earned bachelor's and master's degrees at MIT before receiving a doctorate at Brown University. He married an American woman and spent most of his career with the Cambridge-based consulting firm Arthur D. Little, eventually rising to the position of executive vice president.[20]

In the wake of the Bhopal disaster, Union Carbide hired Arthur D. Little to conduct an investigation of the incident, which Kalelkar led. In matters big or small, consulting firms have often been used to build consensus around a predetermined result desired by their client through reports packaged in the language of objective, technical analysis. Union Carbide's hiring of Arthur D. Little was an effort to enlist American technical credibility on its side against an India unable to martial the same resources.

The Arthur D. Little report asserted that the leak was caused by employee sabotage, a conclusion exculpating Union Carbide. The sabotage theory rendered all other possible contributory factors irrelevant, such as the complications of operating a technology designed for one culture in a completely different culture, or the numerous documented safety lapses. However, the limits to the ability of even a prestigious firm like Arthur D. Little to completely reinterpret the events at Bhopal are suggested by the fact that its report faced skepticism even from the pro-business *Wall Street Journal*.[21]

A further implicit Gandhian critique of MIT was that in contrast to his holism, MIT produced engineers narrowly defined, socialized to work within limits set by an American technological system and American-based capitalism. In these realms, they could see opportunities and exploit them, but outside those realms their vision and power were often much weaker. The careers of the followers of Gandhi who went to MIT in the 1930s and 1940s are particularly poignant, because their experience at MIT opened up new possibilities, but ones that they were largely unable to combine with their association with Gandhi in any generative way. By the 1990s it was a system that put engineers on a path to working in profitable companies, but not to maximize the social returns to India.

There were exceptions. Almitra Patel and Deep Joshi stand as two examples of something approaching Gandhi's citizen-engineer, working outside either the state or the capitalist system. Patel, who as Almitra Sidwha, had been the first Indian women to earn an engineering degree from MIT, continued to work for her family business, Grindwell. However, after her family bought a farm outside Bangalore, their tranquil life was interrupted in 1991 when the municipality of Bangalore began dumping waste alongside the roads to her farm, attracting packs of dogs. Patel became a garbage activist. In 1994, a year when there was a flood-induced plague outbreak in Surat, she and a colleague, J. S. Velu, conducted an all India "clean cities" campaign, driving from city to city in a van, inspecting city dumps, and meeting with city officials on best practices for disposing of wastes while seeking to win publicity for the cause. In 1996 she filed a public interest litigation with the Indian Supreme Court asking that they require India's 300 largest cities to follow set guidelines in solid waste management. In 1998 the Supreme Court appointed a committee, including Patel, to investigate India's solid waste management practices. Their work continues.[22]

Deep Joshi, born in 1947, grew up in the small village of Puriyag in Uttarkhand, one of seven children of farmers. He went to Motilal Nehru Regional Engineering College, in Allahabad, where after earning his degree in mechanical engineering he stayed on as a lecturer. In the early 1970s he won a government scholarship enabling him to study at MIT, where he was expected to earn a doctorate in engineering and return to his college to teach. At MIT Joshi lost interest in the purely technical aspects of engineering and veered toward economics and systems, earning dual master's in mechanical engineering and management with a thesis on nutrition in Guatemala. On his return to India, he worked briefly with Jashwant Krishnayya, an MIT graduate who had mentored Narayana Murthy at his Systems Research Institute in Poona. While on a field assignment in rural Maharashtra, Joshi met two Indian doctors, trained at Johns Hopkins, whose humility and engagement with the people gave him a model for how professionals could work in development. With support first from the Ford Foundation and then other foundations and trusts in India, Joshi went on to form Pradan, an NGO that recruits professionals from schools like the IITs or the IIMs to work in villages to alleviate poverty. Pradan emphasizes the combination of professional skills with empathy—head and heart. Joshi argued that development work was not intellectually less challenging than more high technology work. Pradan's approach of sending workers into villages had resonances with Gandhi's constructive program. In 2009 Joshi was given the Magsaysay Award, sometimes called Asia's Nobel Prize. By 2014 Pradan had grown to almost 400 professionals working with over 270,000 people.[23]

The campaigns of Patel and Joshi have a poignancy to them. Standing outside both the state and capitalism, the forces that both were able to enlist in their causes paled before the magnitude of the challenges they sought to address. Of course, this poignancy echoes Gandhi's efforts to create a system of spinning that would stand outside both the state and capitalism.

M. M. Kunte's hymn of praise to the "art of mechanization" that began this book was an early voice in a chorus that would include more and more people around the world in the succeeding years. But on that May day in 1884, Kunte did not have the last word. After he spoke, M. G. Ranade rose to respond. Ranade, a great and pioneering figure in the Indian nationalist movement, combined a career as a judge in the colonial legal system with efforts to reform Indian society and promote its economic development. Ranade asserted that he was just as eager as Kunte for the spread of machines everywhere. But he denied that machines were the prime cause of "national development," pointing to the fact that India once knew machinery better than the Europeans and that the Portuguese "had not known the art of machinery" when they seized control of western India. Ranade concluded that Kunte's claim "that everything and anything is possible due to promoting the art of machinery is not only wrong but dangerous and full of exaggeration." We, as technological people of varied nationalities, still struggle to grasp that fact today.[24]

NOTES

ACKNOWLEDGMENTS

INDEX

Notes

Abbreviations

CWMG: *The Collected Works of Mahatma Gandhi*, New Delhi, Publications Division Government of India, 1999, 98 volumes. http://www.gandhiserve .org/e/cwmg/cwmg.htm

Introduction

1. "Kunte Vs. Ranade: Or the Wrestling of Two Giants," *Kesari*, June 3, 1884 (in Marathi), translation by S. H. Atre. Kunte's obituary is given in "The Late Mr. M. M. Kunte," *Mahratta*, October 14, 1888. The word "technology" only came to have its current meaning as the useful arts or machines in the early twentieth century. In the nineteenth century, technology meant the discipline of the study of the useful arts or machines. Thus Massachusetts Institute of Technology was not an institute of machines, but an institute for the study of machines. Eric Schatzberg, "Technik Comes to America: Changing Meanings of Technology before 1930," *Technology and Culture* 47, no. 3 (2006): 486–512.

2. "Kunte Vs. Ranade." A taluka is an Indian administrative unit, larger than a city or a town, but smaller than a district.

3. Ibid.

4. Detailing India's technical traditions is beyond the scope of this work, but a sense of these traditions can be gotten through Vibha Tripathi, *Iron Technology and Its Legacy in India* (New Delhi: Rupa, 2008); Giorgio Riello and Tirthankar Roy, *How India Clothed the World: The World of South Asian Textiles, 1500–1850* (Leiden: Brill, 2009); Sven Beckert, *Empire of Cotton: A Global History* (New York: Alfred A. Knopf, 2014), 3–82. Carlyle used the term in his 1829 essay "Signs of the Times," which is reprinted in Thomas Carlyle, *Critical and Miscellaneous Essays*, vol. 2 (London: Chapman and Hall, 1899), 56–82.

5. "Professor M. M. Kunte at Hirabag," *Mahratta*, May 31, 1885. The value for U.S. pig iron production is from *Historical Statistics of the United States, Earliest Times to the Present: Millennial Edition* (New York: Cambridge University Press, 2006), Table Db73.

6. "Sir Richard Temple on Education in India," *Journal of the National Indian Association*, January 1881, 38.

7. "14 Lakh Students Take JEE Exam across 150 Centers in India," *DNA*, April 6, 2014, http://www.dnaindia.com/academy/report-14-lakh-students-take-jee-exam-across-150-centers-in-india-1975800; Sarika Malhotra, "The Dream Factories," *Business Today*, May 12, 2013, http://businesstoday.intoday.in/story/kota-coaching-institutes-brand-iit/1/194170.html.

8. NASSCOM, "Indian IT-BPM Overview," http://www.nasscom.in/indian-itbpo-industry; Patrick Thibodeau, "In a symbolic shift, IBM's India workforce likely exceeds U.S.," *Computer World*, November 29, 2012, http://www.computerworld.com/s/article/9234101/In_a_symbolic_shift_IBM_s_India_workforce_likely_exceeds_U.S.?pageNumber=1; AnnaLee Saxenian, *Silicon Valley's New Immigrant Entrepreneurs* (San Francisco: Public Policy Institute of California, 1999).

9. Raj Varadarajan, interview by author, April 14, 2014, telephone.

10. Before 1980 the title for the commencement programs is *Massachusetts Institute of Technology, Graduation Exercises*. From 1981 on, the title is *Massachusetts Institute of Technology, Commencement Exercises*. These are supplemented by a *Degrees Awarded* document that gives the names of those whose degrees were awarded at times other than the main graduation.

11. My thinking about this study has been shaped by the work of Judith M. Brown, *Windows into the Past: Life Histories and the Historian of South Asia* (Notre Dame, IN: University of Notre Dame Press, 2009); Judith M. Brown "'Life Histories' and the History of Modern South Asia." *The American Historical Review* 114 (June 2009): 587–595.

12. Classic works that discuss nineteenth-century Poona include Stanley A. Wolpert, *Tilak and Gokhale: Revolution and Reform in the Making of Modern India* (Berkeley: University of California Press, 1961); Anil Seal, *The Emergence of Indian Nationalism: Competition and Collaboration in the Later Nineteenth Century* (Cambridge: Cambridge University Press, 1968); Richard I. Cashman, *The Myth of the Lokamanya: Tilak and Mass Politics in Maharashtra* (Berkeley: University of California Press, 1975). Two books that approach economic nationalism in different ways are Bipan Chandra, *The Rise and Growth of Economic Nationalism in India: Economic Policies of Indian National Leadership, 1880–1905*, rev. ed. (New Delhi: Har-Anand Publications, 2010) and Manu Goswami, *Producing India: From Colonial Economy to National Space* (Chicago: University of Chicago Press, 2004).

13. Shahid Amin, "Gandhi as Mahatma: Gorakhpur District, Eastern UP, 1921–1922," in *Selected Subaltern Studies*, ed. Ranajit Guha and Gayatri Spivak (Oxford: Oxford University Press, 1988), 342. Some of the works that emphasize the way Gandhi could speak in a variety of different languages or be seen in a variety of different ways include Lloyd I. Rudolph and Susanne Hoeber Rudolph, *The Modernity of Tradition: Political Development in India* (Chicago: University of Chicago Press, 1967); Benjamin Zachariah, *Developing India: An Intellectual and Social History* (New Delhi: Oxford University Press, 2005), 156–162; Sumit Sarkar, *Modern India: 1885–1947* (Madras: Macmillan, 1983), 181.

14. Goswami, *Producing India*; Gyan Prakash, *Another Reason: Science and the Imagination of Modern India* (Princeton, NJ: Princeton University Press, 1998); David Ludden, "India's Development Regime," in *Colonialism and Culture*, ed. Nicholas Dirks (Ann Arbor: University of Michigan Press, 1992), 247–288.

15. Claude Markovits, "The Tata Paradox," in *Merchants, Traders, and Entrepreneurs: Indian Business in the Colonial Era* (Ranikhet: Permanent Black, 2008), 152–166; Amartya Sen, "The Indian Identity," in *The Argumentative Indian: Writings on Indian History, Culture, and Identity* (New York: Farrar, Straus and Giroux, 2005), 334–356. I am greatly indebted to the work of Dwijendra Tripathi, which has laid a foundation for the study of Indian business history. See especially his *Oxford History of Indian Business* (New Delhi: Oxford University Press, 2004).

16. Vinay Lal, *The Other Indians: A Political and Cultural History of South Asians in America* (Noida: Harper Collins, 2008); Ronald Takaki, *Strangers from a Different Shore: A History of Asian Americans* (Boston: Little, Brown, 1989).

17. Dennis Kux, *India and the United States: Estranged Democracies, 1941–1991* (Washington, DC: National Defense University Press, 1992); H. W. Brands, *India and the United States: The Cold Peace* (Boston: Twayne Publishers, 1990).

18. Charles Maier, *Among Empires: American Ascendancy and Its Predecessors* (Cambridge, MA: Harvard University Press, 2006), 33. My thinking has also been shaped by John Krige, *American Hegemony and the Postwar Reconstruction of Science in Europe* (Cambridge, MA: MIT Press, 2006) and Paul A. Kramer, "Power and Connection: Imperial Histories of the United States in the World," *American Historical Review* 116 (January 2011): 1348–1391.

19. Thomas P. Hughes, *American Genesis: A Century of Invention and Technological Enthusiasm* (New York: Viking, 1989), 249–352; Krige, *American Hegemony and the Postwar Reconstruction of Science*. A different perspective is provided by Michael Adas, *Dominance by Design: Technological Imperatives and America's Civilizing Mission* (Cambridge, MA: Harvard University Press, 2006).

20. Douglas E. Haynes, *Small Town Capitalism in Western India* (Cambridge: Cambridge University Press, 2012); Tirthankar Roy, *Traditional Industry in the Economy of Colonial India* (Cambridge: Cambridge University Press, 1999); Abigail McGowan, *Crafting the Nation in Colonial India* (New York: Palgrave Macmillan, 2009).

21. Mokshagundam Visvesvaraya, *Memoirs of My Working Life* (Bangalore: M. Visvesvaraya, 1951); *Technical and Industrial Education in Bombay: Final Report of the Committee Appointed by Government, 1921–1922* (Bombay: Government Central Press, 1923), Shelfmark V/12763, Asian and African Studies Collection, British Library, London; "Technical Study: A Divided Report," *Times of India*, July 13, 1922. Prakash Tandon describes his father's career as an irrigation engineer in his *Punjabi Century* (London: Chatto and Windus, 1961). Daniel Klingensmith traces the postcolonial history of this civil engineering tradition in his *"One Valley and a Thousand": Dams, Nationalism, and Development* (New Delhi: Oxford University Press, 2007).

1   The Indian Discovery of America

1. "Model Institute of Technology," *Mahratta*, May 11, 1884; "Model Institute of Technology No. II," *Mahratta*, May 25, 1884; "Model Institute of Technology No. III," *Mahratta*, June 1, 1884.

2. "Technical Education," *Mahratta*, July 27, 1884.

3. Michael Adas, *Machines as the Measure of Men: Science, Technology, and Ideologies of Western Dominance* (Ithaca, NY: Cornell University Press, 1989).

4. Dietmar Rothermund, *An Economic History of India: From Pre-Colonial Times to 1991*, 2nd ed. (London: Routledge, 1993), 19–49; Anil Seal, *The Emergence of Indian Nationalism: Competition and Collaboration in the Later Nineteenth Century* (Cambridge: Cambridge University Press, 1968), 1–130; Suresh Chandra Ghosh, *The History of Education in Modern India, 1757–1998*, rev. ed. (Hyderabad: Orient Longman, 2000), 6–91; Daniel R. Headrick, *The Tentacles of Progress: Technology Transfer in the Age of Imperialism, 1850–1940* (New York: Oxford University Press, 1988); David Gilmartin, "Scientific Empire and Imperial Science: Colonialism and Irrigation Technology in the Indus Basin," *Journal of Asian Studies* 53 (November 1994): 1127–1149; David Arnold, *Science, Technology and Medicine in Colonial India* (Cambridge: Cambridge University Press, 2000), 92–128.

5. Bipan Chandra, *The Rise and Growth of Economic Nationalism in India: Economic Policies of Indian National Leadership, 1880–1905*, rev. ed. (New Delhi: Har-Anand Publications, 2010); Manu Goswami, *Producing India: From Colonial Economy to National Space* (Chicago: University of Chicago Press, 2004).

6. C. A. Bayly, *The Birth of the Modern World, 1780–1914* (Malden, MA: Blackwell, 2004); Emily S. Rosenberg, ed., *A World Connecting, 1870–1945* (Cambridge, MA: Harvard University Press, 2012); Jürgen Osterhammel, *The Transformation of the World: A Global History of the Nineteenth Century*, trans. Patrick Camiller (Princeton, NJ: Princeton University Press, 2014); Ronald Findlay and Kevin H. O'Rourke, *Power and Plenty: Trade, War, and the World Economy in the Second Millennium* (Princeton, NJ: Princeton University Press, 2007); Elleke Boehmer, "Global and Textual Webs in an Age of Transnational Capitalism; Or, What Isn't New about Empire," *Postcolonial Studies* 7, no. 1 (2004): 11–26. I take the term informational environment from Richard John, "Recasting the Information Infrastructure for the Industrial Age," in *A Nation Transformed by Information: How Information Has Shaped the United States from Colonial Times to the Present*, ed. Alfred D. Chandler, Jr. and James W. Cortada (New York: Oxford University Press, 2000), 55–105.

7. Balkrishna Govind Gokhale, *Poona in the Eighteenth Century: An Urban History* (Delhi: Oxford University Press, 1988); *Gazetteer of the Bombay Presidency*, vol. 18, part 3 (Poona), (Bombay: Government Central Press, 1885), 266 (population), 313 (craftsmen); Juland Danvers, *Report to the Secretary of State for India in Council on Railways in India for the Year 1862–1863* (London: Her Majesty's Stationary Office, 1863), 23–27; Seal, *Emergence of Indian Nationalism*, 80–81.

8. Gordon Johnson, "Chitpavan Brahmins and Politics in Western India in the Late Nineteenth and Early Twentieth Centuries, in *Elites in South Asia*,

ed. Edmund Leach and S. N. Mukherjee (Cambridge: Cambridge University Press, 1970), 95–118; Richard I. Cashman, *The Myth of the Lokamanya: Tilak and Mass Politics in Maharashtra* (Berkeley: University of California Press, 1975); B. R. Nanda, *Gokhale: The Indian Moderates and the British Raj* (Princeton, NJ: Princeton University Press, 1977); Madhukar J. Jadhav, *The Work of the Sarvajanik Sabha in Bombay Presidency* (Poona: Dastane Ramchandra & Co., 1997).

9. Cashman, *The Myth of the Lokamanya*, 56. Biographical studies of Tilak include N. C. Kelkar, *The Life and Times of Lokamanya Tilak*, trans. D. V. Divekar (New Delhi: Anupama Publications, 1987); Ram Gopal, *Lokamanya Tilak: A Biography* (London: Asia Publishing House, 1956); Stanley A. Wolpert, *Tilak and Gokhale: Revolution and Reform in the Making of Modern India* (Berkeley: University of California Press, 1961).

10. Kelkar, *The Life and Times of Lokamanya Tilak*, 1–115; Gopal, *Lokamanya Tilak*, 1–31; Wolpert, *Tilak and Gokhale*, 13–21.

11. *Gazetteer of the Bombay Presidency*, vol. 18, part 3, 64–65; *Report on Native Papers Published in the Bombay Presidency*, January 6, 1906, 6. The precise role that Tilak played in the papers is not always clear. The papers had no masthead and all its articles and editorials were unsigned. In 1902, the *Mahratta* ran pictures of nine "past and present editors," which included Tilak. The editor most associated with industrial development was M. B. Namjoshi, but the tone and content of the paper did not noticeably change after he died. I am assuming that even if Tilak was not the author of most of the articles and editorials in the *Mahratta*, he was at least in broad sympathy with the sentiments expressed in it. "Past and Present Editors of the Mahratta," *Mahratta*, January 5, 1902. Namjoshi's obituary is given in "The Late Rao Saheb M. B. Namjoshi," *Mahratta*, January 19, 1896.

12. "Our Contemporaries," *Mahratta*, January 2, 1881. Isabel Hofmeyr examines Gandhi's South African editorial career in the context of newspaper exchanges in *Gandhi's Printing Press: Experiments in Slow Reading* (Cambridge, MA: Harvard University Press, 2013).

13. "What the World Says," *Mahratta*, January 2, 1881; "Editorial Notes," *Mahratta*, June 26, 1881.

14. "What the World Tells Us," *Mahratta*, January 23, 1881; "Editorial Notes," *Mahratta*, June 26, 1881; "Editorial Notes," *Mahratta*, April 23, 1882; "Sir W. W. Hunter and the Starving Millions of India," *Mahratta*, June 14, 1891; "Why Some Confectioners Do Not Make Money," *Mahratta*, May 8, 1881. American periodicals were available in India in a variety of ways other than their being sent from an American contact. The Bombay merchant Cooper Madon offered subscriptions to such periodicals as *Scientific American*, *Harper's*, *Scribner's*, or *American Agriculturist*, which could be sent directly from the United States. Bombay also had several lending libraries that would send material out to mofussil areas. In 1889 the Bombay Circulating Library advertised that it had 37 copies of *Scientific American* and 146 copies of *Harper's Monthly Magazine*. It advertised having 1,680 "upcountry" subscribers, which included 224 Hindus, 6 Muslims, and 111 Parsis. "Cooper Madon and Company, Limited, Newspapers and Periodicals," *Bombay Gazette*, May 3, 1886;

"Bombay Circulating Library," *Times of India*, June 19, 1889. Statements about the periodicals cited in the *Mahratta* are based on an examination of every issue published between 1881 and 1912.

15. Various perspectives on nineteenth-century British decline are provided by Martin J. Wiener, *English Culture and the Decline of Industrial Spirit* (Cambridge: Cambridge University Press, 1981); Michael Sanderson, *Education and Economic Decline in Britain, 1870 to the 1990s* (Cambridge: Cambridge University Press, 1999); David Landes, *The Unbound Prometheus: Technological Change and Industrial Development in Western Europe from 1750 to the Present*, 2nd ed. (Cambridge: Cambridge University Press, 2003), 326–358; David Edgerton, *Science, Technology, and British Industrial "Decline," 1870–1970* (Cambridge: Cambridge University Press, 1996); Nicholas Crafts, "Long-Run Growth," in *The Cambridge Economic History of Modern Britain*, vol. 2, *Economic Maturity, 1860–1939*, ed. Roderick Floud and Paul Johnson (Cambridge: Cambridge University Press, 2004), 11–24.

16. "What the World Says," *Mahratta*, January 2, 1881. Although the *Mahratta* attributed the quote to the *Contemporary Review*, I have not been able to find the precise reference in that periodical

17. "The Gokulas of the New World," *Mahratta*, January 9, 1881.

18. "A Bengali in New York," *Mahratta*, November 25, 1883; "A Bengali in New York," *Indian Mirror*, November 16, 1883. This article was reprinted in the Anglo-Indian newspapers, the *Times of India* and the *Bombay Gazette*.

19. "What the World Says," *Mahratta*, May 6, 1883.

20. "Editorial Notes," *Mahratta*, July 17, 1881.

21. "Select Telegrams," *Mahratta*, January 2, 1881; "News and Notes," *Mahratta*, January 15, 1893; "The Indian Tanning Industry," *Mahratta*, November 9, 1902; "News and Notes," *Mahratta*, January 29, 1893; "What the World Says," *Mahratta*, January 7, 1883; "News and Notes," *Mahratta*, December 2, 1894; "Brass and Copper Wares," *Mahratta*, July 22, 1888. In 1883 the Mahratta quoted a Lahore newspaper, "Among the common objects of the bazaar may now be reckoned the empty kerosine tin to be found in every shop, and leading itself to a variety of strange uses." The paper went on to note that the import of kerosene from the United States had been one-fifth the then-current level five years previously. One might well imagine that the tins were the blue Standard Oil tins. "What the World Says," *Mahratta*, September 16, 1883.

22. "Our System of Education: A Defect and a Cure," *Mahratta*, May 15, 1881.

23. "A Few Suggestions to the Young Gaikwar," *Mahratta*, October 30, 1881.

24. "Editorial Notes," *Mahratta*, February 1, 1885.

25. "Mr. Tata's Student Fund," *Mahratta*, March 15, 1891.

26. "Literary Activity in Western India," *Mahratta*, November 2, 1884.

27. "Photophone," *Mahratta*, January 16, 1881; A. Graham Bell, "Selenium and the Photophone," *Nature* 22 (September 23, 1880): 500–503; "The Photophone," *New York Times*, August 30, 1880.

28. "What the World Says," *Mahratta*, October 1, 1882; "What the World Tells Us," *Mahratta*, January 23, 1881; "News and Notes," *Mahratta*, October 30, 1892. In 1889 the *Mahratta* noted that Queen Victoria congratulated Edison on his "grand inventions" by means of a phonograph recording. "Stray Notes," *Mahratta*, October 6, 1889.

29. "Death of M. Pasteur," *Mahratta*, October 6, 1895. The phrase "science has no country" is often attributed to Pasteur without documentation. For one instance of Pasteur referring to it, see "The Pasteur Institute," *Nature* 40 (July 18, 1889): 278.

30. "Editorial Notes," *Mahratta*, January 1, 1882.

31. "Electric Manufactures," *Mahratta*, March 23, 1890.

32. "Will the Natives Be More Enterprising?" *Mahratta*, February 13, 1881.

33. "Industrialism," *Mahratta*, June 5, 1881.

34. "Foreign Travel," *Mahratta*, September 7, 1890; "Maritime Activity of the Ancient Hindus," *Mahratta*, September 21, 1890; "Crossing the Kali Pani," *Mahratta*, December 20, 1891; "Sea Voyage Sanctioned," *Mahratta*, February 26, 1893.

35. "Baboo Amrit Lal Roy and the Boot-Black of New York," *Mahratta*, February 3, 1889.

36. Ibid.

37. "Why Some Confectioners Do Not Make Money," *Mahratta*, May 8, 1881; "Why Some Confectioners Do Not Make Money," *Scientific American*, April 2, 1881, 217. Gunther's approach to modernizing candy sales in drugstores is suggested in "Gunther's Goods for the Drug Trade," *The Pharmaceutical Era* 21 (May 11, 1899): 636.

38. On the development of the Bombay textile industry, see Rajnarayan Chandavarkar, *The Origins of Industrial Capitalism in India: Business Strategies and the Working Classes in Bombay, 1900–1940* (Cambridge: Cambridge University Press, 1994); S. D. Mehta, *The Cotton Mills of India, 1854–1954* (Bombay: Textile Association, 1954).

39. "The Late Mr. M. M. Kunte," *Mahratta*, October 14, 1888. A brief biographical sketch of Kunte is given in Veena Naregal, *Language, Politics, Elites, and the Public Sphere* (New Delhi: Permanent Black, 2001), 228–229.

40. "The Summer Series of Lectures at Poona," *Mahratta*, June 1, 1884; "Editorial Notes," *Mahratta*, October 30, 1887; "Professor M. M. Kunte at Hirabag," *Mahratta*, May 31, 1885.

41. "Editorial Notes," *Mahratta*, June 6, 1886.

42. "Professor M. M. Kunte at Hirabag," *Mahratta*, May 31, 1885; "Economic Reform," *Mahratta*, August 2, 1885.

43. "The Poona Exhibition 1888," *Mahratta*, October 7, 1888. David Arnold discusses nineteenth-century Indian exhibitions in *Everyday Technology: Machines and the Making of India's Modernity* (Chicago: University of Chicago Press, 2013), 24–28.

44. The Poona Exhibition 1888," *Mahratta*, October 7, 1888; "The Poona Exhibition," *Mahratta*, October 21, 1888.

45. "The Poona Exhibition of Native Arts and Manufactures," *Mahratta*, July 1, 1888; "The Poona Exhibition," *Mahratta*, October 7, 1888.

46. "The Poona Exhibition 1888," *Mahratta*, October 7, 1888. William Armstrong was a leading English inventor and industrialist.

47. "The Industrial Association of Western India," *Mahratta*, August 3, 1890; "The Industrial Conference," *Mahratta*, May 18, 1890; "The Industrial Conference," *Mahratta*, August 30, 1891.

48. Mirror, "The Poona Industrial Conferences," *Times of India*, September 14, 1892.

49. David Gostling, "The Poona Industrial Conference and Its Results," *Times of India*, September 16, 1892.

50. "Second Industrial Conference, Poona," *Mahratta*, September 11, 1892; "Stray Jottings," *Mahratta*, September 17, 1893.

51. "The Need of an Industrial School," *Kesari*, April 29, 1884 (in Marathi), translation by S. H. Atre.

52. "Model Institute of Technology," *Mahratta*, May 11, 1884.

53. Samuel C. Prescott, *When MIT Was "Boston Tech"* (Cambridge, MA: Technology Press, 1954), 3–44; Julius A. Stratton and Loretta H. Mannix, *Mind and Hand: The Birth of MIT* (Cambridge, MA: MIT Press, 2005); A. J. Angulo, *William Barton Rogers and the Idea of MIT* (Baltimore: Johns Hopkins University Press, 2009).

54. Charles Eliot, "The New Education: Its Organization," *The Atlantic Monthly*, February 1869, 203–221.

55. Stratton and Mannix, *Mind and Hand*, 113–138, 277–294.

56. Monte A. Calvert, *The Mechanical Engineer in America: Professional Cultures in Conflict* (Baltimore: Johns Hopkins University Press, 1967).

57. Quoted in Prescott, *When MIT Was "Boston Tech,"* 332.

58. "Model Institute of Technology, No. III," *Mahratta*, June 1, 1884.

59. Ibid.

60. Ibid.

61. K. V. Mital, *History of Thomason College of Engineering (1847–1949) on Which Is Founded the University of Roorkee* (Roorkee: University of Roorkee, 1996), 1–31; Deepak Kumar, *Science and the Raj: A Study of British India*, 2nd ed. (New Delhi: Oxford University Press, 2006), 136–148; Headrick, *The Tentacles of Progress*, 315–319.

62. "How to Introduce National Technical Education in India?" *Mahratta*, February 28, 1886 ; "Editorial Notes," *Mahratta*, October 10, 1886; Kumar, *Science and the Raj*, 141.

63. Chandra, *The Rise and Growth of Economic Nationalism in India*, 76–81.

64. Supporters of the Ripon Memorial were unable to raise sufficient funds to establish a technical institute in his name and the money was later folded into that raised for the Victoria Jubilee Technical Institute. "The Ripon Memorial," *Times of India*, April 26, 1886; "Editorial," *Times of India*, August 5, 1886; "The Address of Lord Reay at the Opening of the International Conference on Education," *Mahratta*, June 7, 1885.

65. "Victoria Technical Institute," *Times of India*, January 20, 1887; "Technical Education: Something Will Now Be Done," *Mahratta*, January 23, 1887.

66. "Opening of Victoria Jubilee Technical Institute," *Times of India*, April 12, 1889; "The Opening of the Victoria Jubilee Technical Institute," *Mahratta*, April 14, 1889.

67. Kelkar, *Life and Times of Lokamanya Tilak*, 87. Additional information about Bhat comes from "Keshar Melhar Bhat," Pathfinder File, AC 94, Office of the Registrar, MIT Institute Archives and Special Collections, Cambridge, MA. MIT's records sometimes identified Bhat as Keshar, while he was otherwise known as Keshav.

68. Jadhav, *The Work of the Sarvajanik Sabha in Bombay Presidency*, 92–94; "Editorial Suggestions," *Kesari*, September 13, 1884 (in Marathi), translation by S. H. Atre.

69. Keshav Melhar Bhat, "The Need for Industrial Education in India," *New Ideal* 3 (July-August 1890): 335–336; "A Public Meeting in Poona," *Mahratta*, April 5, 1896; "Report and List of Passengers Taken on Board the Steamer Batavia, July 3, 1882," Passenger Manifests, Commonwealth of Massachusetts Archives, Boston, MA.

70. Bhat, "Need for Industrial Education," 335–336.

71. Ibid.; "Editorial Suggestions," *Kesari*, September 13, 1884; Massachusetts Institute of Technology, *Eighteenth Annual Catalogue, 1882–1883* (Boston: W. J. Schofield, 1882), 18; Massachusetts Institute of Technology, *President's Report, December 12, 1883* (Boston: J. S. Cushing & Co., 1884), 7. Cochrane's background is given in Deloraine Pendre Corey, *Malden Past and Present* (Malden, MA: Malden Mirror, 1899), 69. Cochrane, who was from a Scottish family, went abroad for three years to receive an unspecified business training.

72. Theodore Chateau, "Critical and Historical Notes concerning the Production of Adrianople or Turkey Red and the Theory of This Colour," *The Textile Colourist* 1 (1876): 172–231. The history of Turkey red is also told in Sarah Lowengard, *The Creation of Color in 18th Century Europe* (New York: Columbia University Press, 2006), Gutenberg ebook, http://www.gutenberg-e.org/lowengard/index.html; Anthony S. Travis, *The Rainbow Makers: The Origins of the Synthetic Dyestuffs Industry in Western Europe* (Bethlehem, PA: Lehigh University Press, 1993), 163–203.

73. "Editorial Notes," *Mahratta*, September 14, 1884.

74. "Editorial Notes," *Mahratta*, April 17, 1887.

75. Ibid.; "The Year 1887," *Mahratta*, January 1, 1888. *Swadeshi* is a Sanskrit term meaning "of one's own country." In the late nineteenth century and early twentieth century, it came to mean goods made in India by Indians as part of a technological nationalist campaign. Amartya Sen, "The Indian Identity," in *The Argumentative Indian: Writings on Indian History, Culture, and Identity* (New York: Farrar, Straus and Giroux, 2005), 337.

76. "The Indo-American Dye-House Company Limited," *Mahratta*, September 29, 1889; "Local Notes," *Mahratta*, December 1, 1889.

77. Bhat, "The Need for Industrial Education in India," 332–335.

78. Ibid., 335; "A Public Meeting in Poona," *Mahratta*, April 5, 1896.

79. "Local Notes," *Mahratta*, May 10, 1891.

80. *Bombay Gazette*, June 24 1885; "Art Industries of India," *Mahratta*, June 28, 1885.

81. Industry and Commonsense, "The Industrial Arts," *Bombay Gazette*, June 26, 1885. An example of complaints about American medical degrees can be seen in "Medical Diplomas," *Times of India*, June 7, 1883, and *Times of India*, May 17, 1890.

82. "A Public Meeting in Poona," *Mahratta*, April 5, 1896.

83. "News from the Classes," *Technology Review* 1 (April 1899): 246.

84. Ibid., 246–247.

85. "The Queen's Memorial," *Mahratta*, February 10, 1901. Deepak Kumar discusses this incident in the context of J. N. Tata's offer to fund an institute for scientific research in *Science and the Raj*, 202–205.

86. "The Position of Mahomedans," *Pioneer*, December 13, 1900.

87. "The Queen's Memorial," *Mahratta*, February 10, 1901.

88. "India and Queen Victoria," *Times of London*, February 7, 1901; "Asiatic Society of Bengal," in *Lord Curzon in India* (London: MacMillan and Co., 1906), 526–547.

89. "Technical Education in England," *Mahratta*, November 16, 1902; "Mr. Balfour on Technical Education at Manchester," *Nature* 66, no. 1721 (October 23 1902): 633–634; Myron Pierce, "The Institute and the Commonwealth," *Technology Review* 5 (April 1903): 178.

90. "Technical Education in England," *Mahratta*, November 16, 1902. Gorst's original comments were made in John Gorst, "The Education Bill," *The Nineteenth Century and After: A Monthly Review* 52 (October 1902): 576.

2   American-Made Swadeshi

1. Upendra Nath Roy, "Engineering Education in America," *Modern Review* 11 (May 1912): 481–484.

2. Frank Harris's biography, *Jamsetji Nusserwanji Tata: A Chronicle of His Life*, 2nd ed. (Bombay: Blackie & Son, 1958), has not been surpassed. Other sources include R. M. Lala, *For the Love of India: The Life and Times of Jamsetji Tata* (New Delhi: Penguin, 2004); Dwijendra Tripathi and Makrand Mehta, *Business Houses in Western India: A Study in Entrepreneurial Response, 1850–1956* (New Delhi: South Asia Publications, 1990), 55–75.

3. Eckehard Kulke, *The Parsees in India: A Minority as Agent of Social Change* (New Delhi: Vikas Publishing House, 1978), 32–34; David L. White, *Competition and Collaboration: Parsi Merchants and the English East India Company in 18th Century India* (New Delhi: Munshiram Manoharlal Publishers, 1995); Michael H. Fisher, *Counterflows to Colonialism: Indian Travelers and Settlers in Britain 1600–1857* (Delhi: Permanent Black, 2004, 339–343

4. Harris, *Tata*, 1–28.

5. Ibid., 23–46; Gita Piramal and Margaret Herdeck, *India's Industrialists*, rev. ed. (Boulder, CO: Lynne Rienner Publishers, 1986), 304–306.

6. Harris, *Tata*, 134 (1900 Paris), 29 (1878 Paris); Daniel Headrick, *Tentacles of Progress: Technology Transfer in an Age of Imperialism* (New York: Oxford University Press, 1988), 287 (Chicago, Dusseldorf).

7. Jamsetjee N. Tata, "The Mill Industry," *Times of India*, August 14, 1886; "Editorial Notes," *Mahratta*, August 22, 1886.

8. Jamsetjee N. Tata, *Memorandum Respecting the Growth of Egyptian Cotton in India* (Bombay: Bombay Gazette Steam Printing Works, 1896).

9. "Artesian Wells in India," *Times of India*, December 11, 1899.

10. Harris, *Tata*, 117; "Mr. Tata's Student Fund," *Mahratta*, March 15, 1891; "J. N. Tata's Scheme for Higher Education," *Times of India*, January 29, 1892.

11. Harris, *Tata*, 113–132; "Mr. Tata's Scheme to Establish a Teaching University in India," *Mahratta*, October 2, 1898; "Mr. J. N. Tata's Educational Scheme," *Mahratta*, October 9, 1898; "An Indian Institute of Scientific Research and the Marvelous Munificence of Mr. Tata," *Mahratta*, October 16, 1898; "Mr. Tata's University Scheme," *Times of India*, February 15, 1899. Tata's proposal is discussed in more detail in Deepak Kumar, *Science and the Raj: A Study of British India*, 2nd ed. (New Delhi: Oxford University Press, 2006), 202–206.

12. "Lord Curzon in Bombay," *Times of India*, January 2, 1899.

13. Harris, *Tata*, 142–145; "The Indian Institute of Science," *Modern Review* 13 (January 1913): 114; *Report of the Quinquennial Reviewing Committee of the Indian Institute of Science, Bangalore* (Delhi: Manager of Publication, 1947); "Working of Indian Institute of Science," *Times of India*, July 18, 1936.

14. Vibha Tripathi, *Iron Technology and Its Legacy in India* (New Delhi: Rupa, 2008); Harris, *Tata*, 146–150.

15. Harris, *Tata*, 146–158; R. M. Lala, *The Romance of Tata Steel* (New Delhi: Penguin Viking, 2007), 3–8.

16. "Bombay Merchant Sees the President," *Washington Times*, October 14, 1902; "Society in Washington," *New York Times*, October 15, 1902.

17. Harris, *Tata*, 159–167.

18. Ibid., 165–167; James Howard Bridge, *The Inside Story of Carnegie Steel: A Romance of Millions* (New York: Aldine, 1903), 161.

19. Harris, *Tata*, 166–167. Kennedy's career is discussed in Quentin R. Skrabec, *The Carnegie Boys: The Lieutenants of Andrew Carnegie That Changed America* (Jefferson, NC: McFarland & Co., 2012), 62–63, 213, 221.

20. Harris, *Tata*, 165–166.

21. Lala, *The Romance of Tata Steel*, 10–17; Harris, *Tata*, 169–190; "The Tata Iron Works," *Modern Review* 2 (September 1907): 314–315; "Tata Iron Company: Capital Oversubscribed," *Times of India*, August 17, 1907; "The Tata Iron Works: Progress of the Company," *Times of India*, February 27, 1908.

22. Harris, *Tata*, 191–218; Lala, *The Romance of Tata Steel*, 17–28; "True Swadeshi in India," *The Times*, October 28, 1907; "India's Industrial Growth," *Los Angeles Times*, August 10, 1914; "East India Pig Iron to the United States," *Industrial World*, June 23, 1913, x. Given the small amount of pig iron made in India, it seems unlikely that exports to the United States continued.

23. B. L. Mitter, "The Tata Works and Qualified Indians," *Modern Review* 13 (May 1913): 579; "More Europeans for Jamshedpur." *Modern Review* 36 (September 1924): 354; "State Help Again for Tata Iron and Steel Works," *Modern*

Review 37 (January 1925): 121–122; Lala, *The Romance of Tata Steel*, 68–69. In a study that came to my attention too late to be integrated into this book, Aparajith Ramnath shows the important role that American-trained Indians played at Jamshedpur in the 1920s and 1930s. Aparajith Ramnath, "Engineers in India: Industrialisation, Indianisation, and the State, 1900–1947" (PhD dissertation, Imperial College, London, 2012), 166–179.

24. *Kal*, December 20 and December 28, 1901, cited in *Report on Native Newspapers Published in the Bombay Presidency*, January 11, 1902, 27 (hereafter *RNN Bombay*); *Bombay Samachar*, December 25, 1903, cited in *RNN Bombay*, December 26, 1903, 28–29; *Kesari*, August 2, 1904, cited in *RNN Bombay*, August 6, 1904, 29–30.

25. *Kesari*, August 2, 1904, cited in *RNN Bombay*, August 6, 1904; *Mahratta*, January 20, 1907.

26. *The Paisa Fund Silver Jubilee Number* (Poona: Paisa Fund, 1935), 16–19.

27. T. D. Varshnayi [*sic*], "The Talegaon Glass Works," *Mahratta*, March 12, 1911; Ganesh Sinha, "An Indian Student in America," *Indian People*, January 15, 1905; "Ishwar Das Varshnei," Pathfinder File, AC94, Office of the Register, MIT Institute Archives. Information on Varshnei's arrival in the United States is from Ancestry.com's online database. SS *Coptic*, August 29, 1904, *California, Passenger and Crew Lists, 1882–1959*, Ancestry.com.

28. "Visit of American Members of Society of Chemical Industry to Great Britain," *Electrochemical and Metallurgical Industry* 3 (July 1905): 250–251; Varshnayi [*sic*], "The Talegaon Glass Works."

29. "Late Lala Ishwar Das Varshnei," *All India Glass Manufacturers Bulletin* 14 (1969): xxx; "News from the Classes," *Technology Review* 9 (January 1907): 137.

30. Varshnayi [*sic*], "The Talegaon Glass Works."

31. Ibid.; *Paisa Fund Silver Jubilee Number*, 20–21.

32. *Paisa Fund Silver Jubilee Number*, 21–22; *Kesari*, August 18, 1908, cited in *RNN Bombay*, August 22, 1908, 42; *Kesari*, March 10, 1908, cited in *RNN Bombay*, March 14, 1908, 47; *Mahratta*, June 12, 1910, cited in *RNN Bombay*, June 11, 1910, 17.

33. *Kesari*, August 18, 1908, cited in *RNN Bombay*, August 22, 1908, 42; *Vande Mataram*, January 31, 1909, cited in *RNN Bombay*, February 6, 1909, 19; *Kesari*, September 6, 1910, cited in *RNN Bombay*, September 10, 1910, 23; *Kesari*, March 7, 1911, cited in *RNN Bombay*, March 11, 1911, 8; *Evening Dispatch*, June 21, 1911, cited in *RNN Bombay*, June 24, 1911, 26; *Vinod*, November 22, 1913, cited in *RNN Bombay*, November 29, 1913, 21.

34. Quoted in *Paisa Fund Silver Jubilee Number*, 21–22.

35. *Paisa Fund Silver Jubilee Number*, 23–24; G. P. Ogale, "Glass Industry in India," *Journal of the American Ceramic Society* 5, no. 11 (1922): 295–298.

36. *The Leader*, October 18, 1916; "U. P. Glass Works Bhajoi: His Honour's Visit," *The Leader*, November 26, 1919.

37. "The U.P. Glass Works Limited," *The Leader*, August 12, 1921.

38. *The Leader*, October 18, 1916; "Glass Manufacture in India," *National Glass Budget*, January 26, 1918, 13; "U. P. Glass Works Bhajoi: His Honour's Visit," *The Leader*, November 26, 1919; "The U. P. Glass Works Limited," *The Leader*, August 12, 1921.

39. U.P. Glass Works Bhajoi, *An Appeal Regarding the Urgent Need of Protection to Window Glass Industry* (Bhajoi: U.P. Glass Works, 1935). Document in author's possession. I thank Atul Varshnei for providing me with this document.

40. Dwijendra Tripathi and Makrand Mehta, *Business Houses in Western India: A Study in Entrepreneurial Response, 1850–1956* (New Delhi: South Asia Publications, 1990), 131–146; S. L. Kirloskar, *Cactus and Roses: An Autobiography* (Pune: Macmillan, 2003), 5–9. S. L. Kirloskar gives the 45 rupee a month salary figure, while Tripathi gives the salary as 35 rupees a month.

41. Tripathi and Mehta, *Business Houses in Western India*, 132–133; Kirloskar, *Cactus and Roses*, 7; "Life History of Laxmanrao Kirloskar," *Times of India*, December 6, 1951.

42. Kirloskar, *Cactus and Roses*, 8–9; "Life History of Laxmanrao Kirloskar," *Times of India*, December 6, 1951; "Inventions and Designs," *Times of India*, May 20, 1897.

43. Kirloskar, *Cactus and Roses*, 9, 12. The centrality of foreign journals to Laxmanrao is also affirmed in "Life History of Laxmanrao Kirloskar," *Times of India*, December 6, 1951.

44. Kirloskar, *Cactus and Roses*, 10–16; Tripathi and Mehta, *Business Houses in Western India*, 133–135.

45. "Deccan Agriculture," *Times of India*, August 25, 1909; "Deccan Agriculture," *Times of India*, March 30, 1910. The general question of agricultural improvement in India is discussed in Kumar, *Science and the Raj*, 206–208.

46. Harold A. Mann, "Manufacture of Agricultural Implements in Western India," *Indian Industries and Power* 9 (April 1912): 291–293.

47. Kirloskar, *Cactus and Roses*, 62–63; *Kesari*, September 22, 1908 (translation by Ishwari Kulkarni).

48. Visitor, "The Kirloskar Plough Factory," *Journal of the Mysore Agricultural and Experimental Union* 5 (1923): 19–26.

49. Kirloskar Brothers, untitled Golden Jubilee Brochure (n.p.: Kirloskar Brothers, n.d.), but clearly published in 1961, in author's possession. I thank Mukund Kirloskar for providing me a copy of this brochure.

50. Quoted in "India's Appreciation," *Atlanta Constitution*, October 12, 1927. The editor of the magazine sent a copy to the *Atlanta Constitution* (and one imagines other American newspapers) with "a view to bringing about better acquaintance between our two mutual countries."

51. "Albert Temple of Science," *The Statesman and Friend of India*, January 1, 1878; Pramatha Nath Bose, "Technical and Scientific Education in Bengal," in *Essays and Lectures on the Industrial Development of India and Other Indian Subjects* (Calcutta: W. Newman & Co., 1906), 59–78; Sumit Sarkar, *The Swadeshi Movement in Bengal, 1903–1908* (New Delhi: People's Pub. House, 1973), 109–112. For Bose's biography, see B. P. Radhakrishna, "Pramatha Nath Bose (1855–1934)," *Current Science* 72 (February 10, 1997): 222–224.

52. Sarkar, *The Swadeshi Movement*, 112–114; "Current Comments," *Indian People*, March 16, 1904.

53. Sarkar, *The Swadeshi Movement*, 112–114; "Association for the Advancement of Scientific and Industrial Education of Indians," *Bengalee*, May 5, 1904; "Current Comments," *Indian People*.

54. "Association for the Advancement of Scientific and Industrial Education," *Bengalee*, August 2, 1904; Sarkar, *The Swadeshi Movement*, 113–114; Kumar, *Science and the Raj*, 210–213.

55. Purushottam Nagar, *Lala Lajpat Rai: The Man and His Ideas* (New Delhi: Manohar, 1977).

56. "Association for the Advancement of Scientific and Industrial Education of Indians," *Bengalee*, May 5, 1904.

57. Ibid.

58. The development of complex systems of technology has been a major theme in nineteenth-century American history. See, for example, Thomas P. Hughes, *American Genesis: A Century of Invention and Technological Enthusiasm* (New York: Viking, 1989); Alfred D. Chandler, *The Visible Hand: The Managerial Revolution in American Business* (Cambridge: MA, Harvard University Press, 1977); Louis Galambos, "Technology, Political Economy, and Professionalization: Central Themes of the Organizational Synthesis," *Business History Review* 57 (Winter 1983): 471–493.

59. Sarkar, *The Swadeshi Movement*.

60. "Our Circulation," *Modern Review* 23 (March 1918): 323; R. N. Tagore, "Legumes as Nitrogen Gatherers," *Modern Review* 4 (August 1908): 144–148; D. C. Ahuja and S. Ghosh, "Electricity in the Byproduct Coke Industry of the Tata Iron Works at Jamshedpur," *Modern Review* 37 (March 1925): 312–318.

61. "Association for the Advancement of Scientific and Industrial Education of Indians," *Bengalee*, April 14, 1908; Sarkar, *The Swadeshi Movement*, 114; "Scientific and Industrial Education of Indians," *Bengalee*, September 4, 1904; "Reception of the Scholars of the Indian Industrial and Scientific Association at the Kalighat Temple," *Bengalee*, March 6, 1906.

62. "Association for the Advancement of Scientific and Industrial Education," *Indian People*, October 13, 1904; Keshav Melhar Bhat, "The Need for Industrial Education in India," *New Ideal* 3 (July-August 1890): 332–337; "Value of Technical Institutes," *Mahratta*, June 26, 1904.

63. Tessa Morris-Suzuki, *The Technological Transformation of Japan: From the Seventeenth Century to the Twenty-first Century* (Cambridge: Cambridge University Press, 1994), 71–104; Andrew Gordon, *A Modern History of Japan: From Tokugawa Times to the Present*, 2nd ed. (New York: Oxford University Press, 2009), 93–137.

64. "Japan as a Training School for Indians," *Mahratta*, July 19, 1896; "India House in Japan," *Mahratta*, November 1, 1908; "Studies in Japan," *Mahratta*, December 9, 1900. In the wake of its defeat in the Sino-Japanese War, China looked broadly to Japan as a model for industrialization. Joyman Lee, "Where Imperialism Could Not Reach: Chinese Industrial Policy and Japan, 1900–1930," *Enterprise and Society* 15 (December 2014): 655–71.

65. Rathindranath Tagore, *On the Edges of Time* (Calcutta: Visva-Bharti, 1958), 65–66.

66. "Japan and Indian Students," *Mahratta*, February 9, 1902; Suresh C. Banerje, "Technical Education of Indians: A Few Reflections," *Modern Review* 5 (March 1909): 243–246.

67. *Kesari*, October 21, 1913, cited in *RNN Bombay*, October 25, 1913, 25; "Indian Students Trained Abroad," *Modern Review* 9 (January 1911): 107–109.

68. Sarangadhar Das, "Why Must We Emigrate to the United States of America?" *Modern Review* 10 (July 1911): 69–80; Sarangadhar Das, "Information for Students Intending to Come to the Pacific Coast of the United States," *Modern Review* 10 (December 1911): 602–612.

69. Das, "Why Must We Emigrate," 74.

70. Ibid., 77–78.

71. Ibid., 79–80.

72. Ibid., 77; Krishna Dutta and Andrew Robinson, eds., *Selected Letters of Rabindranath Tagore* (Cambridge: Cambridge University Press, 1997), 422–423.

73. Sarangadhar Das, "Information for Students Intending to Come to the Pacific Coast of the United States," *Modern Review* 10 (December 1911): 602–612.

74. "Information Worth Remembering," *Bulletin of the Hindusthan Association of USA*, August 1913, 6; "Parts from the Annual Report of the General Secretary, December 1912–December 1913," *The Hindusthanee Student* 1 (January 1914): 24–25; "Report of the Eastern Section," *The Hindusthanee Student* 1 (January 1914): 25; "Report of the Western Section," *The Hindusthanee Student* 1 (January 1914), 26; "The Second Annual Convention News," *The Hindusthanee Student* 1 (January 1914): 22.

75. "Some of the Honorary Members of the Hindusthan Association of America," *The Hindusthanee Student* 2 (November 1915): 3; "America's Welcome to Hindusthanee Students," *The Hindusthanee Student* 1 (January 1914): 19.

76. "Dadabhai Naoroji's Message to Our Association," *The Hindusthanee Student* 1 (January 1914): 17; "Some of the Honorary Members of the Hindusthan Association of America," *The Hindusthanee Student* 1 (January 1914): 3; "Parts from the Annual Report of the General Secretary, December 1912–December 1913," *The Hindusthanee Student* 1 (January 1914): 24–25; Krishna Dutta and Andrew Robinson, *Rabindranath Tagore: The Myriad-Minded Man* (London: Bloomsbury Publishing, 1995), 147, 171–173.

77. Dutta and Robinson, *Tagore*, 120–121, 145–149; Rabindranath Tagore, "Our First Duty," *The Hindusthanee Student* 1 (January 1914), 7.

78. "Foreign Students in the United States," *Modern Review* 12 (November 1912): 558; "Foreign Students in the United States," *Los Angeles Times*, October 20, 1912. Paul Kramer provides an introduction to the historical issues involved in international students in the United States in "Is the World Our Campus?: International Students and American Global Power in the Long Twentieth Century," *Diplomatic History* 33 (November 2009): 775–806.

79. Liel Leibovitz and Matthew I. Miller. *Fortunate Sons: The 120 Chinese Boys Who Came to America, Went to School, and Revolutionized an Ancient Civilization* (New York: W. W. Norton, 2011); Stacey Bieler, *"Patriots" or "Traitors"? A History of American-Educated Chinese Students* (Armonk, NY: M. E. Sharpe, 2004), 24–51, 389.

80. "China to Send 2,000 Students to America," *Modern Review* 5 (February 1909): 171–172; Satish Chandra Basu, "China's Student Pilgrims," *Modern Review* 8 (July 1910): 51.

81. The data on Chinese students' majors is from Bieler, *"Patriots" or "Traitors,"* 375–376; Massachusetts Institute of Technology, *Reports of the President and Treasurer for the Year 1919–1920* (Cambridge, MA: Technology Press, 1921), 35.

82. Samuel C. Prescott, *When MIT Was "Boston Tech"* (Cambridge, MA: Technology Press, 1954), 332.

83. Massachusetts Institute of Technology, *President's Report 1883* (Boston: J. S. Cushing, 1884), 16.

84. Massachusetts Institute of Technology, *Reports of the President and Treasurer, January 1910* (Boston: Massachusetts Institute of Technology, 1910), 11–12; Massachusetts Institute of Technology, *Reports of the President and Treasurer, January 1913* (Boston: Massachusetts Institute of Technology, 1913), 17–18.

85. Massachusetts Institute of Technology, *Annual Report of the President and the Treasurer, December 11, 1901* (Boston: Geo. H. Ellis Co., 1902), 11.

86. Massachusetts Institute of Technology, *President's Report, January 1911* (Boston: Massachusetts Institute of Technology, 1911), 10; Massachusetts Institute of Technology, *Reports of the President and Treasurer, January 1913,* 17–18. The reports, issued in January, cover the previous year.

87. Richard B. Maclaurin to Jasper Whiting, December 24, 1910, MIT Office of the President, AC 13, Box 22, Folder 627, MIT Institute Archives and Special Collections, Cambridge, MA (hereafter MIT-IA).

88. Jasper Whiting to Richard B. Maclaurin, June 16, 1911, MIT Office of the President, AC 13, Box 22, Folder 627, MIT-IA.

89. Jasper Whiting to Richard B. Maclaurin, August 7, 1911, MIT Office of the President, AC 13, Box 22, Folder 627, MIT-IA.

90. Vinay Lal, *The Other Indians: A Political and Cultural History of South Asians in America* (Noida: Harper Collins, 2008), 15–38.

91. "Home Rule for India," *Chicago Tribune*, October 3, 1917.

92. Lal, *The Other Indians,* 34–40; Roger Daniels, *Guarding the Golden Door: American Immigration Policy and Immigrants Since 1882* (New York: Hill and Wang, 2004), 41–58.

93. Meeting of the MIT Executive Committee, June 6, 1922, 363, Collection AC 272, Series II, Reel 2, MIT-IA.

94. V. V. Oak, "Opportunities of Studies in America," *Modern Review* 37 (February 1925): 232.

95. *Technical and Industrial Education in the Bombay Presidency: Final Report of the Committee Appointed by Government, 1921–1922* (Bombay: Government Central Press, 1923), British Library; "Technical Study: A Divided Report," *Times of India,* July 13, 1922.

3   Gandhi's Industry

1. Ramachandra Guha, *Gandhi Before India* (New York: Alfred A. Knopf, 2014), 15–81; Judith M. Brown, *Gandhi: Prisoner of Hope* (New Haven, CT: Yale

University Press, 1989), 1–29; David Arnold, *Gandhi* (Harlow: Longman, 2001), 15–43; Pyarelel, *Mahatma Gandhi*, vol. 1, *The Early Phase* (Ahmedabad: Navajivan Publishing House, 1965).

2. Information about Devchand Parekh and his family comes from M. D. Parekh, interview by author, June 20, 2008, Mumbai, India, and Shailaja Kalelkar Parikh, *An Indian Family on the Move: From Princely India to Global Shores* (Ahmedabad: Akshara Prakashan, 2011) as well as numerous discussions with members of the Parekh family. Further background on the princely states is provided by Barbara N. Ramusack, *The Indian Princes and Their States: The New Cambridge History of India* (Cambridge: Cambridge University Press, 2004).

3. *Gazetteer of the Bombay Presidency*, vol. 8, *Kathiawar* (Bombay: Government Central Press, 1884), 457–458; Parikh, *An Indian Family on the Move*, 24–42.

4. Thomas Alfred Walker, ed., *Admissions to Peterhouse or S. Peter's College in the University of Cambridge: A Biographical Register* (Cambridge: Cambridge University Press, 1912), 629.

5. Mohandas K. Gandhi, *An Autobiography: The Story of My Experiments with Truth*, reprint ed. (Boston: Beacon Press, 1957), 42–83.

6. The meaning of "Non-Collegiate Student" was explained to me in a letter from J. Cox, deputy keeper of University Archives, Cambridge University, July 8, 2008. His degree and calling to the bar are documented in Walker, *Admissions to Peterhouse*, 629.

7. Gandhi, *An Autobiography*, 111–112; Pyarelal, *Mahatma Gandhi*, 298. Ramachandra Guha reminds us that we have only Gandhi's later claim of the importance of this event in his life. Guha, *Gandhi Before India*, 68–69.

8. M. D. Parekh, interview by author, June 20, 2008. John Whitaker, the editor of the correspondence of Alfred Marshall, stated to me in a private communication that the very provocative advice Marshall is asserted to have given Parekh (that he was in the wrong place and should go to America instead) was characteristic of Marshall.

9. J. M. Keynes, "Alfred Marshall, 1842–1924," in *Memorials of Alfred Marshall*, ed. A. C. Pigou (London: Macmillan and Co., 1925), 51. Marshall lecturing to students in Parekh's curriculum is asserted by J. Cox, letter to author, July 8, 2008.

10. Marshall to (B. B. Mukerji?), October 22, 1910, in *The Correspondence of Alfred Marshall, Economist*, vol. 3, ed. John K. Whitaker (Cambridge: Cambridge University Press, 1996), 268–269.

11. Marshall to (B. B. Mukerji?), October 22, 1910; Marshall to Manohar Lal, January 28, 1909, in *Memorials of Alfred Marshall*, ed. A. C. Pigou (London: Macmillan, 1925), 457.

12. In a postscript to his 1910 letter, Marshall wrote, "You will of course understand that I know some of the Indians who come to the West do really care to make themselves strong in action: I am very fortunate in counting several such men among my friends. But many more are needed." Marshall to (B. B. Mukerji?), October 22, 1910. One can only speculate if Marshall would have considered Parekh one such friend. A less favorable earlier comment by

Marshall on Indian students is reproduced in Peter Groenewegen, *A Soaring Eagle: Alfred Marshall, 1842–1924* (Aldershot: Edward Elgar, 1995), 607. On the decline in industrial spirit in England, see Martin J. Wiener, *English Culture and the Decline of the Industrial Spirit, 1850–1980* (Cambridge: Cambridge University Press, 1981).

13. Quoted in James Phinney Munroe, *A Life of Francis Amasa Walker* (New York: Henry Holt and Co., 1923), 264. Marshall's visit to America is described in Groenewegen, *A Soaring Eagle*, 193–203.

14. Alfred Marshall, *Principles of Economics*, 9th ed. (London: Macmillan and Co., 1961), 209–210.

15. "Two Hindoos Here," *Washington Post*, July 17, 1893, 6. The *Washington Post* noted with amazement that neither Parekh nor his companion had ever heard of Kipling. The *New York World* noted Parekh's visit to that city, stating that he had stopped into City Hall and met with the acting mayor. "The Rajah in the Hawdah," *New York World*, July 25, 1893. For documentation for Parekh's entry into the United States, SS *Paris*, June 24, 1893, *New York, Passenger Lists, 1820–1957*, Ancestry.com.

16. J. M. Dodd, May 17, 1900. This and other correspondence related to Parekh's appointment at Elphinstone is in the file "Temporary Appointment of Mr. Devcharm [*sic*] Uttamchand Parekh as Professor of History and Political Economy at Elphinstone College," IOR/L/PJ/6/539, File 818, British Library, London, England (hereafter Parekh Appointment).

17. C. J. Lyall to C. H. Hill, August 27, 1900, Parekh Appointment.

18. Telegraph to the Government of India, Home Department, December 15, 1900, Education Department, Vol. 4, 1901 Appointments, Maharashtra State Archives, Mumbai, India.

19. M. K. Gandhi to Devchand Parekh, August 6, 1902, *CWMG* 3:1–2; M. K. Gandhi to D. B. Shukla, November 8, 1902, *CWMG* 3:4. On Gandhi's failure as a lawyer in India, see Guha, *Gandhi Before India*, 150–151.

20. Parikh, *An Indian Family on the Move*, 47–49.

21. Ibid., 49–61; Bhavnagar Chemical Works, *Golden Jubilee Souvenir, 1910 to 1960* (Vartej: Bhavnagar Chemical Works, 1960), 4. I am grateful to Rajesh Mashruwala for providing me with a copy of this document.

22. Gandhi's stay in Jetpur in January 1915, where he is said to have been "put up at Devchandbhai Parekh's," is documented in C. B. Dalal, *Gandhi: 1915–1948: A Detailed Chronology* (New Delhi: Gandhi Peace Foundation, 1971), 1.

23. "Speech at Bagasra on Viramgam Customs Cordon," December 12, 1915, *CWMG* 15:85; "Diary for 1915," *CWMG* 15: 91–122.

24. Shankarlal Banker, *Gandhiji ane Rashtriya Pravrutti* (in Gujarati) (Ahmedabad: Navajivan Prakashan Mandir, 1967), 13–14. I thank Dwijendra Tripathi for finding this source in Ahmedabad and translating this passage for me. An account of that meeting based on this source is M. V. Kamath and V. B. Kher, *The Story of Militant but Non-Violent Trade Unionism* (Ahmedabad: Navajivan Mudranalaya, 1993), 75.

25. "Letter to Fulchand Shah," July 8, 1917, *CWMG* 15: 465. The note to this letter describes Parekh as "a life-long friend of Gandhiji."

26. Sumit Sarkar, *Modern India: 1885–1947* (Madras: Macmillan, 1983), 181.

27. Letter to Premabehn Kantak, February 13, 1932, *CWMG* 55:13.

28. Richard B. Gregg, *Economics of Khaddar* (Madras, 1928), 17, 19, 33. For Gregg's background, see Joseph Kip Kosek, "Richard Gregg, Mohandas Gandhi, and the Strategy of Nonviolence," *Journal of American History* 91 (March 2005): 1318–1348.

29. Scholars disagree as to whether these statements represent immature views, rhetorical flourishes, or the essence of Gandhian thought. M. K. Gandhi, *"Hind Swaraj" and Other Writings: Centenary Edition*, ed. Anthony J. Parel (New Delhi: Cambridge University Press, 2009), 105–109; Dennis Dalton, *Gandhi's Power: Nonviolence in Action* (New York: Columbia University Press, 1993), 16–25; David Hardiman, *Gandhi in His Time and Ours: The Global Legacy of His Ideas* (New York: Columbia University Press, 2003), 67–72.

30. Emma Tarlo, *Clothing Matters: Dress and Identity in India* (Chicago: University of Chicago Press, 1996), 75. However, Tarlo also adds that people interpreted the symbolism behind Gandhi's dress in different ways and Gandhi went to great pains to make his own interpretation clear.

31. Hardiman, *Gandhi in His Time and Ours*, 66–93; Robert J. C. Young, *Postcolonialism: An Historical Introduction* (Oxford: Blackwell, 2001), 317–334. Young discusses Gandhi's sophisticated use of such technologies as railroads, telegraphs, and printing presses in his campaigns.

32. Lloyd I. Rudolph and Susanne Hoeber Rudolph, *The Modernity of Tradition: Political Development in India* (Chicago: University of Chicago Press, 1967), 221–222.

33. Lewis Mumford, *Technics and Civilization* (New York: Harcourt, Brace, 1934), 14; Rudolph and Rudolph, *The Modernity of Tradition*, 222.

34. Rudolph and Rudolph, *The Modernity of Tradition*, 223. The ashram's daily schedule is reprinted in "Satyagraha Ashram," June 14, 1928, *CWMG* 42: 120.

35. Max Weber, *The Protestant Ethic and the "Spirit" of Capitalism and Other Writings* (New York: Penguin Books, 2002); Michael Adas, *Machines as the Measure of Men: Science, Technology, and Ideologies of Western Dominance* (Ithaca, NY: Cornell University Press, 1989), 252–258; Gandhi, *An Autobiography*, 298–299; Dalton, *Gandhi's Power*, ix–xi. Gandhi is typically seen as sui generis, an evaluation which he encouraged through his autobiography. However, at least in the 1880s, there were movements among Brahmins to reform Indian society by promoting the value of manual labor among the higher castes. This can be seen regularly in the newspaper *Mahratta*. See, for example, "Technical Education and the Higher Classes," *Mahratta*, April 1, 1888.

36. "Speech in Reply to Students' Address, Trivandrum," March 13, 1925, *CWMG* 30:409–414.

37. For Gandhi's establishment of the Natal Indian Congress, see Gandhi, *An Autobiography*, 148–151. For Gandhi's early unfavorable impressions of the organization of the Indian National Congress, see ibid., 223–225.

38. Rudolph and Rudolph, *The Modernity of Tradition*, 237–240. For more on Gandhi's reorganization of the Congress, see Brown, *Gandhi: Prisoner of Hope*, 157–158; Rajmohan Gandhi, *Gandhi: The Man, His People, and the Empire* (Berkeley: University of California Press, 2008), 220–221; Gopal Krishna, "The

Development of the Indian National Congress as a Mass Organization, 1918–1923," *Journal of Asian Studies* 25 (May 1966): 413–430.

39. "Speech at Kathiawar Political Conference," March 30, 1929, *CWMG* 45:282.

40. William Thomson, "Electrical Units of Measurement," in *Popular Lectures and Addresses*, vol. 1 (London: Macmillan and Co., 1889), 73.

41. M. K. Gandhi, "An Instructive Table," *Young India* January 12, 1922, 13–14; M. K. Gandhi, "An Instructive Table," *Young India*, September 17, 1925, 322–323. Khadi or khaddar is cloth produced by handspinning and weaving.

42. One of Taylor's points in his famous work was that his principles could be applied in all areas of life. Frederick Winslow Taylor, *The Principles of Scientific Management* (New York: Harper & Brothers, 1913).

43. M. K. Gandhi, "In Andhradesha," *Young India*, May 2, 1929, 137–138.

44. Mahadev Desai, "Weekly Letter," *Young India*, February 3, 1927, 34–35.

45. M. K. Gandhi, "Belgaum Impressions," *Young India*, January 1, 1925, 1–2.

46. M. K. Gandhi, "Cow Protection," *Young India*, April 9, 1925, 122–123; M. K. Gandhi, "All-India Cow Protection Sabha," *Young India*, April 16, 1925, 131; M. K. Gandhi, "A Scheme of Cow Protection," *Young India*, November 5, 1925, 374; V. G. Desai, "The Cow—Mother of Prosperity," *Young India*, February 4, 1926, 50; Thomas Weber, *On the Salt March: The Historiography of Mahatma Gandhi's March to Dandi* (New Delhi: Rupa and Co., 2009), 366; University of California, *The Sixty-Third Commencement, May 12, 1926*, 12. For the cow protection movement before Gandhi, see Sandra B. Freitag, "Sacred Symbol as Mobilizing Ideology: The North Indian Search for a Hindu Community," *Comparative Studies in Society and History* 22, no. 04 (1980): 597–625; C. S. Adcock, "Sacred Cows and Secular History: Cow Protection Debates in Colonial North India," *Comparative Studies of South Asia, Africa, and the Middle East* 30, no. 2 (2010): 297–311.

47. Thomas Weber, *Gandhi as Disciple and Mentor* (Cambridge: Cambridge University Press, 2007), 120–121. Weber makes this point in support of an argument for giving more attention to Gandhi as a social activist.

48. Discussions of Gandhi's program of spinning include Rahul Ramagundam, *Gandhi's Khadi: A History of Contention and Conciliation* (Hyderabad: Orient Longman, 2008); Lisa Trivedi, *Clothing Gandhi's Nation: Homespun and Modern India* (Bloomington: Indiana University Press, 2007); Deepak Kumar, *Science and the Raj*, 2nd ed. (New Delhi: Oxford University Press, 2006), 241–243. Among prominent Gandhi historians, Judith Brown takes his spinning the most seriously. See, for example, Brown, *Gandhi: Prisoner of Hope*, 202–204.

49. Gandhi, *"Hind Swaraj" and Other Writings*, 107.

50. M. K. Gandhi, "My Best Comrade Gone," *Young India*, April 26, 1928, 129–130. Maganlal has largely been neglected by historians, at least partially due to their neglect of Gandhi's manual and technical work. One major exception is Thomas Weber's chapter on Maganlal in his *Gandhi as Disciple and Mentor*. See also C. Shambu Prasad, "Gandhi and Maganlal: Khadi Science and the Gandhian Scientist," in *Mahatma Gandhi and His Contemporaries*, ed. Bindu Puri (Shimla: Indian Institute of Advanced Study, 2001), 228–251.

51. Gandhi, "My Best Comrade Gone," 129.

52. Ibid.

53. Maganlal K. Gandhi, "Handspinning and Handweaving: Its Re-Birth in Gujarat," *Young India*, October 1, 1919; "Circular Letter for Funds for Ashram," July 1, 1917, *CWMG* 15:452; Maganlal K. Gandhi, "Handloom Weaving in Gujarat," *Young India*, October 8, 1919, 6.

54. Gandhi, *An Autobiography*, 489–492; Trivedi, *Clothing Gandhi's Nation*, 7–9; "Letter to Maganlal Gandhi," April-May, 1918, *CWMG* 17:14–15.

55. "The Spinning Wheel," *Young India*, April 7, 1920, 1; "Letter to Maganlal Gandhi," after July 18, 1920, *CWMG* 21: 63.

56. M. K. Gandhi, "Music of the Spinning Wheel," *Young India*, July 21, 1920, 3–4.

57. M. K. Gandhi, "New Spinning Wheel," *Young India*, January 19, 1922, 42.

58. M. K. Gandhi, "Hookworm and Charkha," *Young India*, August 27, 1925, 299; M. K. Gandhi, "Bihar Notes," *Young India*, October 8, 1925, 342.

59. "Moral Basis of Cooperation," September 17,1917, *CWMG* 16:26.

60. M. K. Gandhi, "The Duty of Spinning," *Young India*, February 2, 1921, 35.

61. "Organisation of Spinning," *Young India*, May 4, 1922, 218; "A Model Report," *Young India*, May 3, 1923, 148.

62. Maganlal K. Gandhi, "Khadi Notes," *Young India*, May 3, 1923, 149.

63. M. K. Gandhi, "All India Spinners Association," *Young India*, July 30, 1925, 261–262. In 1926, a year in which Gandhi tried as much as possible to avoid travel and stay at the ashram, his secretary, Mahadev Desai, wrote, "Spinning and prayer are the two things that are more after Gandhiji's heart than anything else, and ever since he began giving more time to the Ashram, he has concentrated his energies on systematising both of them." Mahadev Desai, "Spinning at Sabarmati Ashram," *Young India*, February 11, 1926, 59.

64. "How Should Spinning Be Done," July 11, 1926, *CWMG* 36:24.

65. M. K. Gandhi, "In Far Off Tuticorin," *Young India*, September 30, 1926, 338–339. When groups addressed Gandhi during his tours, he wanted data rather than platitudes. A group of students who reported only 2 percent participation in spinning received praise for at least conducting a precise census. Mahadev Desai, "With Gandhiji in Bengal III," *Young India*, May 28, 1925, 181.

66. M. K. Gandhi, "The Bengal Tour," *Young India*, April 16, 1925, 131.

67. M. K. Gandhi, "Preventible Waste," *Young India*, July 23, 1925, 259.

68. Thompson, *The Making of the English Working Class* (New York: Pantheon Books, 1963), 350–400; David Nasaw, *Schooled to Order: A Social History of Public Schooling in the United States* (New York, 1979).

69. M. K. Gandhi, "Eleven Days' Progress," *Young India*, June 11, 1925, 201.

70. M. K. Gandhi, "Reward of Earnestness," *Young India*, July 29, 1926, 272.

71. M. K. Gandhi, "Khadi Service," *Young India*, September 16, 1926, 328; "Speech to Trainees at Khadi Vidyalaya, Ahmedabad," before April 13, 1928, *CWMG* 41:399–402.

72. "National Week: Public Meeting in Bombay," *Times of India*, April 14, 1920; "National Week," *Times of India*, April 14, 1922.

73. Mahadev Desai, "The National Week at Satyagraha Ashram," *Young India*, April 15, 1926, 139; Devdas M. Gandhi, "National Week at Sabarmati,"

Young India, May 5, 1927, 146–147; Mahadev Desai, "National Week at Satya-grahashram," *Young India*, April 19, 1928, 128.

74. Desai, "The National Week at Satyagraha Ashram," April 15, 1926.

75. On the transformation of European and American holidays, see Robert Muchembled, *Popular Culture and Elite Culture in France, 1400–1750* (Baton Rouge: Louisiana State University Press, 1985); Noah Shusterman, "The Decline of Religious Holidays in Old Regime France (1642–1789)," *French History* 23 (2009): 289–310; Susan G. Davis, "'Making Night Hideous': Christmas Revelry and Public Order in Nineteenth Century Philadelphia," *American Quarterly* 34 (Summer 1982): 185–199.

76. Mahatma Gandhi, *Educational Reconstruction* (Segaon: Hindustani Talmi Sangh, 1939).

77. Benjamin Zachariah, *Developing India: An Intellectual and Social History* (New Delhi: Oxford University Press, 2005), 161–162.

4   From Gujarat to Cambridge

1. "My Notes," March 29, 1925, *CWMG* 31:82–83.

2. Mahatma Gandhi, "Reminiscences of Kathiawar—II," March 8, 1925, *CWMG* 30:360–362.

3. "The National University of Gujarat," *Young India*, November 17, 1920, 8; T. M. Shah, undated resume, copy in author's possession. Shah's time at the Gujarat Vidyapith is also described in Kishorlal G. Mashruwala, "To Reconcile the Irreconciled," in *This Was Sardar: The Commemorative Volume*, ed. G. M. Nandurkar (Ahmedabad: Sardar Vallabhbhai Samark Bhavan, 1974), 177–179.

4. M. K. Gandhi, "National Education," *Young India*, December 10, 1925, 432.

5. "Testimonials of Mr. T. M. Shah," undated recommendation packet provided to me by Shah's son Anant; Anant Shah, interview by author, July 29, 2008, telephone. "Mr. R. P. Paranjape," *Mahratta*, June 25, 1899. That same issue advertised that photoengravings of Paranjape were being sold.

6. Anant Shah, interview, July 29, 2008.

7. The Vidyapith within the general environment of Ahmedabad is discussed in Achyut Yagnik and Suchitra Sheth, *The Shaping of Modern Gujarat: Plurality, Hindutva, and Beyond* (New Delhi: Penguin Books, 2005), 170–176.

8. Samuel C. Prescott, *When MIT Was "Boston Tech, 1861–1916"* (Cambridge, MA: Technology Press, 1954), 247–324; Bruce Sinclair, "Mergers and Acquisitions," in *Becoming MIT: Moments of Decision*, ed. David Kaiser (Cambridge, MA: MIT Press, 2010), 37–57.

9. Massachusetts Institute of Technology, *President's Report, 1893*, 5, 66; Massachusetts Institute of Technology, *Treasurer's Report, 1926*, 5–6.

10. Larry Owens, "Vannevar Bush and the Differential Analyzer: The Text and Context of an Early Computer," *Technology and Culture* 27 (January 1986): 63–95; G. Pascal Zachary, *Endless Frontier: Vannevar Bush, Engineer of the American Century* (New York: The Free Press, 1997), 46–53.

11. James J. Flink, *The Automobile Age* (Cambridge, MA: MIT Press, 1988), 22–55, 232–250; David Nye, *Electrifying America: Social Meanings of a New Technology, 1880–1940* (Cambridge, MA: MIT Press, 1990), 259–286. Data on electrical usage is from *Historical Statistics of the United States, Colonial Times to 1970, Bicentennial Edition* (Washington, DC: Government Printing Office, 1975), 827; Wilbur H. Morrison, *Donald W. Douglas: A Heart with Wings* (Ames: Iowa State University Press, 1991), 30–33.

12. Thomas P. Hughes, *American Genesis: A Century of Invention and Technological Enthusiasm* (New York: Viking, 1989). Hughes has elaborated on his concept of technological systems in *Networks of Power: Electrification in Western Society, 1880–1930* (Baltimore: Johns Hopkins University Press, 1983) and "The Evolution of Large Technological Systems," in *The Social Construction of Technological Systems: New Directions in the Sociology and History of Technology*, ed. Wiebe E. Bijker, Thomas P. Hughes, and Trevor Pinch (Cambridge, MA: MIT Press, 1987), 51–82.

13. Digital copies of these letters between Parekh and Shah are in my possession. I thank Anant Shah for permission to use these letters here. The letters in Gujarati (the letters written after March 7, 1927) were translated by Bharat Bhatt and Neelima Shukla-Bhatt.

14. T. M. Shah to Devchand Parekh, March 6, 1927.

15. In 1912 Har Dayal wrote in the *Modern Review* recommending that students go to the Pacific Coast because "it is not safe for delicate persons of the richer classes to live for a long period in England or Eastern America." Har Dayal, "Education in the West: A Suggestion," *Modern Review* 11 (February 1912): 144. Gandhi's autobiography testifies to the belief that one needed to eat meat in the colder climates of Europe.

16. T. M. Shah to Devchand Parekh, November 15, 1926; February 27, 1927; April 11, 1927; July 17, 1927.

17. Shah to Parekh, March 6, 1927.

18. Shah to Parekh, January 9, 1927; November 22, 1926. On Gandhi's questions to Devchand Parekh, see Krishnadas to Devchand Parekh, February 14, 1928 (digital copy in author's possession). Krishnadas was an assistant to Gandhi, known by his first name only. Rajmohan Gandhi, *Gandhi: The Man, His People, and the Empire* (Berkeley: University of California Press, 2008), 244.

19. Shah to Parekh, November 8, 1926.

20. Shah to Parekh, November 22, 1926.

21. Shah to Parekh, December 26, 1926.

22. Shah to Parekh, December 26, 1927; November 22, 1926.

23. "Advice to Students Proceeding to America," *The Leader*, August 8, 1921; T. M. Shah to Devchand Parekh, April 18, 1927; November 8, 1926. MIT's tuition figure is from Massachusetts Institute of Technology, *President's Report, January 1921* (Cambridge, MA: Massachusetts Institute of Technology, 1921), 11.

24. T. M. Shah to Devchand Parekh, April 18, 1927.

25. Shah to Parekh, August 15, 1927; October 22, 1927.

26. Shah to Parekh, July 28, 1928.

27. Shah to Parekh, March 25, 1927; July 10, 1927, August 8, 1927.

28. Ibid.

29. Shah to Parekh, March 20, 1927; April 18, 1927; April 24, 1927; July 7, 1928.

30. Shah to Parekh, May 15, 1927.

31. Shah to Parekh, April 1, 1928; July 7, 1928.

32. Shah to Parekh, March 14, 1927. The history of the cooperative program is discussed in W. Bernard Carlson, "Academic Entrepreneurship and Engineering Education: Dugald C. Jackson and the MIT-GE Cooperative Engineering Course, 1907–1932," *Technology and Culture* 29 (July 1988): 536–567.

33. Alexander Joseph Krupy from Russia completed the program in 1925, while Hoh Chung Chan from China completed it in 1926, and Ahmed Hassan Halet from Constantinople completed the program in 1928. Massachusetts Institute of Technology, *Graduation Exercises 1925–1928*. S. L. Kirloskar asserted that his cousin Mahdev was denied entry into the cooperative program because he was not a U.S. citizen. Kirloskar, *Cactus and Roses: An Autobiography* (Pune: Macmillian, 2003), 32. T. M. Shah to Devchand Parekh, February 27, 1927.

34. T. M. Shah to Devchand Parekh, January 9, 1927.

35. Shah to Parekh, February 27, 1927.

36. Ibid.; Shah to Parekh, March 1, 1927.

37. Shah to Parekh, March 14, 1927; April 3, 1927; May 1, 1927; May 8, 1927.

38. Shah to Parekh, May 15, 1927.

39. Shah to Parekh, November 15, 1926.

40. Shah to Parekh, March 6, 1927.

41. Ibid.

42. Shah to Parekh, July 10, 1927.

43. Shah to Parekh, February 6, 1928; February 14, 1928.

44. Shah to Parekh, July 28, 1928.

45. "General Electric Sets New Records," *New York Times*, March 30, 1927; Shah to Parekh, July 10, 1927. A company-sponsored history of General Electric during this time is John Winthrop Hammond, *Men and Volts: The Story of General Electric* (Philadelphia: J. B. Lippincott, 1941).

46. Leonard Reich, *The Making of American Industrial Research: Science and Business at GE and Bell, 1876–1926* (Cambridge: Cambridge University Press, 1985); Ronald R. Kline, *Steinmetz: Engineer and Socialist* (Baltimore: Johns Hopkins University Press, 1992).

47. T. M. Shah to Devchand Parekh, July 10, 1927.

48. Shah to Parekh, August 29, 1927.

49. Shah to Parekh, November 13, 1927.

50. Shah to Parekh, March 25, 1928.

51. Shah to Parekh, April 18, 1927.

52. Shah to Parekh, April 18, 1927; November 15, 1926; November 22, 1926.

53. Shah to Parekh, December 20, 1926.

54. T. M. Shah to Devchand Parekh, February 14, 1928; September 17, 1928. The national Hindustan Association of America regularly held national meetings in Boston. Some accounts of these meetings are "Reception Ends

Convention of Hindustan Association," *Boston Globe*, December 29, 1925; "Reception for Indian Students in Boston," *Boston Globe*, April 9, 1927; "Annual Hindustan Association Meeting," *Boston Globe*, December 27, 1930. The Ceylon India Inn, on West 49th Street, was commonly patronized by Indian students when they were in New York. *The Hindustanee Student*, n.s., 2 (September-October 1926), 15.

55. Shah to Parekh, November 8, 1926.

56. "Tech Students Fight Police of Boston and Cambridge," *Boston Globe*, November 5, 1926.

57. T. M. Shah to Devchand Parekh, November 8, 1926.

58. Shah to Parekh, November 15, 1926.

59. Shah to Parekh, November 22, 1926; January 16, 1927.

60. Shah to Parekh, December 20, 1926.

61. Shah to Parekh, November 22, 1926.

62. Shah to Parekh, January 9, 1927.

63. Shah to Parekh, December 18, 1927.

64. Shah to Parekh, January 2, 1927.

65. Shah to Parekh, September 5, 1927; March 20, 1927.

66. Shah to Parekh, July 10, 1927.

67. Shah to Parekh, July 17, 1927.

68. Shah to Parekh, July 21, 1928.

69. "Letter to Devchand Parekh," January 26, 1929, *CWMG* 44:55.

70. T. M. Shah to Devchand Parekh, May 10, 1929.

71. Shah to Parekh, September 5, 1927.

72. Shah to Parekh, April 11, 1927; March 20, 1927.

73. Shah to Parekh, August 15, 1927.

74. Shah to Parekh, February 19, 1928; September 5, 1927.

75. Zachary, *Endless Frontier*; G. Pascal Zachary, "The Godfather," *Wired*, November 1997, http://archive.wired.com/wired/archive/5.11/es_bush.html ?%2520.

76. "Satyagraha Ashram," June 14, 1928, *CWMG* 42:107–110.

77. "Satyagraha Ashram," November 4, 1928, *CWMG* 43:191–192.

78. Saroj Shah, interview by author, February 10, 2009, Ahmedabad, India.

79. Bhavnagar's provision of scholarships for overseas study is mentioned in Pattani's obituary. "Great Work for Bhavnagar: Late Sir P. Pattani," *Times of India*, February 18, 1938.

80. *Report of a Committee Appointed by the Secretary of State for India to Inquire into the System of State Technical Scholarships Established by the Government of India in 1904* (London: His Majesty's Stationery Office, 1913), 9–11.

81. L. M. Krishnan, untitled memoirs in author's possession, 104–105. This is a 247-page document written by Krishnan based on diaries he kept for fifty-two years. I thank Krishnan's daughter, Chitra Viji, for providing me with a copy of this document.

82. Upendra J. Bhatt, "Beloved Brother, Intimate Friend, Brilliant Contemporary," in *Dr. Anant Pandya Commemoration Volume*, ed. Lily Pandya (n.p.:1955), 26.

83. Ibid., 26–27.

84. Ibid., 29–30.

85. Ibid., 31.

86. Ibid., 32–33.

87. Ibid., 33–34.

88. "List or Manifest of Alien Passengers for the United States Immigration Officer at Port of Arrival, S. S. *Aquitania*, Arriving at Port of New York, September 5, 1930," *New York, Passenger Lists, 1820–1957*, Ancestry.com.

89. Mansukhlal D. Parekh, "The effect of pressure on the enthalpy of pure and mixed hydrocarbons" (ScD thesis, MIT, 1940).

90. M. D. Parekh, communication with author.

91. L. M. Krishnan, untitled memoirs.

92. Ibid., 52, 89.

93. Ibid., 89–92.

94. Ibid., 95–96.

95. Ibid., 105–106.

96. Madho Prasad, *A Gandhian Patriarch: A Political and Spiritual Biography of Kaka Kalelkar* (Bombay: Popular Prakashan, 1965), 355–364.

97. "Testimonial to Bal D. Kalelkar," July 5, 1940, *CWMG* 78:391; B. D. Kalelkar, "Potter: Through the Pot's Eyes," in *Incidents of Gandhiji's Life*, ed. Chandrashankar Shukla (Bombay: Vora and Co., 1949), 95. Bal's birth date is given in H. Kothari, ed. *Who's Who in India* (Calcutta: Kothari Publications, 1973), 149.

98. Kalelkar, "Potter: Through the Pot's Eyes," 95–97.

99. Thomas Weber, *On the Salt March: The Historiography of Mahatma Gandhi's March to Dandi* (New Delhi: Rupa, 2009); Bal D. Kalelkar letter to G. D. Birla (unsent draft), April 25, 1940, Birla private papers, New Delhi, India. Further details of the Salt March are provided in Rajmohan Gandhi, *Gandhi: The Man, His People, and the Empire* (Berkeley: University of California Press, 2008), 308–311.

100. B. D. Kalelkar, "A Study of Intake Manifold Design, With Special Emphasis on the Distribution Characteristics of a Six-Cylinder Engine Equipped with a Twin-Carburetor Layout" (PhD dissertation, Cornell University, 1944), unpaginated front matter.

101. Bal D. Kalelkar, "Application for Admission to the Graduate School, Cornell University," February 24, 1941 (in author's possession). I thank Chirag Kalelkar for providing me with a copy of this document.

102. Bal D. Kalelkar letter to G. D. Birla, April 25, 1940.

103. Ibid.

104. Ibid.

105. M. K. Gandhi to G. D. Birla, May 30, 1940, reprinted in G. D. Birla, *Bapu: A Unique Association* (Bombay: Bharatiya Vidya Bhavan, 1977), IV:57.

106. Support from Birla is asserted in Kalelkar, "A Study of Intake Manifold Design."

107. "Testimonial to Bal D. Kalelkar," July 5, 1940, *CWMG* 78:391.

108. In contrast Bal's brother, Satish, faced opposition from both his father and from Gandhi when he went to Oxford in 1935. Shailaja Parikh, *An Indian*

Family on the Move: From Princely India to Global Shores (Ahmedabad: Akshara Prakashan, 2011), 85.

109. SS *President Garfield*, August 18, 1940, *New York, Passenger Lists, 1820–1957*, Ancestry.com.

110. Bal D. Kalelkar, "Stress Concentration in the Eye Section of a Connecting Rod" (master's thesis, MIT, 1941); Kalelkar, "A Study of Intake Manifold Design."

111. Deepak Kalelkar, interview by author, March 5, 2009.

112. Shukla, *Incidents of Gandhiji's Life*, xi; "From Sabarmati," *Indian Express*, January 9, 1967.

113. Mahatma Gandhi to Bal D. Kalelkar, November 3 1944, *CWMG* 85:128.

5   Engineering a Colonial State

1. "Survey Camp," *B. E. College Annual* (5): 1940: 109–110.

2. Anant Shah, interview by author, July 29, 2008, telephone.

3. The letters between Pandya and Frances Siegel are in the Frances Siegel Papers, Radcliffe College Archives, Schlesinger Library, Radcliffe Institute, Harvard University, Cambridge, MA (hereafter Siegel Papers). Siegel's extensive FBI file includes an FBI statement that she was active in the Communist Party between 1930 and 1951. Federal Bureau of Investigation, "Frances Siegel," January 29, 1954, Folder 8.6, Siegel Papers.

4. Anant Pandya to Frances Siegel, September 12, 1933, Folder 3.6, Siegel Papers.

5. Pandya to Siegel, September 24, 1933, Folder 3.6, Siegel Papers.

6. Pandya to Siegel, October 7, 1933, Folder 3.6, Siegel Papers.

7. Pandya to Siegel, September 24, 1933, and October 7, 1933, Folder 3.6, Siegel Papers.

8. Pandya to Siegel, October 16, 1933, Folder 3.6, Siegel Papers.

9. Pandya to Siegel, October 19, 1933, Folder 3.6, and November 11, 1933, Folder 3.7, Siegel Papers.

10. Pandya to Siegel, November 20, 1933, Folder 3.7, Siegel Papers.

11. Pandya to Siegel, December 8, 1933, Folder 3.7, Siegel Papers.

12. Pandya to Siegel, December 10, 1933, Folder 3.7, Siegel Papers.

13. Ibid.

14. Pandya to Siegel, December 14, 1933, Folder 3.7, Siegel Papers.

15. Pandya to Siegel, December 21, 1933 and December 28, 1933, Folder 3.7, Siegel Papers.

16. Ibid.

17. Pandya to Siegel, January 25, 1934; January 8, 1934; January 19, 1934; Folder 3.7, Siegel Papers. The population of Gwalior in 1931 is given in Vishwambhar Prasad Sati and I. K. Mansoori, "Trends of Urbanization in Gwalior Metro-City (India) and Its Environs, *Journal of Environmental Research and Development* 2 (July–September 2007): 86.

18. Pandya to Siegel, January 25, 1934.

19. Ibid.

20. Pandya to Siegel, February 5, 1934, and April 3, 1934, Folder 3.7, Siegel Papers.

21. Pandya to Siegel, February 5, 1934, Folder 3.7, Siegel Papers.

22. "The Fourteenth Annual Dinner," *Journal of the Institution of Engineers (India)* 14 (September 1934): 19–37. The membership figure is given on p. 34. On Willingdon's policies, see Barbara D. Metcalf and Thomas R. Metcalf, *A Concise History of Modern India*, 2nd ed. (Cambridge: Cambridge University Press, 2006), 193; Judith M. Brown, *Modern India: The Origins of an Asian Democracy*, 2nd ed. (Oxford: Oxford University Press, 1994), 285–287.

23. Anant Pandya to Frances Siegel, February 5, 1934, Folder 3.7, Siegel Papers.

24. Upendra J. Bhatt, "Beloved Brother, Intimate Friend, Brilliant Contemporary," in *Dr. Anant Pandya Commemoration Volume*, ed. Lily Pandya (n.p.: 1955), 36–37; R. S. Bhatt, "Anant Pandya: A Biographical Sketch," in *Dr. Anant Pandya Commemoration Volume*, 4.

25. Anant Pandya to Frances Siegel, May 10, 1934, Folder 3.7, Siegel Papers.

26. Pandya to Siegel, July 18, 1934, and July 18, 1935, Folder 3.7, Siegel Papers.

27. Pandya to Siegel, June 15 1935, Folder 3.7, Siegel Papers.

28. Pandya to Siegel, September 18, 1934, Folder 3.7, Siegel Papers. I have gained further details about Shah's family through numerous conversations with Pandya's son, Anand Pandya. Pandya's trip with the Shahs is documented in a series of letters in the Siegel Papers, starting on June 15, 1933 and continuing until August 22, 1933.

29. Pandya to Siegel, January 4, 1935, Folder 3.7, Siegel Papers.

30. Pandya to Siegel, June 15, 1935, Folder 3.7, Siegel Papers.

31. Pandya to Siegel, September 4, 1935, Folder 3.7, Siegel Papers.

32. Upendra Bhatt, "Beloved Brother, Intimate Friend," 37; Pandya to Siegel, October 25, 1935, Folder 3.7, Siegel Papers.

33. R. S. Bhatt, "Anant Pandya: A Biographical Sketch," 4–5; Anant H. Pandya and R. J. Fowler, "Welded Diagonal Grid Framework," *Engineering News-Record*, May 25, 1939, 71–72; Anant Pandya, "Multi-Story Warehouse and Factory, *Civil Engineering* 34 (September 1939): 324–327.

34. R. S. Bhatt, "Anant Pandya: A Biographical Sketch," 5–6; Upendra Bhatt, "Beloved Brother," 38.

35. "Editorial," *Bengal Engineering College Annual* 5 (1940): 4.

36. "Annual Athletic Reports," *Bengal Engineering College Annual* 5 (1940): 111.

37. "Sur le coq," *Bengal Engineering College Annual* 5 (1940): 103–104.

38. "The Lictor (Without Fascism!)," *Bengal Engineering College Annual* 5 (1940): 105.

39. A. H. Pandya, "Address of Welcome," *Bengal Engineering College Annual* 5 (1940): 7–14.

40. *Bengal Engineering College Centenary Souvenir* (Howrah: Bengal Engineering College, 1956), 40, http://www.becollege.org/souvenir/souvenir.asp;

S. R. Sen Gupta, "Dr. Anant Pandya: The Principal of Bengal Engineering College," in *Dr. Anant Pandya Commemoration Volume*, ed. Lily Pandya (n.p.: 1955), 59–60.

41. Gaganvihari L. Mehta, "Dr. Anant Pandya: Some Reminiscences," in *Dr. Anant Pandya Commemoration Volume*, ed. Lily Pandya (n.p.: 1955), 95; John F. Riddick, *The History of British India: A Chronology* (Westport, CT: Praeger, 2006), 309.

42. In April 1930, in the midst of Gandhi's salt satyagraha, which encouraged Indians to make their own salt to avoid paying a salt tax to the British, Gandhi's *Young India* published an article giving instructions on salt manufacture. One of its authors was Mahadevlal Schroff, who had earned a master's degree in chemistry from MIT in 1927. Schroff went on to be one of the leading figures in the development of pharmaceutical education in India. Mahadevlal Shroff and K. G. Mashruvala, "How to Manufacture Salt?" *Young India*, April 24, 1930, 139; Harkishan Singh, *Mahadeva Lal Schroff and the Making of Modern Pharmacy* (Delhi: Vallabh Prakashan, 2005).

43. Shah's remarks are from the brochure *MIT 25th Reunion, Class of 29*, 60–61. I thank Anant Shah for providing me with this document.

44. The letters of recommendation are collected in "Testimonials of Mr. T. M. Shah," in author's possession. This package also contains information about Shah's training and the Tata position. I thank T. M. Shah's son, Anant, for providing me with this document.

45. Ibid.

46. Anant Shah, interview by author, July 29, 2008; "Technology Graduates in Prison in India for Anti-British Activities," *The Tech*, May 13, 1932, 1, 5; Saroj Shah, interview by author, February 10, 2009, Ahmedabad, India.

47. Rudrangshu Mukherjee, *A Century of Trust: The Story of Tata Steel* (New Delhi: Portfolio, 2008), 48, 114; William A. Johnson, *The Steel Industry of India* (Cambridge, MA: Harvard University Press, 1966), 15.

48. *TISCO Review* 10 (August 1942), ix.

49. Judith M. Brown, *Modern India: The Origins of An Asian Democracy*, 2nd ed. (Oxford: Oxford University Press, 1994), 318–325. Francis Hutchins, *India's Revolution: Gandhi and the Quit India Movement* (Cambridge, MA: Harvard University Press, 1973).

50. "Speech at A.I.C.C. Meeting," August 8, 1942, *CWMG* 83:197.

51. Quoted in Hutchins, *India's Revolution*, 282; Brown, *Modern India*, 318–325; Judith M. Brown, *Gandhi: Prisoner of Hope* (New Haven, CT: Yale University Press, 1989), 338–340.

52. P. H. Kutar to the General Manager, August 24, 1942, File L-76, "Labour Movement during 'Quit India' Agitation, 1942," Tata Steel Archives, Jamshedpur, India (hereafter TSA). Although he does not mention Shah, Vinay Bahl also argues that the supervisory personnel were the key actors in the strike. Vinay Bahl, *The Making of the Indian Working Class: A Case of the Tata Iron and Steel Company, 1880–1946* (New Delhi: Sage Publications, 1995), 356–362.

53. "Jamshedpur Labour Situation: Men Arrested at Jamshedpur," File L-139, TSA.

54. "Hartal in the TISCO Works," August 24, 1942, File L-76, "Labour Movement during 'Quit India' Agitation, 1942," TSA.

55. P. H. Kutar to the General Manager, August 24, 1942, 16, File L-76 "Labour Movement during 'Quit India' Agitation, 1942," TSA.

56. The specific plans to bring in government personnel to run the power house (which colonial officials were reluctant to cancel) are outlined in Guthrie Russell to Ardeshir Dalal, September 2, 1942, File L-76, "Labour Movement during 'Quit India' Agitation, 1942," TSA. Anant Pandya would later work for Guthrie Russell.

57. P. H. Kutar to the General Manager, August 24, 1942, 18–19; File L-76 "Labour Movement during 'Quit India' Agitation, 1942," TSA, 19. J. J. Ghandy's background is given in R. M. Lala, *The Romance of Tata Steel* (New Delhi: Penguin Viking, 2007), 68–69.

58. P. H. Kutar to the General Manager, August 24, 1942, 19.

59. Ibid., 19–20.

60. On the broader unrest during the "Quit India," movement, see Brown, *Modern India*, 321–324.

61. "A Brief Record of the Meeting of the TISCO Supervisory Staff held in the Bungalow of Mr. Sethi on Wednesday, the 26th August, 1942," File L-145, TSA.

62. "Address to the Employees of the Tata Iron and Steel Company by Sir Ardeshir Dalal," August 28, 1942, Folder L-135, TSA.

63. Various translated flyers and posters, Folder L-145, TSA.

64. Ardeshir Dalal to J.R.D. Tata, September 2, 1942, File L-138, TSA.

65. Ibid.; "Strike at Jamshedpur, August 1942," August 27, 1942, File L-138, TSA; "Jamshedpur Labour Situation: Men Arrested at Jamshedpur," File L-139, TSA. That September 3 activities are noted in the "Strike at Jamshedpur" document clearly shows that its August 27, 1942 date is incorrect.

66. S. K. Sinha to Sir Frederick James, September 7, 1942, File L-76, "Labour Movement during 'Quit India' Agitation, 1942," TSA; Ardeshir Dalal to Sir Frederick James, September 15, 1942, File L-76, "Labour Movement during 'Quit India' Agitation, 1942," TSA.

67. "Jamshedpur Labour Situation: Men Arrested at Jamshedpur," File L-139, TSA; "Subject: Mr. T. M. Shah," October 16, 1945, File L-120, "Note on Employees arrested and detained or convicted under the D.I. Rules," TSA.

68. Karl L. Wildes, letter to T. M. Shah, July 6, 1945 (in author's possession); Karl L. Wildes, letter to T. M. Shah, October 27, 1947 (in author's possession); Bertha Goodrich, letter to T. M. Shah, January 7, 1947 (in author's possession). I thank Anant Shah for providing me with these documents. *MIT 25th Reunion, Class of 29*, 61; Saroj Shah, interview by author, February 10, 2009, Ahmedabad, India.

69. B. D. Kalelkar, "Potter: Through the Pot's Eyes," in *Incidents of Gandhiji's Life*, ed. Chandrashankar Shukla (Bombay: Vora and Co., 1949), 101; Letter to Kantilal Gandhi, November 16, 1945, *CWMG* 88:340–341.

70. Letter to Bal D. Kalelkar, December 31, 1945, *CWMG* 89:134–135.

71. Umesh Mashruwala, interview by author, July 20, 2009.

6   Tryst with America, Tryst with MIT

1. "MIT (Boston) Past Graduates Meet," *Times of India*, January 9, 1945.

2. "In Far Places," *Technology Review*, April 1945, 372. Figures on American troop strength in India ranging from 170,000 to 250,000 are given in A. F. Steele, "Army's Impression on India Is Only Skin Deep," *Washington Post*, November 11, 1945; Dennis Kux, *India and the United States: Estranged Democracies, 1941–1991* (Washington, DC: National Defense University Press, 1992), 24; *The Last Round-Up*, April 11, 1946, http://cbi-theater-5.home.comcast.net /~cbi-theater-5/roundup/roundup041146.html.

3. Gerhard L. Weinberg, *A World at Arms: A Global History of World War II* (Cambridge: Cambridge University Press, 1994), 321–327; Robert Dallek, *Franklin Roosevelt and American Foreign Policy, 1932–1945* (New York: Oxford University Press, 1979), 328–331.

4. Charles F. Romanus and Riley Sunderland, *Stilwell's Mission to China: United States Army in World War II* (Washington, DC: Office of the Chief of Military History, Department of the Army, 1953), 204; James M. Ehrman, "Ways of War and the American Experience in the China-Burma-India Theater" (PhD dissertation, Kansas State University, 2006).

5. Dallek, *Franklin Roosevelt and American Foreign Policy*, 281–286, 319.

6. Ibid., 325–327.

7. Kenton J. Clymer, *Quest for Freedom: The United States and India's Independence* (New York: Columbia University Press, 1995), 32–34, 48–49.

8. Weinberg, *A World at Arms*, 310–330.

9. "Memorandum by the Assistant Secretary of State (Berle) to the Secretary of State," December 20, 1941, *Foreign Relations of the United States, 1942*, vol. 1, *General, The British Commonwealth, The Far East* (Washington, DC: GPO, 1960), 593–595. "Memorandum of Conversation by the Assistant Chief of the Division of Near Eastern Affairs (Alling)," January 23, 1942, *Foreign Relations of the United States, 1942*, vol. 1, *General, The British Commonwealth, The Far East* (Washington: GPO, 1960), 595–597; "Memorandum of Conversation by the Assistant Secretary of State (Berle)," January 28, 1942, *Foreign Relations of the United States, 1942*, vol. 1, *General, The British Commonwealth, The Far East* (Washington: GPO, 1960), 597–598; "Memorandum by the Assistant Secretary of State (Berle) to President Roosevelt," January 29, 1942, *Foreign Relations of the United States, 1942*, vol. 1, *General, The British Commonwealth, The Far East* (Washington, DC: GPO, 1960), 599.

10. Clymer, *Quest for Freedom*, 58–81; "U.S. War Mission Will Go to India," *New York Times*, March 3, 1942; "Johnson to Be 'Minister,'" *New York Times*, March 17, 1942.

11. Gordon Rentschler to Dean Thresher, April 29, 1942, MIT Office of the President, AC 4, Box 26, Folder 10, MIT Institute Archives and Special Collections, Cambridge, MA (hereafter MIT-IA); "Sees Good Market for Lines in India," *New York Times*, July 14, 1937. For National City Bank's Indian branches, see "National City Bank to Expand on Jan. 1," *New York Times*, December 23, 1926. National City Bank is the progenitor of the current Citigroup.

Durga Bajpai became an important architect in independent India, with such prominent commissions as the Oberoi Hotel in New Delhi as well as the Jehangir Art Gallery and the Tata Institute for Social Sciences, both in Mumbai.

12. Massachusetts Institute of Technology, *Staff and Student Directory, 1942–1943*, 30. The young Mahindra left MIT and graduated from the University of Pennsylvania. K. C. Mahindra's biography is given on the Mahindra website, http://www.mahindra.com/Who-We-Are/How-We-Got -Here/K.C.-Mahindra. The early history of Mahindra and Mahindra is described in Dwijendra Tripathi and Jyoti Jumani, *The Oxford History of Contemporary Indian Business* (New Delhi: Oxford University Press, 2013), 50–52.

13. S. K. Kirpalani, *Fifty Years with the British* (Hyderabad: Orient Longman, 1993), 283, 293.

14. Iswar Saran to K. T. Compton, September 24, 1944, MIT Office of the President, AC 4, Box 193, Folder 2, MIT-IA.

15. Adelaide Lyons, "Why a College Professor Considers Chemistry and Christianity Essential Factors in Developing India's Industry," *World Outlook*, January 1920, 25, 62, 64.

16. Bhatnagar's career is described in Robert S. Anderson, *Nucleus and Nation: Science, International Networks, and Power in India* (Chicago: University of Chicago Press, 2010), 32–37. *The Cosmos of Arthur Holly Compton*, ed. Marjorie Johnston (New York: Alfred A. Knopf, 1967), 39–42.

17. K. T. Compton to Dr. S. S. Bhatnagar, November 27, 1944, MIT Office of the President, AC 4, Box 30, Folder 11, MIT-IA; "New President of the American Association for the Advancement of Science," *Science and Culture* 8 (August 1942): 73.

18. M. N. Saha telegram to K. T. Compton, February 16, 1944; M. N. Saha letter to B. A. Thresher, May 6, 1944; M. N. Saha telegram to K. T. Compton, May 12, 1944, MIT Office of the President, AC4, Box 191, Folder 2, MIT-IA.

19. Sanjoy Bhattacharya and Benjamin Zachariah, "'A Great Destiny': The British Colonial State and the Advertisement of Post-War Reconstruction in India, 1942–1945," *South Asia Research* 19, no. 1 (1999): 71–100.

20. Robert Buderi, *The Invention that Changed the World* (New York: Simon & Schuster, 1996), 35, 114; Bernard Katz, "Archibald Vivian Hill, 26 September 1886–3 June 1977," *Biographical Memoirs of Fellows of the Royal Society* 24 (November 1978): 71–149.

21. A. V. Hill, *A Report to the Government of India on Scientific Research in India* (London: The Royal Society, 1945). Hill's report is discussed in further detail in Deepak Kumar, *Science and the Raj: A Study of British India*, 2nd ed. (New Delhi: Oxford University Press, 2006), 253–259.

22. Hill, *Scientific Research in India*, 17, 28

23. Ibid., 29–30, 53.

24. Ibid., 29–30.

25. "Planning in India," *Times of London*, June 1, 1944. Dalal's obituary is given in "Sir A. Dalal Dead," *Times of India*, October 9, 1949.

26. "Progress in Reconstruction Plans for India," *Times of India*, September 15, 1944.

27. Kim Patrick Sebaly, "The Assistance of Four Nations in the Establishment of the Indian Institutes of Technology, 1945–1970" (PhD dissertation, University of Michigan, 1972), 17–18; "Indian Scientists on Their Visit to the UK and the USA," *Science and Culture* 10 (March 1945): 377.

28. M. D. Parekh to B. A. Thresher, March 30, 1945, MIT Office of the President, AC 4, Box 167, Folder 3, MIT-IA.

29. Sebaly, "The Assistance of Four Nations in the Establishment of the Indian Institutes of Technology," 17–19; *Development of Higher Technical Institutions in India* (Simla: Government of India Press, 1946), 1. Sarkar's obituary is given in "Mr. Sarkar Dead," *Times of India*, January 26, 1953.

30. M. Ahmed to K. T. Compton, July 14, 1945, MIT Office of the President, AC 4, Box 67, Folder 1, MIT-IA; Floyd G. Blair to James R. Killian, July 12, 1945, MIT Office of the President, AC 4, Box 193, Folder 2, MIT-IA; R. N. Kimball to Dean Caldwell, January 8, 1945, MIT Office of the President, AC 4, Box 44, Folder 5, MIT-IA. The composition of the committee is given in *Development of Higher Technical Institutions in India*, 1–2.

31. *Development of Higher Technical Institutions in India*, 2.

32. Ibid., 4–12.

33. Ibid., 6.

34. Massachusetts Institute of Technology, *President's Report*, 1944, 32; Buderi, *The Invention That Changed the World*; Stuart W. Leslie, *The Cold War and American Science: The Military-Industrial-Academic Complex at MIT and Stanford* (New York: Columbia University Press, 1993), 14–25.

35. Benoy Kumar Sarkar, *Education for Industrialization: An Analysis of the Forty Years Work of the Jadavpur College of Engineering and Technology* (Calcutta: Chuckervertty, Chatterjee & Co., 1946); Ananda Lal, Rama Prasad De, and Amrita Sen, *The Lamp in the Lotus: A History of Jadavpur University* (Kolkata: Jadavpur University, 2005).

36. Sarkar, *Education for Industrialization*, 297.

37. Ibid., 298–299. Wallah is an all-purpose Hindi suffix, transferring a noun into a personal identity, often associated with an occupation. Thus a chai wallah is a person who serves tea.

38. Floyd Blair to James Killian, July 12, 1945.

39. Ashoke Nag, "The American Library in Kolkata: A Space Open to All, Say U.S. Consulate General," *The Economic Times*, April 20, 2013; "Information about America: U.S. OWI Library Opened," *Times of India*, November 24, 1944; "American Library in New Delhi Celebrates 60 Years," *Span*, November/December 2006, 54. The OWI and its role is India is discussed more broadly in Allan M. Winkler, *The Politics of Propaganda: The Office of War Information 1942–1945* (New Haven, CT: Yale University Press, 1978); Eric D. Pullen, "'Noise and Flutter,' U.S. Propaganda Strategy and Operation in India During World War II," *Diplomatic History* 34 (April 2010): 275–298.

40. "Information about America," *Times of India*, November 24, 1944.

41. Ibid.

42. Ibid.

43. Amy Schaefer and Flora B. Ludington to Fritz Silber and Trevor Hill, April 2, 1945, New Delhi Mission General Records, 1945, Box 43, Record Group 84, National Archives and Records Administration, College Park, MD (hereafter NARA); "Report on Public Affairs Branches in India," December 9, 1946, New Delhi Mission General Records, 1946, Box 50, Record Group 84, NARA.

44. P. M. Chalmers to Mr. Howard Donovan, July 12, 1944, Bombay Consulate General Records, 1944, Box 112, RG 84, NARA; "To the American Consular Officer in Charge, Bombay, India," August 30, 1944, Bombay Consulate General Records, 1944, Box 112, RG 84, NARA.

45. Massachusetts Institute of Technology, *President's Report* 1944, 30; B. N. Shastri to the Secretary, Office of the Personal Representative of the United States, December 29, 1944, New Delhi Mission General Records, 1945, Box 43, Record Group 84, NARA; "Massachusetts Institute of Technology," *Journal of Scientific and Industrial Research* 3 (June 1945): 569–572; G. N. Banerjee to Ray L. Thurston, May 4, 1944, Bombay Consulate General Records, 1944, Box 112, Record Group 84, NARA.

46. George R. Merrell to Director of Admissions Massachusetts Institute of Technology, March 2, 1945, New Delhi Mission General Records, 1945, Box 43, Record Group 84, NARA; George R. Merrell to Myrl S. Myers, June 27, 1944, Calcutta Consulate General Records, 1944, Box 146, Record Group 84, NARA; George R. Merrell to Roy E. Bower, May 31, 1945, New Delhi Mission, General Records, Box 43, RG 84, NARA; "J. J. Rudra, M.A., Ph.D., MBE," *Current Science* 12 (June 1943): 182.

47. Registrar, Calcutta University to the Consul General for America, Calcutta, May 17, 1944, Calcutta Consulate General Records, Box 146, RG 84, NARA; N. K. Choudhuri to the Consulate General for U.S., February 8, 1945, Calcutta General Records, Box 167, RG 84, NARA; Debabreata Dutt to the American Consulate General, Calcutta, December 12, 1946; Calcutta Consulate General Records, 1946, Box 196, Record Group 84, NARA.

48. Jamshed Patel, recorded discussion with Eric Patel. I thank Eric Patel for sharing this recording.

49. Ibid.

50. W. F. Dickson, "Enclosure No. 1 to Dispatch 313, Dated October 31, 1945," New Delhi Mission General Files, 1945, Box 43, RG 84, NARA; "To the Officer in Charge of the American Mission, New Delhi," November 29, 1945, New Delhi Mission General Files, Box 43, RG 84, NARA.

51. Hill, *Scientific Research in India*, 9–10.

52. *Report of the Selection Board Overseas Scholarships* (Simla: Department of Education, 1945). IOR/V/27/864/8: 1945, British Library.

53. "India Plans Study for Hundreds Here," *New York Times*, April 29, 1945; Benjamin Fine, "India to Send 1,500 of Students Here," *New York Times*, June 18, 1945.

54. R. M. Kimball to M. S. Sundaram, May 18, 1945, MIT Office of the President, AC4, Box 214, Folder 9, MIT-IA; Committee of Stabilization of Enrollment "Report on Foreign Students" to Be Presented to Faculty, April 13,

1945, MIT Faculty Records, AC 1, Series 2, Box 18, Folder "April 13, 1945," MIT-IA.

55. "Report on Foreign Students," 10.

56. Ibid., 7.

57. Ibid., 3.

58. Ibid., 7–8. The "Report on Foreign Students," dealt with this sensitive subject in a deliberately ambiguous manner, stating arguments both for and against the limitation of foreign students, without stating whether these arguments were legitimate or not.

59. Ibid., 6.

60. Ibid.

61. Ibid., 8.

62. Ibid., 4; Massachusetts Institute of Technology, *President's Report*, 1945, 89; MIT Faculty Minutes, AC 1, March 16, 1955, MIT-IA.

63. Massachusetts Institute of Technology, *President's Report*, 1945, 90; "Report on Foreign Students," 11.

64. A. F. Steele, "Army's Impression on India Is Only Skin Deep," *Washington Post*, November 11, 1945.

65. The Karachi *Daily Gateway* article is reproduced in "GI Joe Feels He Is Being Sold Down the River," *Statesville (NC) Record and Landmark*, December 6, 1945.

66. Alfred Wagg, "GIs Beef as U.S. Shares British India Headache," *Chicago Tribune*, December 2, 1945.

67. Ibid.

68. American Consul General, Dispatch 212, December 29, 1945, Calcutta Consulate General Records, 1945, Box 167, RG 84, NARA.

69. Dhawan and Prakash's presence on the *Torrens* is documented in SS *Torrens*, December 7, 1945, *New York, Passenger Lists, 1820–1957*, Ancestry.com.

70. "Passage of Indian Students to US: Priority to Troops," *Times of India*, December 6, 1945.

71. "Memorandum by the Acting Secretary of State to President Truman," June 9, 1945, *Foreign Relations of the United States 1945*, vol. 6, *The British Commonwealth, The Far East* (Washington, DC: GPO, 1969), 287.

72. Harold A. Gould, *Sikhs, Swamis, Students, and Spies: The India Lobby in the United States, 1900–1946* (New Delhi: Sage Publications, 2006), 393–431; Vinay Lal, *The Other Indians: A Political and Cultural History of Indians in America* (Noida: Harper Collins, 2008), 51–52.

73. Charles S. Maier, *Among Empires: American Ascendancy and Its Predecessors* (Cambridge, MA: Harvard University Press, 2006).

7   *High Priests of Nehru's India*

1. Robert J. McMahon, *Cold War on the Periphery: The United States, India, and Pakistan* (New York: Columbia University Press, 1994), 44–64.

2. "Nehru Honored at Reception, Talks to Local Indian Students," *The Tech*, October 25, 1949.

3. "Boston Hails Pandit Nehru During Tour of 3 Colleges," *Boston Globe*, October 22, 1949; " 'We Must Fight Against Stagnation': Prime Minister's Advice to Indians in Massachusetts," *Times of India*, October 24, 1949; "Nehru Honored at Reception."

4. "Draft Constitution of Congress," January 29, 1948, *CWMG* 98: 333–334.

5. A. Vaidyanathan, "The Indian Economy Since Independence," in *The Cambridge Economic History of India*, vol. 2, ed. Dharma Kumar (New Delhi: Orient Longman, 2005), 947–948, 964; Alan Heston, "National Income," in *The Cambridge Economic History of India*, vol. 2, 410–411; *Basic Statistics Relating to the Indian Economy* (New Delhi: Central Statistical Organization, Department of Statistics, Government of India, 1976), 7. The U.S. life expectancy in 1950 is given on the Center for Disease Control's website, http://www.cdc .gov/nchs/data/hus/2010/022.pdf

6. Kim Patrick Sebaly, "The Assistance of Four Nations in the Establishment of the Indian Institutes of Technology, 1945–1970" (PhD dissertation, University of Michigan, 1972), 29–47; S. T. H. Abidi et al., *A Walk through History: 50 Years of IIT Kharagpur* (Kharagpur: Indian Institute of Technology Kharagpur, 2002), 15–72.

7. Abidi, et al., *A Walk through History*, 75; Indian Institute of Technology, Kharagpur, *Annual Report, 1954–1955*, 13–17.

8. Sebaly, "The Assistance of Four Nations."

9. D. D. Kosambi to Norbert Wiener, October 4, 1946, Norbert Wiener papers, MC 22, Box 34B, Folder 1021, MIT Institute Archives and Special Collections, Cambridge, MA (hereafter MIT-IA).

10. *Yearbook of the Universities of the Commonwealth, 1948* (London: G. Bell and Sons, 1948), 1003; Institute of International Education, *Open Doors 1948–1949* (New York: Institute of International Education, 1949), 12.

11. McMahon, *Cold War on the Periphery*, 110–122.

12. Chester Bowles to James Killian, June 5, 1952, MIT Office of the President, AC 4, Box 3, Folder 2, MIT-IA.

13. J. R. Killian Jr. to Chester Bowles, June 16, 1952, MIT Office of the President, AC 4, Box 3, Folder 2, MIT-IA.

14. Ramachandra Guha, *India after Gandhi: The History of the World's Largest Democracy* (New Delhi: Picador, 2007), 201–225; Sunil Khilnani, *The Idea of India* (New York: Farrar, Straus and Giroux, 1997), 61–106.

15. One MIT graduate I do not discuss here who spent his career working for the Indian state is Bal Kalelkar. Kalelkar rose to the position of Deputy Director with the Ministry of Technical Development. He died in 1974.

16. Ramesh Mehta to Frances Siegel, January 27, 1946, Folder 3.9, Frances Siegel Papers, Radcliffe College Archives, Schlesinger Library, Radcliffe Institute, Harvard University, Cambridge, MA.

17. R. S. Bhatt, "Anant Pandya: A Biographical Sketch," in *Dr. Anant Pandya Commemoration Volume*, ed. Lily Pandya (n.p.: 1955), 1–13; "Plan to Double Water Supply of Bombay," *Times of India*, February 10, 1949; "Augmenting Bombay's Water Supply," *Times of India*, April 6, 1952; Ganganvihari L. Mehta,

"Dr. Anant Pandya: Some Reminiscences," in *Dr. Anant Pandya Commemoration Volume*, 98; C. V. S. Rao, "Dr. Anant Pandya: A Good Samaritan," in *Dr. Anant Pandya Commemoration Volume*, 69–70.

18. Sudhir Sen, "Dr. Anant Pandya: An Appreciation," in *Dr. Anant Pandya Commemoration Volume*, 77–78. The relation of the Damodar Valley project to the American Tennessee Valley Authority is discussed in Daniel Klingensmith, *"One Valley and a Thousand": Dams, Nationalism, and Development* (New Delhi: Oxford University Press, 2007).

19. Sudhir Sen, "Dr. Anant Pandya: An Appreciation," 76–80; C. V. S. Rao, "Dr. Anant Pandya: A Good Samaritan," 71.

20. Mastaram H. Pandya, "Anant: His Childhood and Youth," in *Dr. Anant Pandya Commemoration Volume*, 23; "Dr. Anant Hiralal Pandya Dead," *Times of India*, June 3, 1951; Mahendra Pandya, interview by author, July 22, 2009, Bangalore, India.

21. "Current Topics," *Times of India*, June 6, 1951; "Loss of Eminent Engineer," *Times of India*, June 17, 1951.

22. *Kumar*, August 1952 (in Gujarati) in author's possession. I thank Anand Pandya, son of Anant Pandya, for providing me with a copy of this magazine. Many of the articles also appeared (in English) in *Dr. Anant Pandya: Commemoration Volume*. I thank Anand Pandya for providing me with a copy of this volume.

23. Kirit Parikh, interview by author, June 11, 2008. Although I was aware of the *Kumar* tribute to Pandya, Parikh brought it up without prompting by the author.

24. Arvind Panagariya, *India: The Emerging Giant* (New Delhi: Oxford University Press, 2008), 32–33.

25. I use 1951 statistics for India because that year's census allows the calculation of per capita figures. Indian steel production statistics are from *Statistics for Iron and Steel Industry in India* (Ranchi: Hindustan Steel Limited, 1966), 4. *Historical Statistics of the United States, Earliest Times to the Present: Millennial Edition* (New York: Cambridge University Press, 2006), Table Dd399. William A. Johnson, *The Steel Industry of India* (Cambridge, MA: Harvard University Press, 1966), 16–21.

26. Rustam Dastur, interview by author, Kolkata, India, July 28, 2009; "Dr. Minu N. Dastur," *Consulting Engineer*, April 1962, 15–22; Minu Nariman Dastur, "Equilibrium in the Reaction of Hydrogen with Oxygen in Liquid Iron and Iron-Vanadium Alloys" (ScD dissertation, MIT, 1949), 132; "Herman A. Brassert, 86, Dies, Metallurgist Aided Steel Field," *New York Times*, June 19, 1961; "To Expand Plant in India," *New York Times* March 28, 1937; "Reich Contracts Let," *New York Times*, August 3, 1937; "Steel Man Offers Advice for Europe," *New York Times*, August 3, 1947. In addition to the sources cited here, my understanding of M. N. Dastur is based on several days spent with his former colleagues in July 2007 in Kolkata.

27. M. N. Dastur to S. Boothalingam, November 22, 1954, and M. N. Dastur to T. T. Krishnamachari, November 14, 1954, M. N. Dastur & Co., Kolkata, India (hereafter Dastur & Co.); J. R. D. Tata to M. N. Dastur,

January 3, 1955, JRDT-MF-054-JAN-MAR 1955-PG-01, Tata Central Archives, Pune, India; M. N. Dastur, "Nehru and India's Steel Industry," in *Development through Technology: A Symposium on Nehru's Vision*, ed. P. V. Indiresan (Delhi: Indian Institute of Technology, 1989), 115.

28. *Ferromanganese Plant Joda* (Calcutta: M. N. Dastur, n.d.); "Dr. Minu N. Dastur," *Consulting Engineer*, April 1962, 15–20; M. N. Dastur to Morarji Desai, May 15, 1959, Dastur & Co.; Rustom Lalkaka, interview by author, June 15, 2009, telephone.

29. M. N. Dastur to Morarji Desai, May 15, 1959; *Ferromanganese Plant Joda*.

30. Ibid.

31. Dastur to S. Boothalingam, November 22, 1954.

32. Dastur to T. N. Singh, September 2, 1958, Dastur & Co.; Dastur to Pitamber Pant, October 8, 1958, Dastur & Co.

33. Dastur to Pitamber Pant, October 8, 1958.

34. Jawaharlal Nehru to Swaran Singh, August 21, 1958, *Selected Works of Jawaharlal Nehru*, 2nd ser. (New Delhi: Jawaharlal Nehru Memorial Fund, 2011), 43:172.

35. Ibid.

36. Jawaharlal Nehru to Swaran Singh, August 31, 1958, *Selected Works of Jawaharlal Nehru*, 2nd ser. (New Delhi: Jawaharlal Nehru Memorial Fund, 2011), 43:175.

37. Jawaharlal Nehru to Vishnu Sahay, December 3, 1958, *Selected Works of Jawaharlal Nehru*, 2nd ser. (New Delhi: Jawaharlal Nehru Memorial Fund, 2012), 45:545.

38. M. N. Dastur to Secretary, Ministry of Steel, Mines, and Fuel, July 27, 1959, and M. N. Dastur to Sardar Swaran Singh, May 10, 1959, Dastur & Co., Sailendra Nath Ghosh, "M. N. Dastur: A Tribute," *Economic and Political Weekly* 39 (February 21, 2004): 781–783.

39. Dennis Merrill, *Bread and the Ballot: The United States and India's Economic Development, 1947–1963* (Chapel Hill: University of North Carolina Press, 1990), 176–177, 200–201; Robert B. Rakove, *Kennedy, Johnson, and the Nonaligned World* (Cambridge: Cambridge University Press, 2013), 186–188; "Bokaro Steel Project," *The Engineer*, July 12, 1963, 72; "India Drops Request for Steel Mill Aid," *New York Times*, September 12, 1963.

40. M. N. Dastur & Co., *Detailed Project Report on the Establishment of Bokaro Steelworks* (Calcutta: M. N. Dastur & Co., 1963), 51.

41. M. N. Dastur to Jawaharlal Nehru, September 30, 1963, Dastur & Co.

42. "Engineering for Bokaro," *Economic Weekly* 16 (December 5, 1964): 1905–1906.

43. M. N. Dastur & Co., "Steel Plant Design and Engineering Organizations—The Indian Experience," United Nations Industrial Development Organization, Third Interregional Symposium on the Iron and Steel Industry, Brasilia, Brazil, October 1973, 15.

44. Dastur, "Nehru and India's Steel Industry," 115–116.

45. M. N. Dastur & Co., "Detailed Project Report on the Establishment of Bokaro Steelworks," 52; "Bokaro Made in USSR," *Economic Weekly* 16 (October 17, 1964): 1674.

46. Bernard D'Mello, "Soviet Collaboration in Indian Steel Industry, 1954–1984," *Economic and Political Weekly* 23 (March 5, 1988): 473–486.

47. Ibid., 478; "Government Rejects Dastur Plan for Bokaro," *Times of India*, March 24, 1966.

48. "Engineering for Bokaro," 1905–1906.

49. "The Bokaro Colosseum," *Economic and Political Weekly* 3 (March 2, 1968): 380.

50. D'Mello, "Soviet Collaboration," 483.

51. Ghosh, "M. N. Dastur: A Tribute," 781–783; M. N. Dastur & Co., "Steel Plant Design and Engineering Organizations," 15–17, 24; Rustom Lalkaka, interview by author, June 15, 2009.

52. K. P. Mahalingam, "Dastur: A Legend in Steel," *The Hindu*, January 12, 2004; Ghosh, "M. N. Dastur: A Tribute," 781–783.

53. Paul Josephson has written of technology's value as a display of national power in "'Projects of the Century' in Soviet History: Large Scale Technologies from Lenin to Gorbachev," *Technology and Culture* 36 (July 1995): 519–559.

54. Robert S. Anderson, *Nucleus and Nation: Scientists, International Networks, and Power in India* (Chicago: University of Chicago Press, 2010); Itty Abraham, *The Making of the Indian Atomic Bomb: Science, Secrecy, and the Postcolonial State* (London: Zed Books, 1998); George Perkovich, *India's Nuclear Bomb: The Impact on Global Proliferation* (Berkeley: University of California Press, 1999).

55. Jawaharlal Nehru to Baldev Singh, February 29, 1948, *Selected Works of Jawaharlal Nehru*, 2nd ser. (New Delhi: Jawaharlal Nehru Memorial Fund, 1987), 5:420.

56. Perkovich, *India's Nuclear Bomb*, 17–19; Abraham, *The Making of the Indian Atomic Bomb*, 59–61.

57. Quoted in Indira Chowdhury and Ananya Dasgupta, *A Masterful Spirit: Homi J. Bhabha, 1909–1966* (New Delhi: Penguin Books, 2010), 71. Other details about Bhabha's life and his scientific networks are from this book.

58. Abraham, *The Making of the Indian Atomic Bomb*, 59–106; Perkovich, *India's Nuclear Bomb*, 35–37, 64–65.

59. C. V. Sundaram, "Brahm Prakash, 1912–1984," *Biographical Memoirs of the Fellows of the Indian National Science Academy* 16 (1993): 137–151. My understanding of Prakash's career is also based on an interview with his son Arun by telephone, on January 8, 2010, and a meeting with Prakash's wife Rajeshwari and his two daughters, Maneesha and Suman, on March 14, 2010 in Los Angeles, CA.

60. Ibid.; Brahm Prakash, "The Effects of Activators and Alizarin Dyes on the Soap Flotation of Quartz, Cassiterite, and Fluorite" (ScD dissertation, MIT, 1949), 216.

61. Sundaram, "Brahm Prakash," 139–140.

62. Ibid., 140–141.

63. I. M. D. Little, "Atomic Bombay? A Comment on 'The Need for Atomic Energy in Underdeveloped Countries,'" *Economic Weekly* 10 (November 29, 1958): 1483–1486; Perkovich, *India's Nuclear Bomb*, 38; "Misplaced," *Times of India*, March 29, 1961.

64. Perkovich, *India's Nuclear Bomb*, 146–183. India detonated its "peaceful nuclear explosion" on May 18, 1974.

65. Sundaram, "Brahm Prakash," 141.

66. P. V. Manoranjan Rao and P. Radhakrishnan, *A Brief History of Rocketry in ISRO* (Hyderabad: Universities Press, 2012), Kindle edition, chap. 4; A. P. J. Abdul Kalam with Arun Tiwari, *Wings of Fire: An Autobiography* (Hyderabad: Universities Press, 1999), 64–97.

67. Quoted in Deepak Kumar, *Science and the Raj: A Study of British India*, 2nd ed. (New Delhi: Oxford University Press, 2006), 257.

68. Ashok Parthasarathi and Baldev Singh, "Science in India: The First Ten Years," *Economic and Political Weekly* 27 (August 29, 1992): 1852–1858.

69. *Twenty-Five Years of CSIR* (New Delhi: CSIR, 1968); Baldev Singh, "Reviewing the CSIR," *Economic and Political Weekly* 21 (August 23, 1986): 1515.

70. Darshan Singh Bhatia, "Chemical and Physical Effects of Super Voltage Cathode Rays on Amino Acids in Foods and in Aqueous Solutions" (ScD dissertation, MIT, 1950), 184. Another student's account of the voyage to the United States is given by Rajeshwari Chatterjee, who went to the University of Michigan to study electrical engineering. See "A Thousand Streams: A Personal History," chap. 13, http://www.lifescapesmemoirs.net/chatterjee/streams /chatt13.htm. Further information about Bhatia's early life is given by Kavita Chhibber, "A Revered Atlantan," http://www.kavitachhibber.com/main/main .jsp?id=dr_bhatia.

71. Central Food Technological Research Institute Mysore, *Annual Report 1962–1963*, 164; V. Subrahmanyan, Saranya Kumari Reddy, M. N. Moorjani, Gowri Sur, T. R. Doraiswamy, A. N. Sankaran, D. S. Bhatia, and M. Swaminathan, "Supplementary Value of Vegetable-milk Curds in the Diet of Children," *British Journal of Nutrition* 8 (December 1954): 348–352.

72. "Darshan S. Bhatia, Director of Corporate Research and Development Department, The Coca-Cola Corporation," July 1981, Coca-Cola Archives, Atlanta, GA.

73. Brinder Bhatia, interview by author, September 3, 2014, telephone; "Darshan S. Bhatia, Director of Corporate Research and Development Department, The Coca-Cola Corporation"; *International Action to Avert the Impending Protein Crisis: Report to the Economic and Social Council of the Advisory Committee on the Application of Science and Technology to Development* (New York: United Nations, 1968).

74. "Darshan S. Bhatia, Director of Corporate Research and Development Department, The Coca-Cola Corporation"; "Bhatia Heads Corporate Protein Group for the Coca-Cola Company," Coca-Cola news release, December 15, 1972, Coca-Cola Archives, Atlanta, GA; "Protein Beverage Product," *Refresher USA*, January-March 1973, 4–6, 25–27; Derrick Henry, "Darshan Bhatia, 78: Developed Coke Brands," *Atlanta Journal-Constitution*, October 12, 2001.

75. "Laureate Told: No Vacancy," *Times of India*, October 17, 1968; "Brain Drain Issue Mainly Economic, Says Triguna Sen," *Times of India*, November 28, 1968.

76. P. R. Gupta, "Dr. Khorana Comes to India," *Times of India*, November 24, 1968.

77. *Self-Immolation of a Scientist: A Memoir to Dr. Vinod H. Shah, M.Sc. Ph.D.*, ed. Manu Kothari (Ahmedabad: Jayant Shah, 1973[?]), 23. This volume, put together by one of Shah's brothers, contains a wealth of press reports and documents related to Shah's suicide.

78. Jagan Chawla and A. P. Jain, eds., *Whither Indian Science?* (New Delhi: S. Chand & Co., 1973). Chawla's biography is given in *India Who's Who 1980–1981* (New Delhi: INFA Publications, 1981), 311.

79. Chawla and Jain, eds., *Whither Indian Science?*, ix, xvi, 10.

80. Ibid., xviii, xiii.

8 Business Families and MIT

1. John Kenneth Galbraith, *Ambassador's Journal: A Personal Account of the Kennedy Years* (Boston: Houghton Mifflin, 1969), 82. Nehru was not enamored of the United States or Americans, and in contrast to Galbraith's account of this meeting, many high-level encounters between Nehru and U.S. government officials were marked by tensions. It should not be surprising that the son Indira contemplated sending to MIT, Rajiv, went instead to the British Cambridge to study engineering. He returned home with a wife, but without a degree. On Nehru's attitudes towards the United States and his relations with government officials, see Andrew Rotter, *Comrades at Odds: The United States and India, 1947–1964* (Ithaca, NY: Cornell University Press, 2000), 20–25, 89–91.

2. Galbraith, *Ambassador's Journal*, 71. Birla told Galbraith that he had twice read Galbraith's *The Affluent Society*.

3. S. L. Kirloskar, *Cactus & Roses: An Autobiography* (Pune: Macmillan, 2003), 31–33.

4. Ibid., 34–35.

5. Ibid., 36–45.

6. Both Dahanukar's biographies are given in *The India and Pakistan Yearbook and Who's Who* (Bombay: Bennett, Coleman & Co., 1949), 712. "President of Federation," *Times of India*, April 10, 1931. For more on the managing agency system, see Dwijendra Tripathi, *The Oxford History of Indian Business* (New Delhi: Oxford University Press, 2004), 112–113.

7. Dwijendra Tripathi, *The Dynamics of a Tradition: Kasturbhai Lalbhai and His Entrepreneurship* (New Delhi: Manohar Publications, 1981), 1–78.

8. Shrenik Lalbhai, note to author, December 29, 2013. In this note, Lalbhai quotes his father in an interview he gave. Tripathi, *The Dynamics of a Tradition*, 79–127; U.S. Army Air Forces, *Passenger Manifest, November 14, 1944, New York, Passenger Lists, 1820–1957*, Ancestry.com.

9. Kirloskar, *Cactus & Roses*, 88–89.

10. Ibid., 59–60, 130–136.

11. Ibid., 138

12. Ibid., 141–143.

13. Ibid., 166–171; "Indian Made Oil Engines," *Times of India*, February 28, 1955.

14. "Indian-Made Diesel Engines: Foreign Demand," *Times of India*, April 25, 1958; "Big Market for Indian Goods in Rhodesia," *Times of India*, September 17, 1958.

15. Kirloskar, *Cactus & Roses*, 187–193; Jeffrey L. Cruikshank and David B. Sicilia, *The Engine That Could: Seventy-Five Years of Values-Driven Change at Cummins Engine Company* (Boston: Harvard Business School Press, 1997), 224–228.

16. Jack Baranson, *Manufacturing Problems in India: The Cummins Diesel Experience* (Syracuse: Syracuse University Press, 1967), 70–77.

17. Ibid., 98–102.

18. Kirloskar, *Cactus & Roses*, 288–293; "Kirloskar Cummins," *Economic and Political Weekly* 2 (February 18, 1967): 421; "A Second Trombay Power Plant in 2080 Pieces" (Kirloskar advertisement), in "Growth Centres of India: Poona," supplement, *Commerce*, October 25, 1975, 2; "Kirloskar Oil Engines, Speech of the Chairman, Shri S. L. Kirloskar," *Economic and Political Weekly* 4 (August 30, 1969): 1426; "Over to Cummins," *Economic and Political Weekly* 32 (September 6, 1997): 2232–2233.

19. "An MIT Man in Poona," *Fortune*, March 1966, 75.

20. S. L. Kirloskar, "Federation of Indian Chambers of Commerce and Industry: Presidential Address," in *Select Speeches and Writings* (Pune: Kirloskar Press, 1983), 17–29; "PM Defends Plan Size, Taxation, and Food Curbs," *Times of India*, March 13, 1966.

21. "An MIT Man in Poona," 75; "India: Ancient Gods and Modern Methods," *Time*, November 13, 1964, 119–120.

22. Kirloskar, *Cactus & Roses*, 223; "An MIT Man in Poona," 75.

23. Kirloskar, *Cactus & Roses*, 135–136; Cruikshank and Sicilia, *The Engine That Could*, 226; Commerce Research Bureau, "Poona Industrial Complex," in "Industries in and around Poona," supplement, *Commerce*, May 6, 1972, 3–7; Shantanu L. Kirloskar, "How Poona Grew," in "Industries in and around Poona," supplement, *Commerce*, May 6, 1972, 11–14; D. B. Mahatme, "You Can't Create Entrepreneurs," in "Growth Centres of India," supplement, *Commerce*, October 25, 1975, 15, 17.

24. Baba Kalyani, interview by author, February 21, 2009, Pune, India; "City Notes: New Unit for Heavy Forgings," *Times of India*, April 7, 1962; "Poona Industrial Complex," 4; Neelkanth Kalyani, "Largest Forging Shop," in "Industries in and around Poona," supplement, *Commerce*, May 6, 1972, 23; "Poona: An Industrial Profile," in "Growth Centres of India," supplement, *Commerce*, October 25, 1975, 5; "Bharat Forge Company Limited," *Times of India*, December 30, 1976.

25. Kalyani, interview; "Bharat Forge 3.0: The Return of Baba Kalyani Who Changed the Face of Indian Engineering Globally," *Economic Times*, May 18, 2012.

26. Dwijendra Tripathi, *The Oxford History of Indian Business* (New Delhi: Oxford University Press, 2004), 85–86; Medha M. Kudaisya, *The Life and Times of G. D. Birla* (New Delhi: Oxford University Press, 2003), 4–12.

27. Kudaisya, *The Life and Times of G. D. Birla*, 14–44; Tripathi, *Oxford History of Indian Business*, 168.

28. Kudaisya, *The Life and Times of G. D. Birla*, 43–49, 59, 207; Tripathi, *Oxford History of Indian Business*, 168–170.

29. Kudaisya, *The Life and Times G. D. Birla*, 67–71, 163; Jossleyn Hennessy, *India Democracy and Education: A Study of the Work of the Birla Education Trust* (Bombay: Orient Longman, 1955). 68–69.

30. Birla Institute of Technology and Science, *An Improbable Achievement* (New Delhi: Wiley Eastern, 1990), 1; "Pilani—Combination of Gurukul and Varsity," *Modern Review* 97 (November 1955): 414.

31. G. D. Birla to Mahatma Gandhi, March 1, 1932, reprinted in Ghanshyam Das Birla, *Bapu: A Unique Association* (Bombay: Bharatiya Vidya Bhavan, 1977), 1: 175–177.

32. "Mr. G. D. Birla." *Times of India*, July 7, 1945.

33. G. D. Birla to M. O. Mathai, May 15, 1956, Birla Letters, Jawaharlal Nehru Memorial Library, New Delhi. Other details of Birla's trips to the United States are given in Kudaisya, *The Life and Times of G. D. Birla*, 285–286, 332–334.

34. J. R. Killian Jr. to Thomas Bradford Drew, February 9, 1962, MIT Office of the President, AC 134, Box 19, MIT Institute Archives and Special Collections, Cambridge, MA (hereafter MIT-IA); G. D. Birla to James R. Killian Jr., November 11, 1961, File 5, BITS Correspondence, Birla private papers, New Delhi, India (hereafter BITS Correspondence).

35. G. D. Birla to J. A. Stratton, January 11, 1962, MIT Office of the President, AC 134, Box 19, MIT-IA; J. R. Killian to G. D. Birla, January 19, 1962, File 5, BITS Correspondence.

36. G. D. Birla to Thomas Drew, January 29, 1962, File 5, BITS Correspondence. Birla's collaboration with MIT is most fully discussed in Stuart W. Leslie and Robert Kargon, "Exporting MIT: Science, Technology, and Nation-Building in India and Iran," *Osiris* 21 (2006): 118–123.

37. Thomas B. Drew to G. D. Birla, February 12, 1962, and May 15, 1962, File 5, BITS Correspondence.

38. G. D. Birla to James R. Killian Jr., November 11, 1961, File 5, BITS Correspondence.

39. Kudaisya, *The Life and Times of G. D. Birla*, 331–336.

40. Ramachandra Guha, *India after Gandhi: The History of the World's Largest Democracy* (New Delhi: Picador, 2007), 330–344.

41. G. D. Birla to Jawaharlal Nehru, May 25, 1963, G. D. Birla Papers, Jawaharlal Nehru Memorial Library, New Delhi, India.

42. Ibid.

43. Ibid.; Guha, *India after Gandhi*, 338–345.

44. V. Lakshmi Narayanan, Birla Institute of Technology, Report, June 28, 1962, File 5, BITS Correspondence.

45. Thomas B. Drew, "The MIT Plan," March 5, 1963, File 7, BITS Correspondence.

46. Thomas B. Drew to G. D. Birla, February 12, 1964, File 8, BITS Correspondence.

47. Leslie and Kargon, "Exporting MIT: Science, Technology, and Nation-Building in India and Iran," 120–121; Thomas Drew to G. D. Birla, August 14, 1964, File 8, BITS Correspondence,

48. G. D. Birla to Thomas Drew, August 20, 1963, File 8, BITS Correspondence.

49. G. Pascal Zachary, *Endless Frontier: Vannevar Bush Engineer of the American Century* (New York: The Free Press, 1997). One of Zachary's main themes is Bush's contradictory legacy of being committed to the individual while supporting the increased role of the government. MIT President James Killian was Eisenhower's science adviser.

50. "Scheme of the Proposed Birla Institute of Technology, Delhi," File 4, BITS Correspondence; G. D. Birla to S. D. Pande, August 25, 1962, File 4, BITS Correspondence; G. D. Birla to Gordon Brown, May 1, 1966, File 16, BITS Correspondence.

51. *BITS-MIT-Ford Foundation Report, 1964–1974*, Birla Institute of Technology and Science Library, Pilani, India, 2–3, 22.

52. V. Lakshmi Narayanan's CV is included in Thomas Drew to James R. Killian Jr., October 8, 1962, Gordon Brown Papers, MC24, Box 11, Folder 454, MIT-IA; Douglas Ensminger to G. D. Birla, October 21, 1967, MIT Office of the President, AC134, Box 19, MIT-IA; Gordon Brown to G. D. Birla, February 6, 1969, Gordon Brown Papers, MC 24, Box 14, Folder 569, MIT-IA; Harcourt Butler Technological Institute, *Director's Report 1966*, INSTEP Collection, Carnegie Mellon University Archives, Pittsburgh, PA; Sandhya Mitra, interview by author, February 22, 2009, New Delhi, India.

53. Vijay V. Mandke, "BITS Practice School: A Case Study in Industry University Collaboration," in "Practice School: A New Concept in Higher Education," special issue, *Birla Vidhya Vihar Bulletin* 23 (January 1980): 1–73.

54. *BITS-MIT-Ford Foundation Report, 1964–1974*, 4; C. R. Mitra, *Management of Innovation: A Case Study of the Birla Institute of Technology and Science Pilani, India* (Paris: UNESCO International Institute for Educational Planning, 1994), 26–28.

55. James Killian to G. D. Birla, February 13, 1962, File 5, BITS Correspondence.

56. Minhaz Merchant, *Aditya Vikram Birla: A Biography* (New Delhi: Viking, 1997), 77–82; Gita Piramal, *Business Maharajas* (New Delhi: Penguin Books, 1996), 143–145.

57. Merchant, *Aditya Vikram Birla*, 83–112.

58. Ibid., 115–123.

59. Ibid., 107–109.

60. Kudaisya, *The Life and Times of G. D. Birla*, 355–359; Francine R. Frankel, *India's Political Economy 1947–2004*, 2nd ed. (New Delhi: Oxford University Press, 2005), 297–298.

61. Kudaisya, *The Life and Times of G. D. Birla*, 355–359; Frankel, *India's Political Economy*, 298–395.

62. Merchant, *Aditya Vikram Birla*, 129–131, 144–149.

63. Ibid., 222–285.

64. Shailendra Jain, interview by author, September 16, 2014, telephone.

65. Ibid.

66. "Parle: A Journey through Generations," 1–15, n.d., n.p. I thank Ramesh Chauhan with providing me with this undated Parle publication. Ramesh Chauhan, interview by author, June 19, 2008, Mumbai, India; "India's Cola Bottlers Enmeshed in Politics," *New York Times*, May 27, 1980; "Panel to Look into Import Quotas of Soft-Drink Firm," *Times of India*, June 5, 1971; "The Pop, the Fizz, and the Froth: A Hard Look at Soft Drinks," *Times of India*, October 9, 1977; Nikhil Deogun and Jonathan Karp, "For Coke in India, Thums Up is the Real Thing," *Wall Street Journal*, April 29, 1998.

67. B. K. Karanjia, *Godrej: A Hundred Years 1897–1997* (New Delhi: Viking Penguin, 1997); Kaki Hathi, interview with Vrunda Pathare, April 1, 2006, Godrej Archives, Mumbai, India; Adi Godrej, interview by author, June 17, 2008, Mumbai, India; Nadir Godrej, interview by author, June 19, 2008.

68. Almitra Patel, interview by author, July 24, 2014, Montgomeryville, PA.

9   The Roots of IT India

1. Kirit Parikh, interview by author, June 11 2008, New Delhi, India.

2. Martin Campbell-Kelly, William Aspray, Nathan Ensmenger, and Jeffrey R. Yost, *Computer: A History of the Information Machine*, 3rd ed. (Boulder: Westview Press, 2014), 65–85; Kenneth Flamm, *Creating the Computer: Government, Industry, and High Technology* (Washington, DC: Brookings Institution, 1988).

3. Campbell-Kelly et al., 143–152; Kent C. Redmond and Thomas M. Smith, *From Whirlwind to MITRE: The R&D Story of the SAGE Air Defense Computer* (Cambridge, MA: MIT Press, 2000).

4. Kenneth Flamm, *Targeting the Computer: Government Support and International Competition* (Washington, DC: Brookings Institution, 1987), 56; M. Mitchell Waldrop, *The Dream Machine: J. C. R. Licklider and the Revolution and Made Computing Personal* (New York: Viking, 2001), 217–236; "7090 Answers Increased Computation Demands," *The Tech*, February 7, 1962.

5. John von Neumann letter to Homi J. Bhabha, February 3, 1948, reproduced in Indira Chowdhury and Ananya Dasgupta, *A Masterful Spirit: Homi J. Bhabha, 1909–1966* (New Delhi: Penguin Books, 2010), 118–119; R. Narasimhan, "Men, Machines, and Ideas: An Autobiographical Essay," *Current Science* 76 (February 10, 1999): 447–454; R. Narasimhan, interview with Ross Bassett and Indira Chowdhury, March 9, 2007; C. R. Subramanian, *India and the Computer: A Study of Planned Development* (Delhi: Oxford University Press, 1992), 1–37; Dinesh C. Sharma, *The Long Revolution: The Birth and Growth of India's IT Industry* (Noida: Harper Collins, 2009), 1–40.

6. *Electronics in India: Report of the Electronics Committee, February 1966* (Bombay: Electronics Committee, 1966); Subramanian, *India and the Computer,* 1–37.

7. S. K. Bose to M. G. K. Menon, August 7, 1961, TIFR Archives, Mumbai, India; S. R. Sen Gupta to D. G. Carter, May 24, 1957, Box 1, Indian Institute of Technology, Project File, 1953–1966, University of Illinois Archives, Urbana, IL.

8. Lalit Kanodia, interview by author, January 28, 2008 and February 16, 2009, Mumbai, India. The following two paragraphs are also based on these interviews.

9. Nitin Patel, interview by author, August 14, 2008, Cambridge, MA; Ashok Malhotra, interview by author, July 29, 2008, telephone.

10. M. P. Mistri, "Tata Computer Centre, April 29, 1966, Folder "Tata Computer Centre March 1966-July 1967," Box 28 TS/Co/T31/MIS/1, Tata Central Archives, Pune, India (hereafter TCA).

11. "Tata Computer Centre Progress Report No. 3," n.d., Folder "Tata Computer Centre, March 1966-July 1967," Box 28 TS/Co/T31/MIS/1, TCA; Patel, interview.

12. "E. D. P. and the Management Revolution," Box 28 TS/Co/T31/MIS/1, TCA; Dr. Lalit S. Kanodia, "Frontiers of Electronic Data Processing," Box 28 TS/Co/T31/MIS/1, TCA.

13. Ibid.

14. Mistri, "Tata Computer Centre."

15. "Brief Review of Computer Operations TCS/TCC," July 29, 1969, Box 18 TS/CO/T16/FIN/2, TCA; Tata Consultancy Services, Operations Report, 1969, TS/CO/T20/MIS/4, TCA.

16. M. G. K. Menon, "Homi Bhabha and Self-Reliance in Electronics and Information Technology," in *Homi Bhabha and the Computer Revolution,* ed. R. K. Shyamasundar and M. A. Pai (Delhi: Oxford University Press, 2011), 117–118.

17. Patel, interview.

18. "Financial Position of TCC/TCS," undated document, TS/CO/T16/FIN/2 TCA; Lalit Kanodia, "Tata Consultancy Services and Tata Computer Centre," August 1, 1969, TS/CO/T20/MIS/4/ TCA; Patel, interview; "Tata Consultancy Services," two undated brochures in author's possession. I thank Lalit Kanodia for providing me with copies of these brochures.

19. "The Struggle against Automation," *People's Democracy,* March 3, 1968, 9–10; "Man-Eaters on Rampage," *New Age,* November 15, 1964, 15; "Ban Job-Eaters." *New Age,* April 9, 1967, 4.

20. M. Achuthan, "Ram Raj or Computer Raj: National Day against Automation: Feb. 23," *New Age,* February 11, 1968, 3.

21. Naval Tata, "The Inevitability of Automation," *Commerce,* July 20, 1968, 140–141.

22. Patel, interview; B. V. Laud to Tata Services, November 15, 1968, TCS Committee Meetings Finance, TS/CO/T16/FIN/2, TCA.

23. "Officials Confined in Head Office," *Times of India,* April 26, 1967; "Calcutta Electric Supply Corporation Limited," *Times of India,* November 1,

1968; "Calcutta Electric Supply Corporation Limited," *Times of India*, October 13, 1969; F. C. Kohli to The Secretary, Department of Electronics, "ICL 1903 Foreign Exchange Earnings," April 18, 1975, Box No. 19 TS/CO/T16/MIS/4, TCA.

24. F. C. Kohli, "The Managing Director," October 22, 1969, Box No. 18 TS/CO/T16/FIN/2, TCA.

25. F. C. Kohli, interview by author, January 25, 2008, Mumbai, India; F. C. Kohli, "IT takes vision and wisdom," *Tata Review Special Commemorative Issue* 2004, 89–90.

26. Kohli, interview by author, February 18, 2009, Mumbai, India; Kohli, interview, January 25, 2008.

27. F. C. Kohli, "The Managing Director," October 22, 1969, Box No. 18 TS/CO/T16/FIN/2, TCA; Patel, interview; Malhotra, interview.

28. Kanodia, interview, January 28, 2008 and February 16, 2009; L. S. Kanodia, "Tata Consultancy Services and Tata Computer Centre," August 1, 1969, TS/CO/T20/MIS/4, TCA.

29. "Minutes of the TCS Committee Meeting Held in Mr. Agarwala's Chamber on 7th May 1970," TCS Committee Meetings, Finance, March 1968–1970, TS/CO/T16/FIN2 TCA; "Telephone Complaints Swell," *Times of India*, September 1, 1972; "Tata Consultancy Services," *Tata Review* 1974, no. 2, 24–26.

30. "Operations Report 1969," TS/CO/T20/MIS/ 4 TCA.

31. Ibid.

32. "Indian Made Director of U.S. Institute," *Times of India*, December 27, 1972; Kohli, interview, January 25, 2008; Campbell-Kelly et al., *Computer*, 123–133.

33. Folder "Tata Infotech, November 1972-November 1973, July 23, 1973, Box 27, TCA; Kohli, interview, January 25, 2008.

34. Report on TCS Operations (January-December 1976), TS/CO/T16/OPT TCA. The rupee to dollar conversion figure used here and elsewhere in this chapter is from the Reserve Bank of India's website, "Table 154, Exchange Rate of the Indian Rupee," http://rbidocs.rbi.org.in/rdocs/Publications/PDFs /72784.pdf.

35. Report on TCS Operations (January-December 1976)

36. Audrey Mody, "Staying the Course," *Tata Review* 2005, no. 2, 64; N. R. Mody to M. H. Mody, November 14, 1978, TS/CO/T16/MIS/5, TCA. This box contains many clippings from American engineering and business publications.

37. Thomas H. Hopkins to F. C. Kohli, August 6,1975, TS/CO/T16/OPT/7; F. C. Kohli to N. A. Palkiwala, November 18, 1975, TS/CO/T16/OPT/7 TCA; E. S. McCollister to Adi Cooper, December 11, 1975, TS/CO/T16/OPT/7, TCA.

38. F. C. Kohli to the Secretary, Department of Electronics, April 18, 1975, Box 19 TS/CO/T16.MIS/4, TCA.

39. The MRTP Act is discussed in Arvind Panagariya, *India: The Emerging Giant* (New Delhi: Oxford University Press, 2008), 59–62. An example of a Tata

report for MRTP approval is K. C. Mehra to the Deputy Secretary of the Government of India, Ministry of Law, Justice, and Company Affairs, October 3, 1978, TS/CO/MIS/6, TCA. The quote is from M. H. Mody to the Deputy Secretary of the Government of India, Ministry of Law, Justice, and Company Affairs, December 1, 1978, TS/CO/MIS/6, TCA.

40. F. C. Kohli to T. S. Natarajan, October 29, 1975, Folder "TCS General No. 2 Jan 1976 Oct 1977," TS/CO/T16/MIS/5, TCA; S. Ramadorai, *The TCS Story. . . . and Beyond* (New Delhi: Portfolio Penguin, 2011), 81.

41. "TCS Scales New Heights Abroad," *Tata Review* 1978, no. 1, 21–28.

42. Kohli, interview, January 25, 2008. The speech is reprinted in F. C. Kohli, *The IT Revolution in India: Selected Speeches and Writings* (New Delhi: Rupa and Co., 2005), 3–6.

43. Kohli, interview, January 25, 2008; F. C. Kohli and C. Iyer, "Software Development for Self-Reliance in Computers," in *SEARCC 76: Proceedings of the IFIP Regional Conference, Singapore, 6–9 September 1976* (Amsterdam: North Holland Publishing Co., 1977), 615–620.

44. Ibid., 616.

45. Ibid., 618.

46. "TCS Trains European Computer Professionals," *Tata Review* 1977, no. 4, 24.

47. "Evening Courses in Computers," *Times of India*, November 6, 1968; "Datamatics Institute of Management," *Times of India*, May 10, 1972; Kanodia, interview, January 28, 2008 and February 16, 2009.

48. "India's Leading Data Processing Company," *Times of India*, March 12, 1981; Kanodia, interview, January 28, 2008 and February 16, 2009.

49. "India's Leading Data Processing Company," *Times of India*, March 12, 1981; "Promoting Export of Computer Software," *Times of India*, April 24, 1981; Kanodia, interview, January 28, 2008 and February 16, 2009.

50. "Grand Mansion with Imported Bricks," *Times of India*, August 17, 1989; "Datamatics Consultants," *Times of India*, October 19, 1989; Kanodia, interview.

51. Narendra Patni, interview by author, January 23, 2008, Mumbai, India.

52. Ibid.; Robert Weisman, "At the Center of a Cultural Shift," *Boston Globe*, May 25, 2004.

53. Sharma, *The Long Revolution*, 262–264; Patni, interview.

54. Narayana Murthy, interview by author, January 20, 2006.

55. Murthy, interview; Jashwant Krishnayya, interview by author, January 31, 2008.

56. Sharma, *The Long Revolution*, 264–265.

57. Ibid., 303–326; G. K. Gurumurthy, "Parley on Software in India," *India Abroad*, November 13, 1987, 29, ProQuest; Lavanya Mandavilli, "Offshore Software Development: Selling a New Idea to American Business," *India Currents*, November 30, 1988, 18, ProQuest; "Grand Mansion with Imported Bricks," *Times of India*.

58. Reproduced in Subramanian, *India and the Computer*, 154.

10   From India to Silicon Valley

1. "Statement by the President Concerning an Educational Consortium to Aid in Developing a Technical Institute in India," November 11, 1961, *Public Papers of the Presidents of the United States: John F. Kennedy, 1961* (Washington, DC: Government Printing Office, 1962), 1:712; "U.S. Colleges Aid India Technology," *New York Times,* November 11, 1961; *Kanpur Indo-American Program Final Report* (Newton, MA: Educational Development Center, 1972), A4, B1; "The Most Chaotic Education Anywhere," *Nature* 308 (April 12, 1984): 593.

2. Roger L. Geiger, *Research and Relevant Knowledge: American Research Universities since World War II* (New York: Oxford University Press, 1993): 30–91; Bruce Seely, "The Other Re-engineering of Engineering Education, 1900–1965," *Journal of Engineering Education* 88 (July 1999): 285–294; Stuart W. Leslie, *The Cold War and American Science: The Military Industrial Complex at MIT and Stanford* (New York: Columbia University Press, 1993), 14–43; Massachusetts Institute of Technology, *President's Report, October 1948,* 21.

3. Karl L. Wildes and Nilo A. Lindgren, *A Century of Electrical Engineering and Computer Science at MIT, 1882–1982* (Cambridge, MA: MIT Press, 1985), 310–318; G. S. Brown, "Educating Electrical Engineers to Exploit Science," *Electrical Engineering* 74 (February 1955):110–115; Gordon S. Brown, "The Modern Engineer Should Be Educated as a Scientist," *Journal of Engineering Education* 43 (December 1952): 274–281.

4. *Report of the Reviewing Committee of the Indian Institute of Technology, Kharagpur* (New Delhi: Ministry of Scientific Research and Cultural Affairs, 1959), 10, 13.

5. Dean G. S. Brown, Dean H. L. Hazen, Professors N. C. Dahl, C. H. Norris, and L. D. Smullin, "Status of the 'Indian Project,'" November 21, 1960, Gordon Brown Papers, MC 24, Box 5, Folder 218, MIT Institute Archives and Special Collections, Cambridge, MA (hereafter MIT-IA).

6. Ibid.; P. K. Kelkar, unpublished reminiscences in author's possession, 24; Robert S. Green to Norman C. Dahl, March 28, 1961, Gordon Brown Papers, MC, 24, Box 5, Folder 211, MIT-IA. Further details on MIT's involvement with IIT Kanpur are given in Stuart W. Leslie and Robert Kargon, "Exporting MIT: Science, Technology, and Nation-Building in India and Iran," *Osiris* 21 (2006): 110–130; Ross Bassett, "Aligning India in the Cold War Era: Indian Technical Elites, the Indian Institute of Technology at Kanpur, and Computing in India and the United States," *Technology and Culture* 50 (October 2009): 783–810.

7. Max F. Millikan to Jay Stratton and Gordon Brown, "The Kanpur Proposal," August 26, 1960, MIT Office of the President, AC 134, Box 74, Folder 8, MIT-IA.

8. Charles Montrie to Gordon Brown, March 13, 1961, Gordon Brown Papers, MC 24, Box 5, Folder 211, MIT-IA.

9. Robert S. Green to Norman C. Dahl, March 28, 1961.

10. For specific information on Kelkar, I have relied on S. Ranganathan, "Uncanny Confidence: An Obituary of P. K. Kelkar," *Current Science* 60 (February 10, 1991): 185–186; "The Passing of a Legend," *Indian Institute of Technology Kanpur Alumni Association Newsletter*, January 1991, 1–2; Arawind Parasnis, "IITK, Kelkar and I: Reminiscences of a Bygone Era," undated document in author's possession, as well as lengthy discussions with Arawind Parasnis (one of the first professors at IIT Kanpur). P. K. Kelkar, unpublished reminiscence in author's possession, 23. The volume Kelkar read was *Recent Advances in the Engineering Sciences: Their Impact on Engineering Education, Proceedings of the Conference on Science and Technology for Deans of Engineering* (New York: McGraw-Hill, 1958).

11. Leslie and Kargon, "Exporting MIT," 114–115.

12. John Kenneth Galbraith, *Ambassador's Journal: A Personal Account of the Kennedy Years* (Boston: Houghton Mifflin, 1969), 411–412; *Kanpur Indo-American Program Final Report*.

13. This observation is based on discussions with many IIT Kanpur and Bombay faculty members and students.

14. Dennis Kux, *India and the United States: Estranged Democracies* (Washington, DC: National Defense University Press, 1992), 289–312; H. W. Brands, *India and the United States: The Cold Peace* (Boston: Twayne Publishers, 1990), 122–138; George Perkovich, *India's Nuclear Bomb: The Impact on Global Proliferation* (Berkeley: University of California Press, 1999), 161–166. The 1972 state of Indo-American relations are alluded to in *Kanpur Indo-American Program Final Report*, iii–iv.

15. Indian Institute of Technology, Bombay, *Final Report of the Curriculum Committee. Part I: General* (Bombay, 1972), 5, 8. Suhas Sukhatme, interview by author, January 28, 2008, Mumbai, India. Sukhatme later became the director of IIT Bombay and wrote the "brain drain" studies discussed later in this chapter.

16. The numbers of the IIT's student populations are based on annual reports of each institution from the late 1960s. The figures for education expenditures are from George Tobias and Robert S. Queener, *India's Manpower Strategy Revisited, 1947–1967* (Bombay: N. M. Tripathi Private Ltd., 1968), 60, 120–121.

17. Indian Institute of Technology, Madras, *Information Bulletin, Academic Session 1973-74*, 24–25; *Basic Statistics Relating to the Indian Economy, 1950–1951 to 1972–1973* (New Delhi: Central Statistical Organization, 1976), 10. A history of IIT Bombay noted that in the 1960s, tuition accounted for only 2 percent of the Institute's revenue. Rohit Manchanda, *Monastary, Sanctuary, Laboratory: 50 Years of IIT Bombay* (Delhi: Macmillan, 2008), 109.

18. The following paragraphs are based on many years of discussions with IIT students and faculty.

19. All-India Council on Technical Education, *Technical Education in Independent India, 1947–1997* (New Delhi: All-India Council on Technical Education, 1999), 85–86.

20. A. D. King, "Elite Education and the Economy: IIT Entrance, 1965–1970," *Economic and Political Weekly* 5 (August 29, 1970): 1463–1472.

21. Ibid. IIT Kanpur's 1988–1989 *Annual Report* gives a vivid picture of this, showing the All-India Rankings of the students placed into each major. Indian Institute of Technology Kanpur, *29th Annual Report, 1988–1989*, IV-7-8.

22. Indian Institute of Technology, Kharagpur, *Seventh Convocation, 10 March 1962*; Indian Institute of Technology Bombay, *Eighth Annual Report, 1965–1966*, iii; Indian Institute of Technology Kanpur, *Fifteenth Annual Report, 1974–1975*, 46–47. For IIT Kharagpur's 1962 graduates, I matched the names in the convocation list with MIT graduates, contacted some graduates, and matched some graduates up with people who had earned graduate degrees in the United States.

23. Institute of International Education, *Open Doors, 1964* (New York: Institute of International Education, 1964), 30; Institute of International Education, *Open Doors, 1969* (New York: Institute of International Education, 1969), 38.

24. Tapan Kumar Gupta, interview by author, August 14, 2013, telephone; Tapan Kumar Gupta, "Sintering of Zinc Oxide" (ScD dissertation, MIT, 1967), 124.

25. Pramud Rawat, interview by author, January 22, 2009, telephone.

26. Ani Chitaley, interview by author, 2013, telephone; Satya Atluri, interview by author, September 24, 2013, telephone.

27. Thomas Kailath, interview by author, May 18, 2007, telephone; Tekla S. Perry, "Medal of Honor: Thomas S. Kailath," *IEEE Spectrum*, May 2007, 44–47; G. S. Krishnayya and J. M. Kumarappa, *Going to USA: A Guidebook for Students and Other Visitors* (Bombay: Tata Institute for Social Sciences, 1952).

28. Bassett, "Aligning India in the Cold War Era."

29. Massachusetts Institute of Technology, *Report of the President, 1967*, 513–514; *Employment Outlook for Engineers, 1969–1979* (New Delhi: Institute of Applied Manpower Research, 1969), 34; Institute of International Education, *Open Doors 1967*, 23.

30. On MIT's policy of expecting foreign students to pay for at least the first semester of their education, see P. M. Chalmers to J. R. Killian, October 26, 1948, MIT Office of the President, AC 4, Box 50, Folder 1, MIT-IA. Several Indians who went to MIT described to me being admitted without any financial assistance and then writing to MIT and subsequently receiving assistance.

31. B. V. Bhoota to G. E. Brown, April 29, 1964, and P. E. Chalmers to U. J. Bhatt, May 1, 1964, both Gordon Brown Papers, MC 24, Box 5, Folder 214, Box 5, MIT-IA.

32. Ibid.

33. This comment is based on many interviews with Indian graduates of the IITs. The most formal system of distributing applicants across American graduate schools seems to have been developed at IIT Madras.

34. I looked at the dissertations of all the Indian students who earned MIT doctorates. Up through the early 1970s, the dissertations typically included a brief biography. The acknowledgments typically provide a sense of who funded the work and the students' networks at MIT.

35. Sandhya Mitra, interview by author, February 22, 2009, New Delhi.

36. D. D. Kosambi to Lawrence Arguimbau, January 19, 1949, Folder 11, Carton 4, and D. D. Kosambi to Lawrence Arguimbau, April 25, 1949, Folder 12, Carton 4, both Blake-Clapp-Arguimbau Family Papers, Massachusetts Historical Society, Boston.

37. "In Memoriam: Jamshed R. Patel," *Synchrotron Radiation News* 20 (2007): 44; Jamshed R. Patel, "The Influence of Stress and High Pressure on Martensitic Transformation" (ScD dissertation, MIT, 1954); Susan Furland, interview by author, July 17, 2013, telephone; Eric Patel, interview by author, August 20, 2013, telephone.

38. Vijay Gopal Paranjpe, "Equilibrium Relationships in the Iron-Nitrogen System" (ScD dissertation, MIT, 1949). This paragraph is based on an analysis of the acknowledgments section of every available doctoral dissertation by an Indian between 1950 and 1971. The Mansfield amendment, passed in 1969, and continuing on in intent, if not in law, put limits on funding of research by the Department of Defense, resulting in more university research being funded by the National Science Foundation. Daniel Kevles, *The Physicists: The History of a Scientific Community in Modern America* (New York: Alfred A. Knopf, 1977), 414–415.

39. "Avinash C. Singhal, CV, May 5, 1965, *Kanpur Indo-American Program Records, 1961–1972*, AC 334, Box 5, Folder "IIT/K Prospective Indian Faculty," MIT-IA; Anil Malhotra, *A Passion to Build: India's Quest for Offshore Technology, a Memoir* (n.p., lulu.com, 2007), 47–49; Ashok Malhotra, interview by author, July 29, 2008, telephone.

40. Kailath, interview by author, May 18, 2007; Thomas Kailath, "Optimum Receivers for Randomly Varying Channels," *Information Theory: Papers Read at a Symposium on 'Information Theory' held at the Royal Institution, London, August 29th to September 2nd 1960*, ed. Colin Cherry (London: Butterworths, 1961), 109–122. In a later interview, Kailath asserted that he did not believe that he received permanent residency status until after 1965. Kailath, interview by author, January 25, 2015.

41. Vinay Lal, *The Other Americans: A Political and Cultural History of South Asians in America* (Noida: Harper Collins, 2008), 53–54; "The Gold Drain and the Other Drain," *Economic and Political Weekly* 1 (September 24, 1966): 229; Carl T. Rowan, "Dangers in the Brain Drain," *Los Angeles Times*, September 9, 1966. The number of Indian professionals migrating is from U.S. Senator Walter Mondale, who made unsuccessful efforts to stop skilled migration from developing countries. For Indian migration to the United States considered in a broader context, see Devesh Kapur, *Diaspora, Development and Democracy: The Domestic Impact of International Migration from India* (Princeton, NJ: Princeton University Press, 2010).

42. This idea has been expressed to me in numerous interviews. It is also described in S. P. Sukhatme, *The Real Brain Drain* (Bombay: Orient Longman, 1994), 32–36.

43. Tapan Kumar Gupta, interview by author.

44. David A. Hounshell, "The Evolution of Industrial Research in the United States," in *Engines of Innovation: U.S. Industrial Research at the End of an*

Era, ed. Richard S. Rosenbloom and William J. Spencer (Boston: Harvard Business School Press, 1996), 13–85.

45. Praveen Chaudhari, interview by author, May 21, 2008, Yorktown Heights, NY.

46. Praveen Chaudhari, "The Effects of Irradiation with Protons on the Electrical Properties and Defect Configurations of the Compound Bi_2Te_3" (ScD dissertation, MIT, 1965).

47. Chaudhari, interview; Supratik Guha, Robert Rosenberg, and Jochen Mannhart, "Praveen Chaudhari," *Physics Today*, April 2010, 64–65.

48. National Science Foundation, *Women and Minorities in Science and Engineering* (Washington, DC: National Science Foundation, 1982), 31; Patricia E. White, *Women and Minorities in Science and Engineering: An Update* (Washington, DC: National Science Foundation, 1992), 137–139; P. P. Parikh, S. P. Sukhatme, "Women Engineers in India," *Economic and Political Weekly* 39 (January 10–16 2004): 193–201. The MIT commencement programs do not identify the sex of degree recipients, so this analysis is based on names, supplemented by other information.

49. Uma Chowdhry, interview by author, March 13, 2015, telephone; Uma Chowdhry, "Defect Detection in Single Crystals and Polycrystals of CoO" (PhD dissertation, MIT, 1976), 129.

50. Chowdhry, interview.

51. Ibid.; Chowdhry, "Defect Detection," 129.

52. Chowdhry interview; Uma Chowdhry, CV, March 2014. I thank Dr. Chowdhry for providing me with this document.

53. Suhas Patil, interview by author July 9, 2010, Mountain View, CA; Avinash Singhal interview by author, October 18, 2013, telephone; Sanjay Amin, interview by author, November 5, 2013, telephone.

54. Except where noted, the following paragraphs are based on B. C. Jain, interview by author, October 4, 2013, telephone.

55. "Is Studying to Electrify India," *Washington Post*, January 2, 1938; "Mr. Nanubhai Amin, June 27, 1991," document in author's possession; *Jyoti Jagat*, special issue, 1999. I thank Amin's son, Rahul, for providing me with these documents.

56. T. E. R. Simhan, "Baroda Unit Taps Solar Power," *Times of India*, July 28, 1981.

57. Kailath, interview, May 18, 2007; Anurag Kumar, "Introductory Remarks on Prof. Thomas Kailath, Plenary Speaker, International Workshop on Information Theory, Bangalore, India, May 2002" (in author's possession). IIT Kanpur's part of the program is described in Indian Institute of Technology Kanpur, *Fifteenth Annual Report, 1974–75*, 60–64. For the Joint Services Program in the United States, see Leslie, *The Cold War and American Science*, 25–26.

58. "Bid to Reverse Brain Drain," *Times of India*, June 14, 1985; "PM woos Indian Scientists in U.S.," *Times of India*, June 16, 1985; Praveen Chaudhari, interview by author; Satya Atluri, interview by author. Rajiv Gandhi's interest in technology is described in Dinesh Sharma, *The Long Revolution:*

The Birth and Growth of India's IT Industry (Noida: Harper Collins, 2009), 124–173.

59. AnnaLee Saxenian, *Regional Advantage: Culture and Competition in Silicon Valley and Route 128* (Cambridge, MA: Harvard University Press, 1994).

60. Thomas Kailath, interview by author, January 25, 2015; S. Perry, "Medal of Honor: Thomas Kailath," 44–47.

61. Suhas Patil, interview by author, July 9, 2010.

62. Ibid.; Suhas S. Patil and Terry A. Walsh, "A Programmable Logic Approach for VLSI," *IEEE Transactions on Computers* 28 (September 1979): 601; "New Stock Listings," *Wall Street Journal,* June 19, 1989; Francis Assisi, "Cirrus Logic on Fast Track," *India-West,* October 26, 1990, ProQuest Ethnic Newswatch.

63. Mohan Ramakrishna, "Inspiring TIE Seminar: Silicon Valley Entrepreneurs Share Expertise at Historic Two-Day Seminar," *India Currents Magazine,* May 31, 1994, ProQuest Ethnic Newswatch; Arvind Kumar, "TIE Seminar Focus: Marketing," *India Currents Magazine,* April 30, 1995, ProQuest Ethnic Newswatch; Richard Springer, "Sold-Out Event Offers Networking: TIE Rises to a New TIEr," *India-West,* May 17, 1996, ProQuest Ethnic Newswatch.

64. Vivek Ranadive, interview by author, July 13, 2010, Palo Alto, CA.

65. Ibid. An exposition of the Teknekron parent's philosophy is given in Harvey E. Wagner, "The Open Corporation," *California Management Review* 33 (Summer 1991): 46–60.

66. Vivek Ranadive, *The Power of Now: How Winning Companies Sense and Respond to Change Using Real-Time Technology* (New York: McGraw-Hill, 1999); "Teknekron Scores Big at Salomon Brothers, Winning Phase One of Fulcrum Project," *Sell-Side Technology,* September 11, 1989, http://www.waterstechnology .com/sell-side-technology/feature/1624547/teknekron-scores-big-at-salomon -brothers-winning-phase-one-of-fulcrum-project; Lawrence M. Fisher, "Reuters Buying Teknekron," *New York Times,* December 18, 1993.

67. Malcolm Gladwell, "How David Beats Goliath: When Underdogs Break the Rules," *New Yorker,* May 11, 2009, 40–49; Peter Delevett, "Software to Hardwood," *San Jose Mercury News,* May 30, 2013; Ryan D'Agostino, "The Man Who Knows Everything," *Esquire,* February 2012.

68. S. P. Sukhatme, *The Real Brain Drain,* 12.

69. Suhas P. Sukhatme, "Obituary: Pandurang Vasudeo Sukhatme," *The Statistician* 47 (1998): 389–390; Manju Sheth, "Movers and Shakers in Medicine: Dr. Vikas Sukhatme," *Lokvani,* June 20, 2012, http://www.lokvani.com /lokvani/article.php?article_id=8232; Pace University, Office of the Provost, "Pace University Names New Provost," January 18, 2012, http://www.pace .edu/provost/about-office-provost.

Conclusion

1. Rohit Manchanda, *Monastery, Sanctuary, Laboratory: 50 Years of IIT-Bombay* (New Delhi: Macmillan, 2008), 392; Gokul Rajaram, "GRE Days from 1994," in *Indian Institute of Technology, Golden Jubilee Celebrations, January 17–18, 2003*, 68–69; P. V. Indiresan and N. C. Nigam, "The Indian Institutes of Technology: Excellence in Peril," in *Higher Education Reform in India: Experience and Perspectives*, ed. Suma Chitnis and Philip G. Altbach (New Delhi: Sage, 1993), 334–364.

2. For two different perspectives on India's liberalization, see Arvind Panagariya, *India: The Emerging Giant* (New Delhi Oxford University Press, 2008) and Atul Kohli, *Poverty Amid Plenty in the New India* (Cambridge: Cambridge University Press, 2012).

3. "Editorial Notes," *Mahratta*, June 6, 1886.

4. Gary Gereffi, Vivek Wadhwa, Ben Rissing, and Ryan Ong, "Getting the Numbers Right: International Engineering Education in the United States, China, and India," *Journal of Engineering Education* 97 (January 2008): 13–25.

5. Evaluserve, "India Emerging as Preferred Career Destination for IITians," April 14, 2008, http://www.clubofamsterdam.com/contentarticles/48%20India/Evaluserve%20Article%20India%20Emerging%20as%20Preferred%20Career%20Destination%20for%20IITians%20April14-2008.pdf.

6. I used LinkedIn and other online sources to track the locations of the 107 Indian engineering graduates. (I could not identify the location of ten.)

7. Institute of International Education, *Open Doors, 1991/1992* (New York: Institute of International Education, 1992), 20; Institute of International Education, *Open Doors 2002* (New York: Institute of International Education, 2002), 8; Institute of International Education, *Open Doors 2014* (New York: Institute of International Education, 2014), 4. Neil G. Ruiz, "The Geography of Foreign Students in U.S. Higher Education: Origins and Destinations" (Washington, DC: Brookings Institution, 2014).

8. U.S. Citizenship and Immigration Services, "Characteristics of H-1B Specialty Occupation Workers for Fiscal Year 2012," http://www.uscis.gov/sites/default/files/USCIS/Resources/Reports%20and%20Studies/H-1B/h1b-fy-12-characteristics.pdf; Rama Lakshmi, "For Indian Engineers, H1-B Visa Is the Key to Career Growth," *Washington Post*, June 28, 2013.

9. AnnaLee Saxenian, *Silicon Valley's New Immigrant Entrepreneurs* (San Francisco: Public Policy Institute of California, 1999).

10. Glenn Rifkin, "Amar G. Bose, Acoustic Engineer and Inventor, Dies at 83," *New York Times*, July 12, 2013; Elizabeth M. Hoeffell, Sonya Rastogi, Myoung Oak Kim, and Hasan Shahid, "The Asian Population: 2010," United States Census Bureau, March 2012, 15, http://www.census.gov/prod/cen2010/briefs/c2010br-11.pdf; Clive Thompson, "How Khan Academy Is Changing the Rules of Education," *Wired*, July 15, 2011, http://www.wired.com/2011/07/ff_khan/all/.

11. Kume Kunitake, comp., *The Iwakura Embassy, 1871–1873* (Princeton, NJ: Princeton University Press, 2002), I:382; Massachusetts Institute of Technology, *President's Report 1875*, 203.

12. For the size of the Chinese student delegation at MIT in the early twentieth century, see Massachusetts Institute of Technology, *President's Report 1914*, 52. "Transcript of Premier Zhu Rongji's Speech at MIT, April 15, 1999," http://newsoffice.mit.edu/1999/zhufull. The history of Tsinghua University is discussed in Joel Andreas, *Rise of the Red Engineers: The Cultural Revolution and the Origins of China's New Class* (Stanford, CA: Stanford University Press, 2009).

13. This analysis is based on MIT's *President's Report 2011–2012* and the Institute of International Education's 2014 report, *Open Doors*. See http://web.mit.edu/annualreports/pres12/; http://www.iie.org/Research-and-Publications/Open-Doors/Data/International-Students/Leading-Places-of-Origin/2012–14.

14. Zuoyue Wang makes this an analogous point based on his study of Chinese scientists during the Cold War. See Zuoyue Wang, "Transnational Science during the Cold War: The Case of Chinese/American Scientists," *ISIS* 101 (June 2010): 367–377. Michael J. Boylan, "Foreigners DOT Stony Brook Campus," *New York Times*, April 16, 1972. A 2014 study showed that foreign students contributed $27 billion to the American economy in the academic year 2013–2014. See http://www.nafsa.org/Explore_International_Education/Impact/Data_And_Statistics/The_International_Student_Economic_Value_Tool/.

15. Amartya Sen, "Quality of Life: India Vs. China," *New York Review of Books*, May 12, 2011; Jean Drèze and Amartya Sen, *An Uncertain Glory: India and Its Contradictions* (London: Penguin Books, 2014); Gardiner Harris, "Poor Sanitation in India May Afflict Well-Fed Children with Malnutrition," *New York Times*, July 13, 2014.

16. Quoted in Ramachandra Guha, ed., *Makers of Modern India* (Cambridge, MA: Harvard University Press, 2011), 279.

17. Judith M. Brown, *Nehru: A Political Life* (New Haven, CT: Yale University Press, 2003), 235–236.

18. M. K. Gandhi, *"Hind Swaraj" and Other Writings*, ed. Anthony J. Parel (Cambridge: Cambridge University Press, 2009), 106–109.

19. Edward Broughton, "The Bhopal Disaster and Its Aftermath: A Review," *Environmental Health* 4 (2005), http://www.ehjournal.net/content/4/1/6; Dominique Lapierre and Javier Moro, *It Was Five Past Midnight in Bhopal*, rev. ed. (New Delhi: Full Circle Publishing, 2009), 218; T. R. Chouhan, *Bhopal: The Inside Story* (New York: The Apex Press, 1994).

20. Ashok Kalelkar, interview by author, September 11, 2007, telephone Kalelkar's life and career are also discussed in Shailaja Kalelkar Parikh, *An Indian Family on the Move: From Princely India to Global Shores* (Ahmedabad: Akshara Prakashan, 2011), 211–213.

21. Ashok S. Kalelkar, "Investigation of Large-Magnitude Accidents: Bhopal as Case Study" (paper presented at The Institution of Chemical Engineers Conference on Preventing Major Chemical Accidents, London, May 1988), http://storage.dow.com.edgesuite.net/dow.com/Bhopal/casestdy.pdf; Laurie Hays and Richard Koenig, "How Union Carbide Fleshed Out Its Theory of Sabotage at Bhopal," *Wall Street Journal*, July 7, 1988.

22. Almitra Patel, interview by author, July 24, 2014, Montgomeryville, PA; Asha Sridhar, "A Woman's Campaign to Keep Waste from Ending Up in Landfills," *The Hindu*, August 15, 2013.

23. Deep Joshi, interview by author, August 30, 2014, telephone; "Indian Activist Deep Joshi Wins Magsaysay Award," *Times of India*, August 3, 2009; "Questions and Answers: Deep Joshi," *Wall Street Journal*, August 10, 2009, http://www.wsj.com/articles/SB124990333366219177.

24. "Kunte Vs. Ranade: Or the Wrestling of Two Giants," *Kesari*, June 3, 1884 (in Marathi), translation by S. H. Atre. Ranade's career is discussed in Richard P. Tucker, *Ranade and the Roots of Indian Nationalism* (Chicago: University of Chicago Press, 1972).

Acknowledgments

THIS IS A PROJECT that would have been impossible without the help of many people, most of whom were unknown to me as I started. It is a source of great joy after twelve years of work to be able to finally thank them publicly.

This book began as a study of the Indian Institute of Technology at Kanpur and only later turned into its present form. For help in that stage of the project, I thank Dorothy Dahl, M. A. Pai, Barun Banerjee, Ashok Mittal, and the former director of IIT Kanpur, Sanjay Dhande. In the earliest days of this project, I had the good fortune of meeting Indira Chowdhury, who provided invaluable advice and introductions. Shail Jain and Yogesh Andlay provided me with important contacts at IIT Delhi. Juzer Vasi and Devang Khakhar were gracious hosts to me at IIT Bombay. Jiten Divgi, Chirag Kalelkar, Raj Mashruwala, Anand Pandya, Anant Shah, Rusheed Wadia, and Kirit Yajnik have provided me abundant material, contacts, advice, and friendship over the years of this project. I thank Ravi Meattle and his wife Mona for the hospitality they showed in New Delhi. While I have benefited from the work of many scholars of Indian history, I have had the good fortune to enjoy the personal generosity of Ramachandra Guha, Deepak Kumar, and Dwijendra Tripathi.

Bharat Bhatt and Neelima Shukla-Bhatt provided me with careful translations of the letters of T. M. Shah as well as comments on drafts of my chapters. I am grateful to Professor Shyam Atre for translations of *Kesari* articles. Special thanks are due to Ishwari Kulkarni, who spent a week with me in front of a microfilm machine in Cambridge, England, finding and translating articles in the *Kesari*.

Many librarians and archivists have provided essential assistance. I am deeply grateful to Nora Murphy and the staff at the MIT Institute Archives, and to Rajendra Prasad Narla and staff at the Tata Central Archives. I thank Jenny Shah for allowing me access to the Tata Steel Archives and Vrunda Pathare for access to the Godrej Archives. The libraries at Harvard, IIT Kanpur, Delhi, Bombay, Madras, and Kharagpur, BITS Pilani, as well as the archives cell at the Indian Institute of Science in Bangalore and the Jawaharlal Nehru Memorial Library in New Delhi all provided me essential access to materials. I thank both them and their staffs. The staff of the Inter-Library Loan Department at North Carolina State University was especially helpful in tracking down a variety of difficult-to-find articles. Dr. Pragnya Ram provided helpful materials about Aditya Birla's life and career. I thank Mr. Basant Kumar Birla and Mrs. Sarala Birla for permission to see and use papers in their private collection. I thank Rustam Dastur for providing me with access to materials from Dasturco and the staff at Dasturco for the time they spent helping me to understand the life and career of M. N. Dastur. I thank Anant Shah, Chitra Viji, and Anand Pandya for permission to use materials from their fathers in this book.

The Godrej family and Vrunda Pathare generously invited me to give the Godrej Lecture in Business History in 2009. I had the opportunity to present parts of my research as lectures at various institutions in India and have my hosts and audiences to thank at the Indian Institute of Technology at Kanpur, the National Chemical Laboratory, the Observer Foundation, and the Indian Institute of Science. At an early state of my work, I had the good fortune to participate in Gary Downey's workshop at Virginia Tech on "Locating Engineers" and am grateful to Gary and the conference participants for their encouragement. I first presented the material from this project at a workshop on Science and Technology and National Identity and I acknowledge Ann Johnson and Carol Harrison for organizing that meeting and seeing the papers into print. Portions of this work have been presented before audiences at the Society for the History of Technology, the Social Science History Conference, the

Business History Conference, the Computer History Museum, the University of South Carolina, Stanford University, MIT, UC Irvine, and North Carolina State University. The audiences at each of these venues provided constructive feedback. Satya Atluri was my gracious host at Irvine and shared his insights with me. I thank Sheila Jasanoff for a number of helpful conversations.

This book is based on work supported by the National Science Foundation under Grant No. 0450808. Any opinions, findings, and conclusions or recommendations expressed in this book are those of the author and do not necessarily reflect the views of the National Science Foundation. I thank the NSF as well as the American taxpayers for their support. Ron Rainger and Fred Kronz provided essential help at the NSF. At North Carolina State, Amanda Tueting, Missy Seate, and Paula Braswell played vital roles in research administration.

I have the good fortune to be a member of the department of history at North Carolina State University, whose faculty and staff have been a great source of support. I particularly thank LaTonya Tucker, Norene Miller, and Courtney Hamilton, who have provided invaluable assistance with good cheer. I thank also the taxpayers of the state of North Carolina for their commitment to supporting great public research universities.

I have had the remarkable privilege of having David Gilmartin as the occupant of the office next to mine. He read many drafts of chapters and the penultimate draft of the entire manuscript. He has given me many impromptu lessons in South Asian history, doing much to educate a newcomer to the field. I thank him for his seemingly limitless patience and generosity.

Rosalind Williams at MIT has been a great encouragement over the years, and I thank her for many helpful conversations as well as her comments on a draft of this manuscript. I also thank Bill Leslie for his comments on a draft of this manuscript. Two anonymous readers for Harvard University Press made helpful suggestions for improving the manuscript.

My editor, Sharmila Sen, has done much to shape this manuscript over the years. She has provided the perfect combination of advice, encouragement, and prodding. Without her, this book would not be in front of you. I thank Heather Hughes for her help in shepherding this manuscript into print. I thank also Angela Piliouras and Karen Brogno for their work in the copyediting and production process.

The saddest job is acknowledging the help of those who live only in memory. Arawind Parasnis, Manubhai Parekh, and R. N. Narasimhan each gave me invaluable assistance in the early days of my work. It is impossible to register my sorrow at being unable to personally thank my department head, Jonathan Ocko, for his help. His belief in me and this project was a great source of strength. Working in a difficult financial environment and at great personal sacrifice, Jonathan spent innumerable hours and displayed great entrepreneurial creativity in making sure that I, as well as the other members of our department, had the resources necessary to do our job. At the same time he worked indefatigably to ensure that our department maintained a collegial and humane environment.

Many others provided help and I thank them collectively while I apologize for not being able to name each here. Space prevents me from individually thanking the scores of people I interviewed for this project. Although I could not include every story in this book, those people who shared their (and their parents') stories provided me with an understanding essential to writing this book.

I thank my sister Beth and her husband Rob for their support, which included sending me old copies of MIT's *Technology Review* and providing me with a friendly place to stay during my research trips. While my parents, Alvira and Knox, also live only in memory, my debt to them is one that cannot be repaid. They instilled in me a love of learning, and of history especially, and gave me the freedom to pursue a course that was my own.

I would finally like to thank my family. I could not have written this book without the support and love of my wife Debbie. She shared my enthusiasm for the project and graciously welcomed it into our family. She has read drafts and endured my long hours in my study and away from home, taking care of many things in my absence. My daughter Anna has also endured my absences and I thank her both for her patience and for the many occasions she gave me to put my work aside. While working on this book over the last twelve years has been a pleasure, it cannot compare with the pleasure of seeing a first-grader mature into an independent young woman. I thank both Debbie and Anna for their sustaining love.

While many have contributed and worked to improve this manuscript, the responsibility for any errors is mine alone.

Index